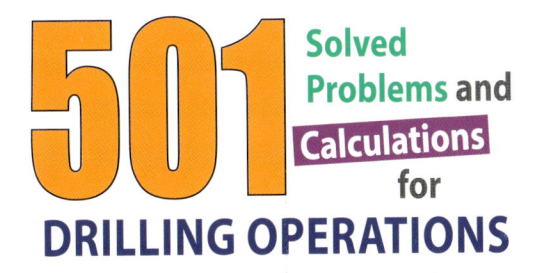

501 Solved Problems and Calculations for DRILLING OPERATIONS

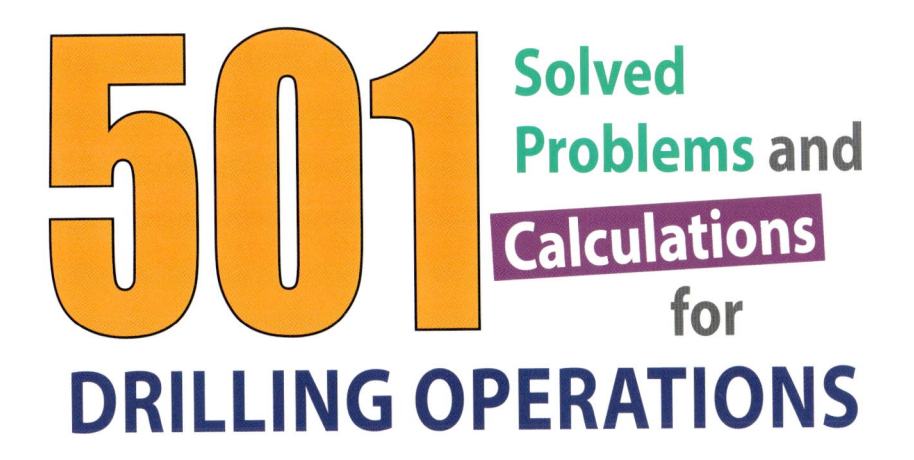

501 Solved Problems and Calculations for DRILLING OPERATIONS

Robello Samuel

σQuadrant
sigmaquadrant.com

A SigmaQuadrant Engineering Publication
Houston/Beijing/Chennai

DISCLAIMER

While both authors and the publisher have used their best efforts in preparing and producing the book, they make no representations or warranties with respect to the accuracy or completeness of the contents of this book and specifically disclaim any implied warranties of merchantability or fitness for a particular purpose. No warranty may be created or extended by marketing or sales representatives or in print or online sales and marketing materials. The advice and strategies contained herein are the opinions of the authors and may not be suitable for your situation. You should consult with the proper professional where appropriate. Neither the publisher nor the authors shall be held liable for any loss of profit or any other commercial damages, including but not limited to special, incidental, consequential, or any other damage.

This publication or any part thereof may not be copied, reproduced, stored in a physical or electronic retrieval system, or transmitted in any form by any means, electronic, mechanical, photocopying, scanning, recording, or otherwise, except as permitted under Section 107 or 108 of the 1976 United States Copyright Act, without either: (1) the prior written permission of the publisher, or (2) authorization through payment of the appropriate per-copy fee to the Copyright Clearance Center, 222 Rosewood Drive, Danvers, Massachusetts, 01923, (978) 750-8400, fax (978) 646-8600, or at www.copyright.com.

501 Solved Problems and Calculations for Drilling Operations

Copyright © 2015 by Sigmaquadrant LLC, Houston, Texas. All rights reserved.
No part of his publication may be reproduced or transmitted in any form without the prior written permission of the publisher.
HOUSTON, TX:
Sigmaquadrant.com
11306 Dawnheath Dr
Cypress, TX 77433
Director: **Cynthia Samuel**
Managing Editor: **Cynthia Samuel**
Production Editor: **Hubert Daniel**
Senior Design Editor: **Balaji Srinivasan**

Library of Congress Cataloging-in-Publication Data
Samuel, Robello.
501 Solved Problems and Calculations for Drilling Operations / Robello Samuel.
p. cm.
Includes bibliographical references and index.

ISBN: 0-9906836-1-3 | ISBN-13: 978-0-9906836-1-2
10 9 8 7 6 5 4 3 2 1
1. Drilling Operations —Equipment and supplies. 2. Oil well drilling—Equipment and supplies. 3. Oil well drilling. 4. Gas well drilling. I. Title.
Printed in the United States of America
Printed on acid-free paper.

Text design and composition by: Kryon Publishing Services (P) Ltd., Chennai, India.
www.kryonpublishing.com

...The mathematical beauty of Euler's formula $e^{i\pi} + 1 = 0$, draped with transcendental, imaginary and real numbers and three basic arithmetic operations provides an enigmatic union between the human intellect and The Creator.

Ollebor Leumas

To
Cynthia, Nishanth and Sharon

Contents

From the Author's Pen   viii

Acknowledgments   ix

About the Author   x

1. **Rig Equipment** (Problem 1 to 50)   1
2. **Mud Pumps** (Problem 51 to 85)   43
3. **Well Path Design** (Problem 86 to 152)   73
4. **Pressure Calculations** (Problem 153 to 175)   131
5. **Mud Weighting** (Problem 176 to 219)   147
6. **Fluids** (Problem 220 to 258)   181
7. **Hydraulics** (Problem 259 to 297)   209
8. **Drillbit Hydraulics** (Problem 298 to 325)   269
9. **Drilling Tools** (Problem 326 to 344)   295
10. **Hole Cleaning** (Problem 345 to 380)   317
11. **Tubulars** (Problem 381 to 449)   347
12. **Tubular Wear** (Problem 450 to 458)   395
13. **Drilling Operations** (Problem 459 to 473)   413
14. **Cementing** (Problem 474 to 496)   429
15. **Offshore** (Problem 497 to 519)   449
16. **Well Cost** (Problem 520 to 534)   467

Appendix: Useful Conversion Factors   481

Bibliogrpahy   491

Index   509

From the Author's Pen

This book is an expanded and corrected version of the author's "Formulas and Calculation for Drilling Operations – Edition 1" book. I believe this new book is the most comprehensive practical handbook with calculations and solved problems for drilling operations. This central premise of this book is easy to use step-by-step calculations which can be used by students, lecturers, drilling engineers, consultants, software programmers, operational managers, and researchers. The readers are advised to refer to the books in the bibliography as well as the author's other books such as *Horizontal Drilling Engineering, Drilling Optimization and Positive Displacement Motor* (SigmaQuadrant.com) for the underlying theoretical construct.

We took every effort to correct errors and a work with many equations, numbers and calculations, errors may be there due to ignorance. We appreciate your comments and every effort will be taken to correct them if any in the next version.

Houston

Acknowledgments

I would like to thank Dr. Hubert (Kannan) Daniel of John Hopkins University, and Kryon Publishing Services (P) Ltd., Chennai for helping us at various stages of the book.

My heartfelt appreciation goes to my family, friends, and students for their support and encouragement.

Every possible effort has been taken to acknowledge and give appropriate credits in using the copyrighted materials. Should there be any omission, we sincerely apologize for the mistake and suitable corrections will be made at the first possible update.

About the Author

Dr. Robello Samuel is a Chief Technical Advisor and a Halliburton Technology Fellow in the Drilling division of Halliburton and has been working with Halliburton since 1998. He is currently a technical and engineering lead for well engineering and provides guidance and direction on drilling and cross disciplinary projects of major significance. He makes decisions and recommendations that are authoritative and have far reaching impact on research and scientific activities of the company and he serves as a corporate resource, providing technical direction and advice to management in long-range planning for new or projected areas of drilling. He has more than 30 years of multi-disciplinary experience in domestic and international oil/gas drilling and completion operations, management, consulting, software development and teaching and one year of creditable work experience in a manufacturing industry. His skills include practical and theoretical background in onshore and offshore well engineering, design, cost estimates, supervision of drilling and completion operations, personnel and technical review; project management; and creative establishment of project relationships through partnering and innovation. He has been an educator in the USA, Venezuela, Mexico, China, and India and an adjunct Professor at the University of Houston and Texas Tech University, Lubbock for the past 12 years,

About the Author

teaching drilling and well completions and complex well architecture courses.

Dr. Samuel has written or co-written more than 140 journal articles, conference papers, and technical articles. He has given several graduate seminars at various universities and keynote speeches at forums and conferences. Dr. Samuel has been the recipient of numerous awards including the SPE Regional Drilling Engineering Award and the "CEO for A Day (Halliburton)" award. He is presently serving as a review chairman on several journals and professional committees. He has also worked at Oil and Natural Gas Corporation (ONGC), India from 1983 to 1992 as a drilling engineer.

Dr. Samuel, a Society of Petroleum Engineer Distinguished Lecturer, holds B.S. and M.S. (mechanical engineering) degrees from The University of Madurai (Madurai, India) and College of Engineering, Guindy, Anna University (Chennai, India), and M.S. and Ph.D. (petroleum engineering) degrees from Tulsa University. He is also the author of nine drilling books and a forthcoming book *Applied Drilling Engineering Optimization*. Dr. Samuel can be reached via e-mail at robellos@hotmail.com, on the web www.oilwelltechnology.com or by phone at (832)275-8810.

CHAPTER 1

RIG EQUIPMENT

This chapter focuses on the different basic calculations involved in rig equipment and associated calculations are listed below:
- Rig capacity
- Engine calculations
- Rotary torque, HP
- Block line work and capacity
- Offshore environmental forces

The mud pumps will be covered in Chapter 2.

1. Problem 1.1

What two basic components make up a bottom hole assembly (BHA)?
Solution:
Drill collars and Heavy weight drill pipe.

2. Problem 1.2

Explain the three basic elements to drill a well (Fig. 1.1).
Solution:
Weight on bit (WOB), Bit rotation (N), and Flow rate (Q).

3. Problem 1.3

A cross section of a Kelly used in a drillstring may be all of the following except—
   a. square.
   b. round.

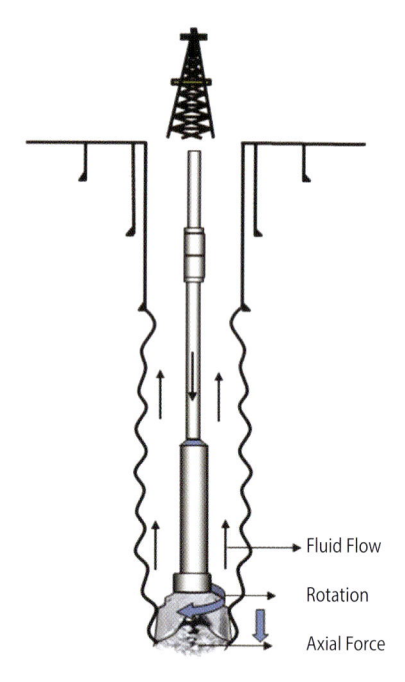

Figure 1.1   Basic elements needed to drill a well

Rig Equipment

   c. elliptical.
   d. six-sided.

Solution:

A Kelly may be all of the following except <u>round</u>.

4. Problem 1.4

Which of the following correctly traces the path of the drilling mud?
a. Mud tank, standpipe hose, annulus, swivel, bit, drill string, mud tank.
b. Swivel, drill string, mud tank, standpipe hose, annulus, bit, mud tank.
c. Mud tank, standpipe hose, swivel, drill string, bit, annulus, mud tank.
d. Mud tank, standpipe hose, swivel, bit, drill string, bit, annulus, mud tank.

Solution:

c. Mud tank, standpipe hose, swivel, drill string, bit, annulus, mud pit

5. Problem 1.5

What are the hardware systems that make up a rotary drilling rig?

Solution:

The hardware systems that make up a rotary drilling rig are:
a. power generation system,
b. hoisting system,
c. drilling fluid circulation system,
d. rotary system,
e. well blowout control systems, and
f. drilling data acquisition and monitoring system.

6. Problem 1.6

Functions of the swivel include—
a. providing a path for the drilling mud to the drill string.
b. providing the power to make the top drive to rotate.
c. supporting the weight of the drill stem.

Solution:
   a. providing a path for the drilling mud to the drill string.

7. Problem 1.7
What is a stand of pipe and a standpipe?
Solution:
Stand of pipe consists of 2 or 3 or single pipes.
Standpipe is a pipe attached to the rig to flow the drilling fluid to the drill hose.

8. Problem 1.8
Horsepower Calculation
Mechanical horsepower (HP) is given as

$$HP = \frac{\text{Force (lbf)} \times \text{distance (ft)}}{550 \times \text{time (sec)}}$$

It is also given as

$$HP = \frac{\text{Force (lbf)} \times \text{velocity (ft/min)}}{33{,}000}$$

Hydraulic horsepower (HHP) is given as

$$HHP = \frac{\text{Flow rate (gpm)} \times \text{pressure (psi)}}{1714}$$

Rotating horsepower is given as

$$HP = \frac{\text{Torque (ft-lbf)} \times \text{speed (rpm)}}{5252}$$

One horsepower = 550 foot-pounds/second = 33,000 ft-lbf/minute, and 33,000 ft-lbf / 6.2832 ft-lbf = 5252.

Other conversion factors are 1 horsepower = 0.0007457 megawatts = 0.7457 kilowatts = 745.7 watts.

9. Problem 1.9

The torque on a motor shaft follows a sinusoidal pattern with a maximum amplitude of 8000 ft-lbf. The torque will always be positive through the cycle. Shaft speed is 200 rpm. Calculate the motor horsepower.

Solution:

Horsepower is given as

$$HP = \frac{2\pi NT}{3300}$$

The torque is calculated using the sinusoidal pattern:

$$T = 2\int_0^\pi 8000 \sin x\, dx$$

$$T = 2[8000 \cos x]_0^\pi$$

Substituting the limits,

$$T = 16{,}000[\cos \pi - \cos 0]_0^\pi = 32{,}000 \text{ ft-lbf}$$

$$HP = \frac{2\pi \times 200 \times 32{,}000}{33{,}000} = 1219 \text{ hp.}$$

10. Problem 1.10

Overall Efficiency of Engines Calculations

The overall efficiency of power generating systems may be defined as

$$\text{Efficiency (\%)} = \frac{\text{output power}}{\text{input power}} \times 100 \text{ or}$$

$$\eta_o = 100 \frac{P_o}{P_i}$$

The output power of an engine is

$$P_o = \frac{2\pi NT}{33{,}000},$$

where

T = output torque in ft-lbs.

N = engine rotary speed in revolution per minute (rpm).
P_o = output power in horsepower, hp.

The input power is expressed as

$$P_i = \frac{Q_f H}{2545},$$

where

Q_f = rate of fuel consumption in lbm/hr.
H = fuel heating value in BTU/lb.
P_i = input power in horsepower, hp.

Fuel consumption can be given as:

$$Q_f = 48.46 \frac{NT}{\eta H} \text{ lb/hr}$$

11. Problem 1.11

Energy Transfer Calculations

Efficiency transfer from the diesel engines to the hoisting system can be given as in Figure 1.2. Due to interrelated equipment, various efficiencies can be used:

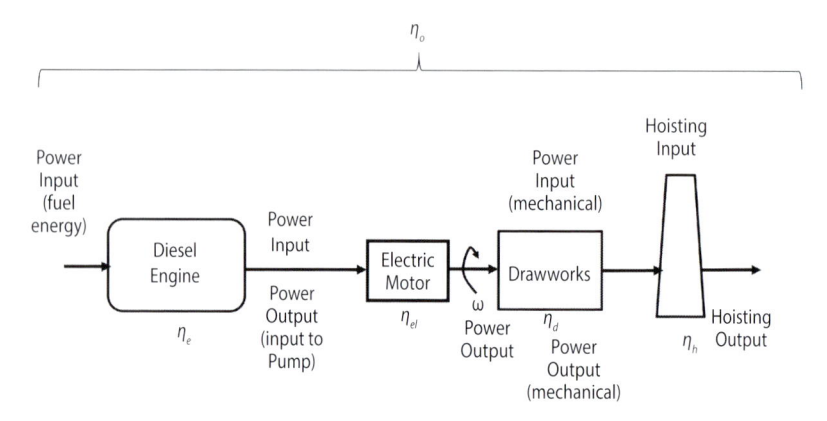

Figure 1.2 Energy transfer

- Engine efficiency, η_e
- Electric motor efficiency, η_{el}
- Drawworks mechanical efficiency, η_m
- Hoisting efficiency, η_h
- Overall efficiency $\eta_o = \eta_e \times \eta_{el} \times \eta_m \times \eta_h$

12. Problem 1.12

A diesel engine's output torque is estimated to be 1870 ft-lbf at 1100 rpm. Determine the output power and efficiency of the engine if the fuel consumption rate is 30 gal/hr. The heating value of diesel oil is 19,000 BTU/lbm.

Solution:

Fuel consumption = 30 (gal/hr) × 7.2 (lbm/gal) = 216 lbm/hr

The output power of an engine is

$$P_o = \frac{2\pi NT}{33{,}000} = \frac{2\pi \times 1100 \times 1870}{33{,}000} = 391.7 \text{ hp}$$

The input power is

$$P_i = \frac{Q_f H}{2545} = \frac{216 \times 19{,}000}{2545} = 1612.6 \text{ hp}$$

Efficiency is given as

$$\eta = 100 \frac{P_o}{P_i} = \frac{391.7}{1612.6} \times 100 = 24.3\%$$

13. Problem 1.13

A drilling rig has three diesel engines for generating the rig power requirement. Determine the total daily fuel consumption for an average engine running at a speed of 900 rpm, an average output torque of 1610 ft-lb, and an engine efficiency of 40%. The heating value of diesel oil is 19,000 BTU/lbm.

Solution:

Given: $N = 900$ rpm, $\eta = 40\%$, $H = 19{,}000$ BTU/lb, and $T = 1610$ ft-lbs

Fuel consumption is

$$Q_f = 48.5 \left(\frac{1610 \times 900}{40 \times 19{,}000} \right) = 92.5 \text{ lb/hr}$$

or

$$Q_f = \frac{92.5 \text{ lb}}{\text{hr}} \times \frac{24 \text{ hr}}{\text{day}} \times \frac{\text{gal}}{7.2 \text{ lb}} = 308 \text{ gal/day}$$

For the three engines, $Q_f(\text{total}) = 308 \times 3 = 924$ gal/day

14. Problem 1.14

Determine the fuel cost (dollars/day) to run an engine at 1800 rpm with 3000 ft-lb of output torque. The engine efficiency at the above rotary speed is 30%. The cost of diesel oil is $1.05/gal, weight is 7.14 lb/gal, and the heating value is 19,000 BTU/lb.

Solution:

The output power for an engine is given as

$$P_o = \frac{2\pi NT}{33{,}000} = \frac{2\pi \times 1800 \times 3000}{33{,}000} = 1028.16 \text{ hp}$$

The input power is expressed in terms of the rate of fuel consumption, Q_f, and the fuel heating value, H.

$$P_i = \frac{Q_f H}{2545} = \frac{Q_f \times 19{,}000}{2545} \text{ hp}$$

Hence,

$$Q_f = \left(\frac{1028.16 \times 2545}{0.30 \times 19{,}000} \right) = 460 \text{ lb/hr}$$

or

$$Q_f = \frac{460 \text{ lb}}{\text{hr}} \times \frac{24 \text{ hr}}{\text{day}} \times \frac{\text{gal}}{7.2 \text{ lb}} = 1533 \text{ gal/day}$$

Cost = 1533 × 1.05 = $1610/day

15. Problem 1.15

Engine Speed = 1500 rpm
Engine Torque = 1800 ft-lbf
Fuel Consumption = 30 gal/hr
Calculate

i. output power of the engine
ii. input power
iii. overall efficiency of the engine

Solution:

The output power developed by the engine is

$$P_o = \frac{2\pi \times 1500 \times 1800}{33{,}000} = 514 \text{ hp}$$

The input power is

$$P_i = \frac{216 \times 19{,}000 \times 779}{33{,}000 \times 60} = 1615 \text{ hp}$$

The overall efficiency of the engine is

$$\eta_o = \frac{514}{1615} \times 100 = 31.83\%$$

16. Problem 1.16

Blocks and Drilling Line

The efficiency of block and tackle system is measured by

$$\eta = \frac{\text{power output}}{\text{power input}} = \frac{P_0}{P_i} = \frac{F_h v_{tb}}{F_f v_f}$$

The output power is defined as

$$P_0 = F_h v_{tb},$$

where

F_h = load hoisted in pounds (buoyed weight of the string + travelling block, compensator, etc.).
v_{tb} = traveling block velocity.

Input power from the drawworks to the fast line is given by

$$P_i = F_f v_f,$$

where

F_f = load in fast line.
v_f = fast line speed.

The relationship between the travelling block speed and the fast line speed is

$$v_{tb} = \frac{v_f}{n}$$

Therefore,

$$\eta = \frac{F_h}{n F_f},$$

where

n = the number of lines strung between the crown block and traveling block.

17. Problem 1.17
Derrick Load Calculations
Static derrick load is calculated as

$$F_s = \frac{n+2}{n} \times F_h$$

Dynamic fast line load is calculated as

$$F_f = \frac{F_h}{En}$$

Dynamic derrick load is given by

$$F_f + F_h + F_{dl}$$

Derrick load is given as

$$F_d = \left(\frac{1+E+En}{En}\right) \times F_h$$

Maximum equivalent derrick load is given as

$$F_{de} = \left(\frac{n+4}{n}\right) \times F_h$$

Derrick efficiency factor is calculated as

$$E_d = \left(\frac{F_d}{F_{de}}\right) = \frac{E(n+1)+1}{E(n+4)},$$

where
F_{dl} = the dead line load.

18. Problem 1.18
Block Efficiency Factor Calculation
Overall block efficiency factor is given as

$$E = \frac{(\mu^n - 1)}{\mu^s n(\mu - 1)},$$

where
μ = the friction factor, ~1.04.
n = number of rolling sheaves (usually $s = n$).

A simplified overall block efficiency factor can be given as

$$E = 0.9787^n$$

19. Problem 1.19
Block Line Strength Calculation

The safety factor (SF) is calculated as

$$SF = \frac{\text{Breaking strength of rope}}{\text{fast line load}}$$

A minimum safety factor of 3.5 is recommended for drilling, and 2.5 is recommended for casing running and fishing operations.

20. Problem 1.20

A rotary rig that can handle triples is equipped with 1200 hp drawworks. The efficiency of the hoisting system is 81%. Determine the time it takes to trip one stand of pipe at a hook load of 300,000 lbs.

Solution:

The efficiency is

$$\eta = \frac{\text{power output}}{\text{power input}} = \frac{P_0}{P_i} = 0.81 = \frac{P_0}{1200}$$

$$P_0 = 927 \text{ hp}$$

But, output power $P_0 = v_{tb} F_h$

Therefore,

$$v_{tb} = \frac{P_0}{F_h} = \frac{927 \text{ hp} \times 33{,}000 \left(\frac{\text{ft-lb}}{\text{min}} / \text{hp} \right)}{300{,}000} = 102 \text{ fpm}$$

and

$$t = \frac{\text{Length}}{v_{tb}} = \frac{90 \text{ ft}}{102 \text{ ft/min}} = 0.84 \text{ min} = 50.5 \text{ sec}$$

21. Problem 1.21

Determine the required drawworks horsepower for a maximum hook load of 400,000 lbs and traveling block velocity of 2 ft/sec. Hoist efficiency is 80% and drawworks efficiency is 70%.

Solution:

Hoisting efficiency is

$$\eta_h = \frac{\text{power output}}{\text{power input}} = \frac{P_0}{P_i} = 0.80 = \frac{2 \times 60 \times 400,000}{P_i \times 33,000}$$

$P_i = 1455$ hp

Output of the drawworks is the input to the hoisting system. So, the drawworks efficiency is

$$\eta_d = \frac{\text{power output}}{\text{power input}} = \frac{P_0}{P_i} = 0.70 = \frac{1455}{P_i}$$

Therefore, the input horsepower for the drawworks = 2080 hp

22. Problem 1.22

A trip is to be made from a depth of 20,000 ft. Determine the minimum time at which the first stand can be tripped out of the hole. Assume the following data:

- Rig: can handle triples
- Drawworks: 1000 hp, efficiency = 75%
- Drilling lines: 12; hoist efficiency = 80%
- Drill pipe: effective weight = 14 lb/ft
- Drill collar: effective weight = 90 lb/ft, length = 1000 ft
- Other suspended loads = 30,000 lbs

Solution:

Well depth at which trip out is made = 20,000 ft
Drill collar length = 1000 ft
Drillpipe length = 20,000 − 1000 = 19,000 ft

Total load at the hook = 30,000 + 14 × (20,000 − 1000) + 90 × 1000
= 386,000 lbf

Fast line load

$$F_f = \frac{F_h}{n\eta} = \frac{386{,}000}{12 \times 0.8} = 40{,}208 \text{ lb}$$

The fast line speed

$$v_f = \frac{1000 \times 0.75 \times 33{,}000}{40{,}208} = 615.5$$

Time to pull the first stand $= \dfrac{90 \times 12}{615.5} = 1.75 \text{ min}$

$$v_{tb} = \frac{P_0}{F_h} = \frac{1000 \text{ hp} \times 0.75 \times 33{,}000 \frac{\text{ft} - \text{lb}}{\text{min}} / \text{hp} \times 0.8}{386{,}000} = 51.28 \text{ fpm}$$

$$t = \frac{L_s}{v_{tb}} = \frac{90 \text{ ft}}{51.28 \text{ ft/min}} = 1.75 \text{ min}$$

23. Problem 1.23

Using the following data
Drill pipe: Effective weight = 16.77 ppf
Drill collar: Effective weight = 94.6 ppf
Length = 300 ft
Depth of the well: 9000 ft
Drilling lines: 10 lines
Hook is hoisted at a velocity of 40 ft/min

i. Calculate the velocity of the fast line.
ii. Calculate line pull at the drawworks, assuming 2.2% frictional losses per line.
iii. Calculate the horsepower of drawworks.

Rig Equipment

Solution:
Hook is hoisted at a velocity of 40 fpm
Velocity of the fast line = v_f = 40 × 10 = 400 fpm
Hook load = F_h = 94.6 × 300 + 16.77 × (9000–300) = 174,279 lbf
2.2% frictional losses per line results in 78% efficiency:

$$F_f = \frac{174{,}279}{10 \times 0.78} = 22{,}343 \text{ lbf}$$

Horsepower of drawworks = $\dfrac{22{,}343 \times 400}{33{,}000}$ = 271 hp

24. Problem 1.24

A drilling rig has 10 lines strung between crown and traveling blocks. 9 ⅝" – 47 ppf casing operation is planned for a depth of 10,000 ft. Neglecting the buoyancy effects, calculate the equivalent mast load, derrick efficiency factor. If an overpull margin of 30% casing weight for stuck conditions is allowed, calculate the equivalent mast load and derrick efficiency factor.

Solution:
Total hook load = F_h = 47 × 10,000 = 470,000 lbf
Deadline load is calculated assuming an efficiency factor of 81%.

$$F_d = \left(\frac{1 + 0.81 + 0.81 \times 10}{0.81 \times 10}\right) \times 470{,}000 = 575{,}025 \text{ lbf}$$

The equivalent derrick load and derrick efficiency factor can be calculated, respectively, as follows:

$$F_{de} = \left(\frac{n+4}{n}\right) \times F_h = \left(\frac{10+4}{10}\right) \times 470{,}000 = 658{,}000$$

$$E_d = \frac{E(n+1)+1}{E(n+4)} = \frac{0.81(10+1)+1}{0.81(10+4)} = 87.4\%$$

or

$$E_d = \frac{658{,}000}{805{,}035} = 87.4\%$$

When an overpull margin of 30% used the hook load to be handled is 1.3 × 470,000 = 611,000 lbf, but the derrick efficiency factor does not change as it is independent of the hook load.

25. Problem 1.25

Calculate the derrick load and the equivalent derrick load using the following data:

   i. Drawworks input power: 500 hp
   ii. Hook load needed to be lifted: 400 kips
   iii. Number of lines strung: 12

Solution:

The derrick load is

$$F_d = \left(\frac{1+E+En}{En}\right) W = \left(\frac{1+0.77+0.77 \times 12}{0.77 \times 12}\right) 400{,}000 = 476.6 \text{ kips}$$

The equivalent derrick load is given as

$$F_{de} = \left(\frac{n+4}{n}\right) W = \left(\frac{16}{12}\right) 400{,}000 = 533.3 \text{ kips}$$

Derrick efficiency is

$$\eta_d = \frac{476.6}{533.3} \times 100 = 89.39\%$$

26. Problem 1.26

A well is being drilled with a drawworks that is capable of providing a maximum input power of 1000 hp. Maximum load expected to hoist is 250,000 lbf. Twelve lines are strung between the traveling block and crown block, and the deadline is anchored to a derrick leg. The drilling line has a nominal breaking strength of 51,200 lbf.

   i. Calculate the dynamic tension in the fast line.

ii. Calculate the maximum hook horsepower available.
iii. Calculate the maximum hoisting speed.
iv. Calculate the derrick load when upward motion is impending.
v. Calculate the maximum equivalent derrick load.
vi. Calculate the equivalent derrick efficiency factor.
vii. What is the maximum force that can be applied to free a stuck pipe that has a strength of 400,000 lbf? Assume safety factors of 2.0 for the derrick, drillpipe, and drilling line.

Solution:

i. The efficiency for 12 lines = 0.782
The dynamic fast line tension is

$$F_f = \frac{F_h}{E n} = \frac{250{,}000}{0.782 \times 12} = 26{,}641 \text{ lbf}$$

ii. Maximum hook horsepower available = P_h = 0.782 × 1000 = 782 hp

iii. The maximum hoisting speed is

$$v_b = \frac{0.782 \times 782 \times 33{,}000}{250{,}000} = 80.7 \text{ fpm}$$

iv. The maximum derrick load is

$$F_d = \left(\frac{1 + 0.782 + 0.782 \times 12}{0.782 \times 12}\right) 250{,}000 = 297{,}474 \text{ lbf}$$

v. The maximum equivalent derrick load is

$$F_{de} = \left(\frac{12 + 4}{12}\right) 250{,}000 = 333{,}333 \text{ lbf}$$

vi. The maximum equivalent derrick efficiency factor is

$$E_d = \left(\frac{297{,}474}{333{,}333}\right) = 89\%$$

vii. Checking for pipe, rope, and derrick failure

based on derrick: $\left(\dfrac{333{,}333}{2} \times \dfrac{12}{16}\right) = 125{,}000$ lbf (lowest)

based on pipe: $\left(\dfrac{400{,}000}{2}\right) = 200{,}000$ lbf

based on rope: $\left(\dfrac{51{,}200}{2} \times 0.782 \times 12\right) = 240{,}230$ lbf

27. Problem 1.27

Calculate the block efficiency for a block with 12 lines strung using simplified method as well as with a friction factor of 1.04.

Solution:

A simplified overall block efficiency factor is given as

$$E = 0.9787^n$$

$$E = 77.23\%$$

The overall block efficiency factor is

$$E = \dfrac{(\mu^n - 1)}{\mu^s n(\mu - 1)} = \dfrac{(1.04^{12} - 1)}{1.04^{12} \times 12\,(12-1)} = 0.782$$

$$E = 78.2\%$$

28. Problem 1.28

Given the following data:
Target well depth = 26,000 ft
Drillpipe effective weight = 20 lbs/ft
Drillcollar effective weight = 100 lbs/ft
Maximum length of drillcollar required = 1000 ft
Additional suspended loads to drill string = 60,000 lbs
Number of lines strung between crown block and traveling block
=10 lines
Drawwork maximum input power = 2000 hp

Rig Equipment

Drawwork efficiency = 75%
Rig can handle triples (93 ft)
Time to handle a stand once out of hole = 1 min
Drilling cost = $15,200/day

a. Determine cost of tripping out from 26,000 ft. Assume that maximum allowed speed of pulling pipe out is that of pulling the first stand.
b. Determine cost of tripping out with the maximum allowed speed of the first stand up to 20,000 ft and thereafter 1.25 ft/min. Use time to handle a drillcollar stand = 2 minutes.
c. If rig can handle quadruples (124 ft), what would be the cost in part a?

Solution:

A. Given: $n = 10$, $P_i = 2000$ $\eta_{dw} = 0.75$
Maximum length of the drill collar required = 1000 ft
Total hook load = DP weight + DC weight + other hanging weight
= 25,000 × 20 + 1000 × 100 + 60,000 = 660 kips
Drawworks output is $P_o = 2000 \times 0.75 = 1500$ hp

$$V_{tb} = \frac{P_o}{F_h} = \frac{0.81 \times 1500 \text{ hp} \times 33,000 \frac{\text{ft}-\text{lb}}{\text{min}}/\text{hp}}{660,000} = 60.75 \text{ fpm}$$

$$t = \frac{L_s}{V_{tb}} = \frac{93 \text{ ft}}{60.75 \text{ ft/min}} = 1.53 \text{ min} = 93 \text{ sec}$$

Total number of stands = $\frac{26,000}{93} = 280$ stands

Total time consumed = 280 × 1.53 + 280 × 1 = 11.80 hrs

Total cost = $\frac{15,200}{24} \times 11.80 = \$7474$

B. If variable speed is used, the total cost will be $ 6968.78

C. Time equals

$$t = \frac{L_s}{V_{tb}} = \frac{124 \text{ ft}}{60.75 \text{ ft/min}} = 2.04 \text{ min} = 122.5 \text{ sec}$$

$$\text{Total number of stands} = \frac{26{,}000}{124} = 210 \text{ stands}$$

Total time consumed = 210 × 2.04 + 210 × 1 = 10.64 hrs
Total cost = $6738.66
Savings of $738.33 is realized for this particular trip.

For 26,000 ft, with the given efficiency and day rate, pulling out in quadruples may not save much compared to the cost of the rig to handle stands with four singles.

29. Problem 1.29

A well is to be drilled to a depth of 20,000 ft. The effective weight of drillpipe is 20 lbs/ft, and that of the drill collar is 120 lbs/ft. If the traveling blocks and swivel are rated at a maximum capacity of 200 tons, determine the required drawworks horsepower to trip at 10 ft/sec from a depth of 19,000 ft. The drill collar length is 1000 ft, and the hoisting system has 10 lines strung between traveling block and crown block (efficiency = 80%).

Solution:

Drill pipe length = 18,000 ft
Drill collar length = 1000 ft
Hook load = 18,000 × 20 + 1000 × 120 = 480,000 lb

$$P_o \text{ for hoist system} = \frac{480{,}000 \times (10 \text{ ft/sec} \times 60 \text{ sec/min})}{33{,}000} = 8727 \text{ hp}$$

$P_i = 8727/0.8 = 10{,}909$ hp

30. Problem 1.30

Using the following data calculate the safety factor for a 1.5-inch block line with a breaking strength of 228,000 lbf. Assume a block efficiency of 82% for 12 lines in the sheave.

Mud weight = 9.9 ppg
Casing length = 9500 ft, 29 ppf
Block weight = 35 kips

Solution:

Buoyed weight of the casing $= \left(1 - \dfrac{9.9}{65.5}\right) \times 29 = 24.62$ ppf

Total weight of the casing $= 24.62 \times 9500 = 233{,}890$ lb
Total hook load $= 233{,}890 + 35{,}000 = 268{,}890$ lb
The fast line is

$$F_f = \dfrac{F_h}{E_n} = \dfrac{268{,}890}{0.82 \times 12} = 27{,}326 \text{ lbf}$$

The safety factor is

$$SF = \dfrac{\text{Breaking strength of rope}}{\text{fast line load}} = \dfrac{228{,}000}{27{,}326} = 8.3$$

31. Problem 1.31
Ton-Miles (TM) Calculations
Round-trip ton miles is calculated as follows:

$$T_f = \dfrac{L_h(L_s + L_h)W_{dp} + 4L_h\left(W_b + \dfrac{1}{2}W_1 + \dfrac{1}{2}W_2 + \dfrac{1}{2}W_3\right)}{10{,}560{,}000} \text{ TM,}$$

where
L_h = measured depth of the hole or trip depth, ft.
L_s = length of the stand, ft.
W_{dp} = buoyed weight of the drillpipe per foot, ppf.
W_b = weight of the block, hook, etc., lbs.
W_1 = excess weight of drill collar in mud, lbs (buoyed drill collar weight in mud—buoyed drillpipe weight of same length in mud).
W_2 = excess weight of heavy weight pipe in mud, lbs (buoyed heavy weight pipe in mud–buoyed drillpipe weight of same length in mud).
W_3 = excess weight of miscellaneous drilling tools in mud, lbs (buoyed miscellaneous drilling tools in mud—buoyed drillpipe weight of same length in mud).

32. Problem 1.32
Drilling Ton-Miles Calculations
When a hole is drilled only once without any reaming
$$T_d = 2(T_{i+1} - T_i)$$
When a hole is drilled with one time reaming
$$T_d = 3(T_{i+1} - T_i)$$
When a hole is drilled with two times reaming
$$T_d = 4(T_{i+1} - T_i),$$
where
T_i = round trip ton-mile calculated, from depth i.

Coring Ton-Miles Calculations
When a hole is drilled only once without any reaming,
$$T_d = 2(T_{i+1} - T_i)$$

Casing Ton-Miles Calculations
$$T_f = \frac{1}{2}\left(\frac{L_h(L_s + L_h)W_c + 4W_b L_h}{10{,}560{,}000}\right) \text{TM},$$

where
W_c = buoyed weight of the casing per foot, ppf.

33. Problem 1.33
Calculate the ton-mile for a round trip from 8000 ft with 450 ft of drill collar with a weight of 83 ppf. The weight of the drillpipe is 19.5 ppf. The mud density is 9.6 ppg. Assume the stand length to be 93 ft and the total block weight is 40,000 lbs.
Solution:
Using the following data:
- L_h = 8000 ft
- L_s = 90 ft

- W_b = 40,000 lbs

Buoyancy factor = $\left(1 - \dfrac{9.6}{65.5}\right) = 0.853$

W_{dp} = 19.5 × 0.853 = 16.64 ppf
Buoyed weight of drill collar – buoyed weight of drillpipe
= 0.853 × 450 (83–19.5) = 24,387 lbs
Substituting the values

$$T_f = \dfrac{8000(93 + 8000)16.64 + 4 \times 8000(40,000 + 24,387)}{10,560,000} = 297 \text{ TM}$$

34. Problem 1.34

Compute the ton-miles required to drill from 15,000 to 15,900 ft for the following conditions:

Drillpipe = 5 in × 19.5 ppf
Drill collars = 6 ½ in × 2 ½ in – 660 ft
Heavy weight drillpipe = 5 in × 3 in × 50 ppf – 270 ft
Expected bit performance = 300 ft/ bit
Mud used is 12.5 ppg for 15,000 ft; 12.7 ppg for 15,300 ft; 12.9 ppg for 15,600 ft; and 13.0 ppg for 15, 900 ft.
Traveling block weight = 40 kips
Average length of the stand = 93 ft
Additionally, a 15-ft core was cut from 15,900 ft

Solution:

Total Operations:

T_1: Drilling from 15,000 ft to 15,300 ft with mud weight 12.5 ppg

T_2: Since the bit's performance is 350 ft, it can drill another 50 ft with 12.7 ppg mud

T_3: Pull out the bit/change the bit and run in to 15,350 for drilling with 12.7 ppg mud

T_4: While running in reaming one time from 15,097 to 15,127 ft with 12.7 ppg mud

T_5: Drilling from 15,350 ft to 15,600 ft with mud weight 12.7 ppg

T_6: Pull out the bit/change the bit with core bit and run in to 15,600 for coring with a mud weight of 12.7 ppg

T_7: Drilling from 15,600 ft to 15,900 ft with mud weight 13 ppg

T_8: Round trip from 15,900 ft to run the core barrel in 13 ppg mud

T_9: Coring from 15,900 ft to 15,915 ft with 13 ppg mud

1. Round trip ton-mile from 15,300 ft with 12.7 ppg:

$$T_f = \frac{L_h(L_s + L_h)W_{dp} + 4L_h\left(W_b + \frac{1}{2}W_1 + \frac{1}{2}W_2 + \frac{1}{2}W_3\right)}{10,560,000}$$

$$BF = \left(1 - \frac{12.5}{65.5}\right) = 0.809$$

Effective weight of the drill collar

$$= \left(\frac{\frac{\pi}{4} \times (6.5^2 - 2.5^2)}{0.2945}\right) \times BF = 77.67 \text{ ppf}$$

Effective weight of drill pipe = $19.5 \times BF = 15.78$ ppf
Effective weight of heavy weight drill pipe = $50 \times BF = 40.45$ ppf

$$T_{15,300} = \frac{15,300(93+15,300)15.78 + 4}{10,560,000}$$

$$\times \frac{15,300\left(40,000 + \frac{1}{2}6663 + \frac{1}{2}40,857\right)}{10,560,000} = 722 \text{ TM}$$

$$T_{15,000} = \frac{15,000(93+15,000)15.78 + 4}{10,560,000}$$

$$\times \frac{15,000\left(40,000 + \frac{1}{2}6663 + \frac{1}{2}40,857\right)}{10,560,000} = 701 \text{ TM}$$

1. Drilling from 15,000 ft to 15,300 ft with mud weight 12.5 ppg:
Drilling ton-mile for 15,000 ft to 15,300 ft = 2(722 − 701)= T_1
= 42 TM

$T_{15,350} = 723$ TM; $T_{15,300} = 720$ TM

2. Drilling another 50 ft to 15,350 ft with 12.7 ppg mud:
   Drilling to 15,350 ft with 12.7 ppg = 2(723 − 720) T_2 = 7 TM
3. Round trip with 12.7 ppg mud:
   Round trip ton-mile from 15,350 ft with mud weight 12.7 ppg
   $T_3 = 723$
4. While running in reaming one time from 15,097 to 15,127 ft with 12.7 ppg mud:
   Reaming ton-mile T_4 = 6 TM
5. Drilling from 15,350 to 15,600 ft with mud weight 12.7 ppg:

$$T_{15,600} = 741 \text{ TM}; T_{15,350} = 723 \text{ TM}$$

   Drilling ton-mile = 2(741 − 723) = T_5 = 35 TM
6. Pull out the bit/change the bit with core bit and run in to 15,600 for coring with a mud weight of 12.7 ppg:
   Round trip ton-mile from 15,600 ft T_6 = 741 TM

$$T_{15,900} = 760 \text{ TM}; T_{15,600} = 741 \text{ TM}$$

7. Drilling from 15,600 ft to 15,900 ft with mud weight 13 ppg:
   Drilling from 15,600 ft to 15,900 ft = 2(760 − 741) = T_7 = 39 TM
8. Round trip from 15,900 ft to run the core barrel in 13 ppg mud:
   Round trip from 15,900 ft for coring is

   Round trip ton-mile T_8 = 760 TM

9. Coring from 15,900 ft to 15,915 ft with 13 ppg mud:
   Coring from 15,600 to 15,615 ft with mud weight 13 ppg is

$$T_{15,915} = 760.4 \text{ TM}; T_{15,900} = 760.3 \text{ TM}$$

   Coring ton-mile = 2(760.4 − 760.3) = T_9 = 0.2 TM
   Total ton miles to drill, ream and core from 15,000 to 15,915 = 2353 TM

35. Problem 1.35

Determine the casing ton-miles for a casing run of 7 inch 29 ppf casing to 9,500 ft in a mud of 9.9 ppg. Travelling block weight is 35 kips.

Solution:

Buoyed weight of the casing $= \left(1 - \dfrac{9.9}{65.5}\right) \times 29 = 24.62$ ppf

Assuming 40 ft of casing, casing ton-miles can be calculated as

$$T_f = \dfrac{1}{2}\left(\dfrac{9500(40+9500) \times 24.62 + 4 \times 35{,}000 \times 9500}{10{,}560{,}000}\right) = 169 \text{ ton-miles}$$

36. Problem 1.36

Crown Block Capacity

The crown block capacity required to handle the net static hook-load capacity can be calculated using the following formula:

$$R_c = \dfrac{(H_L + S)(n+2)}{n},$$

where

R_c = required crown block rating (lbs).
H_L = net static hook load capacity (lbs).
S = effective weight of suspended equipment (lbs).
n = number of lines strung to the traveling block.

37. Problem 1.37

What minimum drawworks horsepower is required to drill a well using 10,000 ft of 4-½" OD 16.6# drillpipe and 50,000 lbs of drill collars? Efficiency is 65%.

Solution:

Drill string weight (air weight) = 50,000 lbs + (10,000 ft) (16.6 lb/ft)
= 216,000 lbs

$$\text{Hook horsepower} = \dfrac{(216{,}000 \text{ lbs})(100 \text{ ft/min})}{33{,}000} = 655 \text{ hp}$$

$$\text{Required minimum drawworks horsepower rating} = \dfrac{655}{0.65} = 1007 \text{ hp}$$

38. Problem 1.38
Line Pull Efficiency Factor Calculations

Hoisting engines should have a horsepower rating for intermittent service equal to the required drawworks horsepower rating divided by 85% efficiency. Drawworks also have a line pull rating efficiency depending on the number of lines strung between crown block and traveling block (see Table 1.1).

39. Problem 1.39

What line pull is required to handle a 500,000 lb casing load with 10 lines strung?
Solution:

$$\text{Line pull} = \frac{500{,}000}{(10)(0.810)} = 61{,}728 \text{ lbs}$$

40. Problem 1.40
Rotary Power Calculations

Rotary horsepower can be calculated as follows:

$$H_{rp} = \frac{2\pi NT}{33{,}000},$$

where
N = rotary table speed (RPM).
T = torque (ft-lbs).

Table 1.1 Line pull efficiency factor

No. of lines	Efficiency factor
6	0.874
8	0.842
10	0.811
12	0.782

41. Problem 1.41

The drillpipe must transmit rotating power to the bottom hole assembly (BHA) and the bit. The following example illustrates the calculation of the horsepower that the drillpipe can transmit without torsion failure. For example, if a drillpipe has a maximum recommended make-up torque of 20,000 ft-lbs, then what is the rotary horsepower that can be transmitted at 100 rpm?

Solution:

Rotary horsepower is given as

$$H_{rp} = \frac{(2\pi \times 100 \times 20{,}000)}{33{,}000} = 381 \text{ hp}$$

The empirical relationship to estimate the rotary horsepower requirements is

$$H_{rp} = F \times N,$$

where
F = torque factor, ft-lbf.
N = rotary speed (RPM).

The torque factor (F) is generally estimated as follows:

- $F = 1.5$ to 1.75 for shallow holes less than 10,000 ft with light drill string.
- $F = 1.75$ to 2.0 for 10,000 – 15,000 ft wells with average conditions.
- $F = 2.0$ to 2.25 for deep holes with heavy drill string.
- $F = 2.0$ to 3.0 for high-torque.

The above empirical estimates are subject to many variables but have proved to be reasonable estimates of rotary requirements. However, for highly deviated wells, torque/H_{rp} requirements must be closely calculated using available computer software programs.

42. Problem 1.42

Derive the equations to calculate the length, L, in feet of the drill line in a drawworks reel drum (Fig. 1.3) with the following dimensions:

Rig Equipment

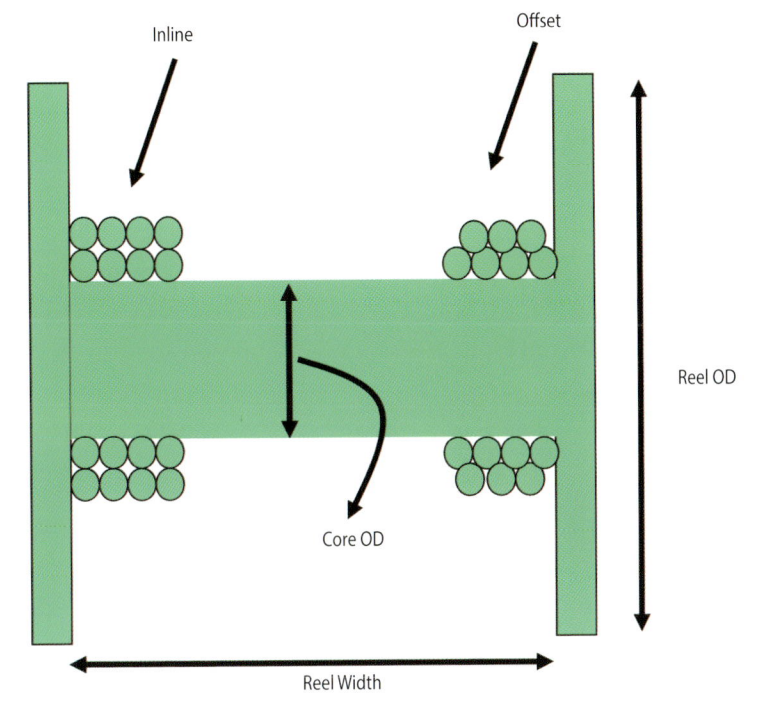

Figure 1.3 Drawworks reel drum

- Reel diameter: D_R in.
- Reel width: W in.
- Core diameter: D_c in.
- Diameter of the drill line: d in.
- Wrapping types: offset and inline

A drilling engineer is planning to estimate the number of times the drill line can be slipped before cutting. Establish an equation to estimate the number of times 200 ft of drill line can be slipped. The initial drill line length on the drum is m% of the drawworks drum capacity. The maximum drill line length is not to exceed n% of the drawworks drum capacity.

Solution:

Inline:

first layer = $L_1 = \dfrac{\pi}{12}(D_c + d)\dfrac{W}{d}$ ft second layer = $L_2 = \dfrac{\pi}{12}(D_c + 3d)\dfrac{W}{d}$ ft

nth layer = $L_n = \dfrac{\pi}{12}[D_c + (2n-1)d]\dfrac{W}{d}$ ft,

where
n = total number of layers on the drum.

$$L_T = L_1 + L_2 \ldots + L_n = \dfrac{\pi}{12}[D_c + d + D_c + 3d \ldots + D_c + (2n-1)d]\dfrac{W}{d}$$
$$= \dfrac{\pi}{12}\dfrac{W}{d}(nD_c + n^2 d)\, \text{ft}$$

From the geometry, the number of laps is

$$n = \dfrac{D_r - D_c}{2d}$$

Offset:

$$h' = \sqrt{d^2 - \left(\dfrac{d}{2}\right)^2} = \dfrac{\sqrt{3}}{2}d$$

first layer = $L_1 = \dfrac{\pi}{12}(D_c + d)\dfrac{W}{d}$ ft;

second layer = $L_2 = \dfrac{\pi}{12}(D_c + d + \sqrt{3}d)\left(\dfrac{W}{d} - 1\right)$ ft

3rd layer = $L_3 = \dfrac{\pi}{12}(D_c + d + 2\sqrt{3}d)\dfrac{W}{d}$ ft

nth layer = $L_n = \dfrac{\pi}{12}[D_c + d + (n-1)\sqrt{3}d]\left(\dfrac{W}{d} - f(n)\right)$ ft,
where
n = total number of layers on the drum.
$f(n) = 0$ if n is odd.
$f(n) = 1$ if n is even.

$$L_T = L_1 + L_2 \ldots + L_n = \dfrac{\pi}{12}\left[(D_c + d)\dfrac{W}{d} + (D_c + d + \sqrt{3}d)\left(\dfrac{W}{d} - 1\right) \ldots \right.$$
$$\left. + \dfrac{\pi}{12}[D_c + d + (n-1)\sqrt{3}d]\left(\dfrac{W}{d} - f(n)\right)\right]\text{ft}$$

$f(n) = 0$ if n is odd and $f(n) = 1$ if n is even

Adding and simplifying,

$$1+2+3... = \frac{(n-1)n}{2}; 1+3+5... = (n-2)^2$$

$$L_T = \frac{\pi}{12}\frac{W}{d}\left(nd + nD_c + \frac{(n-1)n}{2}\sqrt{3}d\right)$$

$$-\frac{\pi}{12}\left[(n-2)d + (n-2)D_c + (n-2)^2\sqrt{3}d + \left(d + D_c + (n-1)\sqrt{3}d\right)f(n)\right] \text{ft}$$

From the geometry,

$$\frac{D_r - D_c}{2} = d + (n-1)\frac{\sqrt{3}}{2}d$$

Therefore,

$$n = (D_r - D_c - 2d)\frac{1}{d\sqrt{3}} + 1$$

Slip and cut:
Usually three or four times the drill line is slipped into the drum before cutting and discarding certain amount of drill line from the drum so that the lap points are removed.

Number of slips = $\left(\frac{(n-m) \times L_T}{200 \times 100}\right)$ (expressed in integer)

43. Problem 1.43

Determine the total time it takes to make a trip at 11,700 ft depth.
Rig: Can handle triples
Drill Pipe: Length = 10,700 ft; Effective weight = 15 lb/ft
Drill collar: Effective weight = 80 lb/ft
Maximum available drawworks horsepower = 1200 hp
Drawworks consists of 12 lines strung between crown block and traveling block
Block and tackle efficiency = 85%
Drawworks efficiency = 75%
Rig crew can break a joint or connect a joint in average time of 18 seconds.

Assume average stand length = 90 ft
Solution:
Given: $n = 12$, $\eta = 0.85$, $P_i = 1200$ hp $\eta_{dw} = 0.75$
Total hook load = DP weight + DC weight + other hanging weight
= 10,700 × 15 + 1000 × 80 = 240,500 lbs

Drawworks output is
$P_o = 2000 \times 0.75 = 1500$ hp

$$v_{tb} = \frac{P_o}{F_h} = \frac{0.85 \times 1500 \text{ hp} \times 33,000 \frac{\text{ft} - \text{lb}}{\text{min}}/\text{hp}}{240,500} = 175 \text{ fpm}$$

$$t = \frac{L_s}{v_{tb}} = \frac{90 \text{ ft}}{175 \text{ ft/min}} = 0.51 \text{ min} = 31 \text{ sec}$$

Total number of stands = $\frac{11,700}{90} = 130$ stands

Total time consumed = $130 \times 0.51 + 130 \times \frac{18}{60} \approx 106$ min

44. Problem 1.44

You are given the following data:
Drawworks output horsepower = 800 hp
Trip to be made from 20,000 ft well depth
Effective drill pipe weight = 15 lbs/ft
Effective drill collar weight = 80 lbs/ft
Maximum required weight on drill bit is 80,000 lbs, between 15,000 and 20,000 ft well depth
Hoist efficiency = 85%
Drawworks efficiency = 80%

i. Determine the amount of time it takes to trip out of hole, if the speed is maintained at a value equal to the maximum speed that can be obtained for pulling the first stand of pipe. Assume rig crew can handle a stand in average of 20 seconds. Rig can handle doubles.

Rig Equipment

ii. Determine the total cost of diesel oil used in tripping out.
Assume diesel engine efficiency = 40%
Drawworks efficiency = 80%
Diesel oil heating value = 18,000 BTU/lb
Weight = 7.2 lbs/gal
Cost = $1/gal.

Solution:

Given: $n = 12$, $\eta = 0.85$, $P_i = 1200$ hp $\eta_{dw} = 0.75$

Assuming air weight the 15,000 ft to 20,000 the drillcollar length needed with safety factor 1 is 80,000/80 = 1000 ft

Assuming air weight

Length of the drillcollar needed = 80,000/80 = 1000 ft

So the length of the drillpipe = 20,000 − 1000 = 19,000 ft

Total hook load = DP weight + DC weight + other hanging weight
= 19,000 × 15 + 1000 × 80 = 365,000 lbs.

Drawworks output is $P_o = 800$ hp

$$V_{tb} = \frac{P_o}{F_h} = \frac{800 \text{ hp} \times 33,000 \frac{\text{ft}-\text{lb}}{\text{min}}/\text{hp}}{365,000} = 72.32 \text{ fpm}$$

$$t = \frac{L_s}{V_{tb}} = \frac{60 \text{ ft}}{72.32 \text{ ft/min}} = 0.8296 \text{ min} = 50 \text{ sec}$$

Total number of stands = $\frac{20,000}{60}$ = 334 stands

Total time consumed = 334 × 0.83 + 334 × $\frac{18}{60}$ ≈ 377 min

P_i to drawworks = 800/0.8 = 1000 hp
P_i to diesel engine = 1000/0.4 = 2500 hp

$$2500 = \frac{Q_f \times H}{2545}$$

Q_f = 354 lb/hr

Weight of diesel for one trip = 354 × 377/60 = 2227 lb = 2227/7.2
= 310 gal

Cost for 1 trip = $310

45. Problem 1.45

Given the following information:
Total well depth = 20,000 ft
Drillcollar weight = 150 lb/ft
Drillcollar total length = 1000 ft
Drillpipe weight = 20 lb/ft
Average drag force on drill string while being tripped out at 60 ft/mm = 3.5 lb/ft
Mechanical efficiency between power generating engine output and drawworks output = 70%
Ten lines strung between crown and traveling blocks
Diesel engine mechanical efficiency = 50%
Diesel oil heating value = 20,000 BTU/lb
Diesel oil weight = 7.2 lb/gal
Diesel cost = $2.00/gal
Determine engine fuel cost ($/day)

Solution:

$$\text{Input } H_{HP} = \frac{QP}{1714} = \frac{360 \times 3400}{1714 \times 0.85} = 840 \text{ hp}$$

Or

$$\frac{QP}{1714} = \frac{324 \times 3400}{1714 \times 0.85 \times 0.9} = 840 \text{ hp}$$

$$H_{HP} = \frac{840}{0.55} = 1527 \text{ hp}$$

$$Q_f = \frac{1527 \times 2545 \times 24}{20,000 \times 0.5 \times 7.2} = 1177 \text{ gal/day}$$

Consider the following:
Engine input (fuel energy) → engine output (input to pump) → mechanical pump output (input to hydraulic) → hydraulic output.

Overall efficiency can also be used to find the engine input to the pump output horsepower.

46. Problem 1.46
Offshore Vessel Calculations
Offshore vessels' sides and the types of vessel motions encountered are shown in Figure 1.4 and Figure 1.5, respectively.

Draft is the depth of the vessel in the water, whereas the freeboard is the distance above the water. They are shown in Figure 1.6.

Environmental Forces
Environmental forces are shown in Figure 1.7. These forces include the following:

- Wind force
- Wave force
- Current force

Wind force is given by

$$F_w = 0.00338 \, V_w^2 \, C_h \, C_s \, A,$$

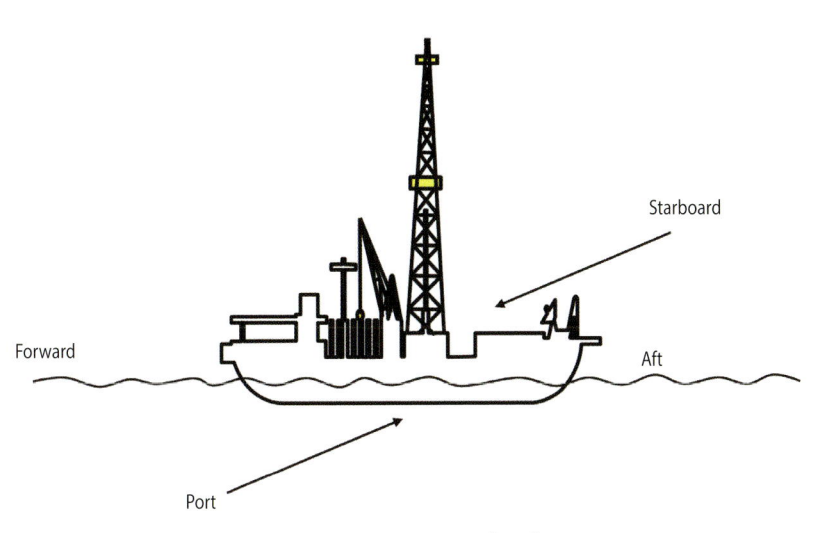

Figure 1.4 Vessel sides

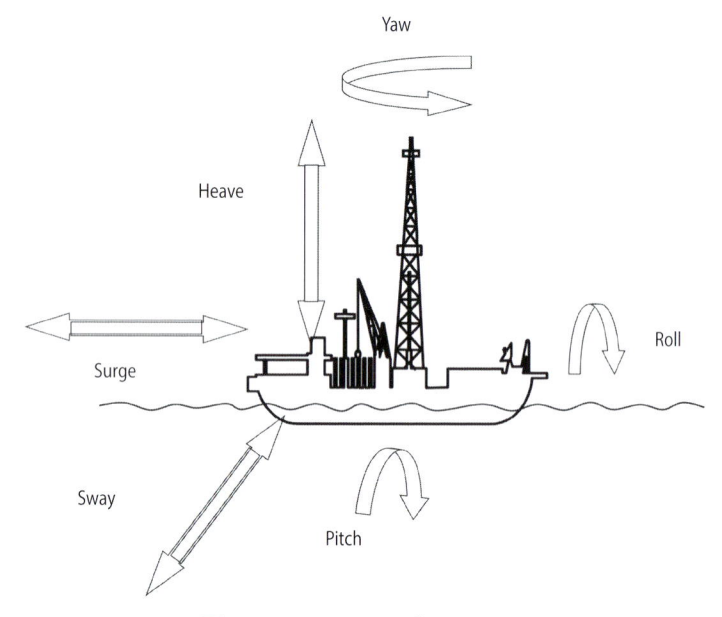

Figure 1.5 Vessel motion

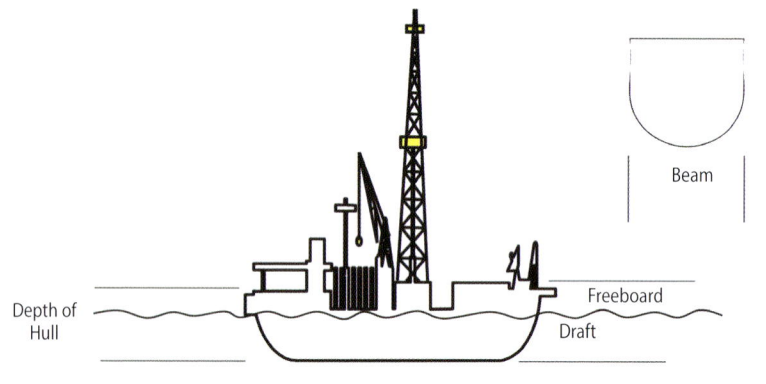

Figure 1.6 Draft and freeboard

where
F_A = wind force, lbf.
V_A = wind velocity, knots.
C_s = shape coefficient.
C_h = height coefficient.
A = projected area of all exposed surfaces, sq. ft.

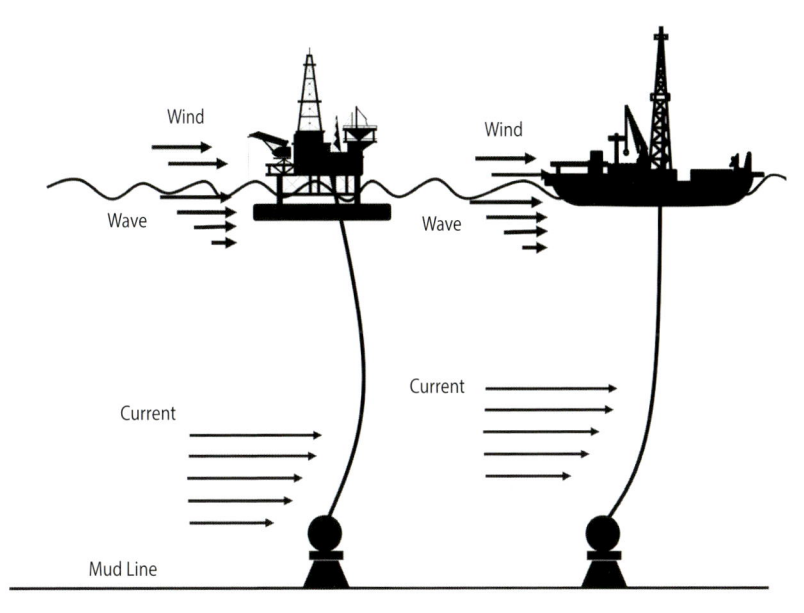

Figure 1.7   Environmental forces

Table 1.2   Shape coefficients

Shape	Cs
Cylindrical shapes	0.5
Hull (surface type)	1.0
Deck house	1.0
Isolated structural shapes (cranes, beams, etc.)	1.0
Under deck areas (smooth surfaces)	1.0
Under deck areas (exposed beams, girders)	1.3
Rig derrick (each face)	1.25

Shape coefficients can be estimated from Table 1.2, and height coefficients can be estimated from Table 1.3.

Current force is calculated from

$$F_c = g_c V_c^2 C_s A,$$

where
F_c = current drag force, lbf.

Table 1.3 Height coefficients

From to (ft)	Ch
0–50	1.0
50–100	1.1
100–150	1.2
150–200	1.3
200–250	1.37
250–300	1.43
300–350	1.48
350–400	1.52
400–450	1.56
450–500	1.6

V_c = current velocity, ft/sec.
C_s = drag coefficient same as wind coefficient.
A = projected area of all exposed surfaces, sq. ft.

The following are the facts regarding the nautical mile:

- One nautical mile is one minute of latitude
- A speed of one nautical mile per hour is termed the Knot
- Determined by latitude (not longitude)
- 1 min of latitude = 1 nautical mile – 6076 ft
- 1 degree of latitude = 60 NM

Wave force is calculated for various conditions.

Bow forces are under different conditions, depending on the wave period.

$$\text{If } A > 0.332\sqrt{L}$$

$$F_{bow} = \frac{0.273 H^2 B^2 L}{T^4}$$

$$\text{If } A < 0.332\sqrt{L}$$

$$F_{bow} = \frac{0.273H^2B^2L}{\left(0.664\sqrt{L} - A\right)^4},$$

where
F_{bow} = bow force, lbf.
  A = wave period, sec.
  L = vessel length, ft.
  H = significant wave height, ft.
  B = vessel beam length, ft.

Beam forces are as follows:

If $T > 0.642\sqrt{B+2D}$

$$F_{beam} = \frac{2.10H^2B^2L}{A^4}$$

If $A < 0.642\sqrt{B+2D}$

$$F_{beam} = \frac{2.10H^2B^2L}{\left(1.28\sqrt{B+2D} - A\right)^4}$$

where
F_{beam} = bow force, lbf.
  A = wave period, sec.
  L = vessel length, ft.
  H = significant wave height, ft.
  B = vessel beam length, ft.
  D = vessel draft, ft.

47. Problem 1.47
Riser Angle Calculations
Riser angle is measured relative to vertical. Riser angles are measured with respect to, x and y axis and the resultant riser angle is given as

Exact equation:

$$\theta = \tan^{-1}\sqrt{\tan^2\theta_x + \tan^2\theta_y}$$

The following is an approximate equation:

$$\theta \cong \sqrt{\theta_x^2 + \theta_y^2},$$

where
θ_x = riser angle in x-direction, deg.
θ_y = riser angle in y-direction, deg.
θ = resultant angle, deg.

48. Problem 1.48

Calculate the percentage error in using the approximate equation to calculate the resultant riser angle when the riser angle in X and Y directions are 4° and 5°, respectively.

Solution:

Exact calculation for riser angle:

$$\theta = \tan^{-1}\sqrt{\tan^2\theta_x + \tan^2\theta_y} = \tan^{-1}\sqrt{\tan^2 4 + \tan^2 5} = 6.39^0$$

Approximate calculation for riser angle is

$$\theta \cong \sqrt{\theta_x^2 + \theta_y^2} = \sqrt{4^2 + 5^2} = 6.40^0$$

Percentage error = $\dfrac{6.39^0 - 6.40^0}{6.39^0} = 0.2\%$

49. Problem 1.49

With the following information regarding the floater, calculate the wind and current forces.

Floater details:

- Draft: 45 ft
- Freeboard: 50 ft

- Length: 450 ft
- Width: 80 ft

Substructure: 40 ft ×25 ft

Rig Derrick:
- Bottom section: 40 ft × 90 ft height and 20 ft width at the top
- Top section: 20 ft × 20 ft height and 10 ft width at the top

Heliport Truss: 200 ft² area of wind path

Wind velocity: 50 mph (towards port side perpendicular to floater)

Current velocity: 3 ft/sec (towards port side perpendicular to floater)

Solution:

Rig wind force = $0.00338 \times 43.45^2 \times 1.1 \times 1.25$

$$\times \left[\frac{1}{2}(20+10)20 + \frac{1}{2}(40+20)90\right] = 26{,}322 \text{ lbf}$$

Helipad = $0.00338 \times 43.45^2 \times 1.1 \times 1.25 \times 200 = 1754$ lbf

Hull = $0.00338 \times 43.45^2 \times 1 \times 1 \times 450 \times 50 = 143{,}575$ lbf

Total wind force = 180 kips

Current force = $1 \times 1 \times 3^2 \times 450 \times 45 = 182{,}250$ lbf (smooth surface assumed)

50. Problem 1.50

What are the four modes of vibration?

Solution:

Axial, torsional, lateral, and coupled.

CHAPTER 2

MUD PUMPS

This chapter focuses on the different basic calculations involved in mud pumps and other related mud pump operations.

51. Problem 2.1
Linear Capacity Calculation

Capacities of the pipe, annular capacity, and annular volume can be calculated using the following equations:

The linear capacity of the pipe is

$$C_i = \frac{A_i}{808.5} \text{ bbl/ft},$$

where

A_i = cross-sectional area of the inside pipe in square inches and equals $0.7854 \times D_i^2$.

D_i = inside diameter of the pipe, in.

Annular linear capacity against the pipe is

$$C_o = \frac{A_o}{808.5} \text{ bbl/ft},$$

where

A_o = a cross-sectional area of the annulus in square inches
$= 0.7854 \times (D_h^2 - D_o^2)$

D_o = the outside side diameter of the pipe, in.

D_h = the diameter of the hole or the inside diameter of the casing against the pipe, in.

52. Problem 2.2
Volume Capacity Calculation

Volume capacity is

$$V = C_i \times L \text{ bbl},$$

where
L = the length of the pipe, ft.

Annular volume capacity is

$$V = C_o \times L \text{ bbl}$$

Mud Pumps

53. Problem 2.3
Displacement Calculation
Displacement of the pipe based on the thickness of the pipe
Open-ended displacement volume of the pipe is

$$V_o = \frac{0.7854\left(D_o^2 - D_i^2\right)}{808.5} \text{ bbl/ft}$$

Displacement volume = $V_o \times L$ bbl
Close-ended displacement volume of the pipe is

$$V_c = \frac{0.7854\left(D_o^2\right)}{808.5} \text{ bbl/ft}$$

Displacement volume = $V_c \times L$ bbl

54. Problem 2.4
Calculate the drillpipe capacity, open-end displacement, closed end displacement, annular volume, and total volume for the following condition: 5000 ft of 5" drillpipe with an inside diameter of 4.276" inside a hole of 8 ½".

Solution:
Linear capacity of pipe is

$$C_i = \frac{A_i}{808.5} = \frac{0.7854 \times D_i^2}{808.5} = \frac{0.7854 \times 4.276^2}{808.5} = 0.017762 \text{ bbl/ft}$$

Pipe volume capacity = $0.017762 \times 5000 = 88.81$ bbl
Open-end displacement of pipe is

$$V_o = \frac{0.7854\left(D_o^2 - D_i^2\right)}{808.5} = \frac{0.7854\left(5^2 - 4.276^2\right)}{808.5} = 0.006524 \text{ bbl/ft}$$

Close-end displacement of pipe, is

$$V_c = \frac{0.7854\left(D_o^2\right)}{808.5} = \frac{0.7854\left(5^2\right)}{808.5} = 0.024286 \text{ bbl/ft}$$

Annular volume is

$$V = C_o \times L = \frac{A_o}{808.5} \times L = \frac{0.7854}{808.5} \times (D_h^2 - D_o^2) \times L$$

$$= \frac{0.7854}{808.5} \times (8.5^2 - 5^2) \times 5000 = 229.5 \text{ bbl}$$

Total volume = Pipe volume + annular volume = 88.81 + 229.50 = 318.31 bbl

55. Problem 2.5

Horsepower Calculation

Mechanical horsepower is given as

$$HP = \frac{\text{Force (lbf)} \times \text{distance (ft)}}{550 \times \text{time (sec)}}$$

It is also given as

$$HP = \frac{\text{Force (lbf)} \times \text{velocity (ft/min)}}{33{,}000}$$

Hydraulic horsepower is given as

$$HHP = \frac{\text{Flow rate (gpm)} \times \text{pressure (psi)}}{1714}$$

Rotating horsepower is given as

$$HP = \frac{\text{Torque (ft-lbf)} \times \text{speed (rpm)}}{5252}.$$

One horsepower = 550 foot-pounds/sec = 33,000 ft-lbf/min and 33,000 ft-lbf / 6.2832 ft-lbf = 5252.

Other conversion factors are 1 horsepower = 0.0007457 megawatts = 0.7457 kilowatts = 745.7 watts.

56. Problem 2.6

The torque on a motor shaft follows a sinusoidal pattern with a maximum amplitude of 8000 ft-lbf. The torque will always be positive through the cycle. Shaft speed is 200 rpm. Calculate the motor horsepower.

Mud Pumps

Solution:

Horsepower is given as

$$HP = \frac{2\pi NT}{3300}$$

The torque is calculated using the sinusoidal pattern:

$$T = 2\int_0^\pi 8000\sin x\, dx$$

$$T = 2\big[8000\cos x\big]_0^\pi$$

Substituting the limits,

$$T = 16{,}000[\cos\pi - \cos 0]_0^\pi = 32{,}000 \text{ ft-lbf}$$

$$HP = \frac{2\pi \times 200 \times 32{,}000}{33{,}000} = 1219 \text{ hp}$$

57. Problem 2.7
Mud Pump Calculations

The following are calculations for theoretical volume of fluid displaced. For single-acting pump,

$$V_t = \left(\frac{\pi}{4} D_L^2 L_s\right) N_c$$

For double-acting pump,

$$V_t = \frac{\pi}{4} N_c L_s \left(2D_L^2 - D_r^2\right)$$

Actual flowrate is

$$Q_a = Q_t \eta_v$$

Pump hydraulic horsepower is

$$HHP_p = \frac{P_p Q}{1714},$$

where
D_L = liner or piston diameter, in.
L_S = stroke length, in.
N_C = number of cylinders, 2 for duplex and 3 for triplex.
D_r = rod diameter, in.
η_v = volumetric efficiency.

58. Problem 2.8
Volumetric Efficiency Calculations

$$\eta_v = \frac{Q_a}{\text{Displacement Volume} \times \text{Speed}} \times 100,$$

where
Q_a = actual flowrate.

$$\eta_v = \frac{Q_t - \Delta Q}{Q_t} \times 100,$$

where
ΔQ = leakage losses.
Q_t = theoretical flowrate = displacement volume × pump speed.

59. Problem 2.9
Pump Factor Calculations

The pump factor is the pump displacement per cycle and is given as F_d in bbl/stroke or gal/stroke.

Duplex pump factor is given by

$$2\left(V_{fs} + V_{bs}\right) = PF_d$$

$$PF_d = \frac{\pi}{2} L_s \left(2D_L^2 - D_r^2\right)$$

Triplex pump factor is given by

$$3V_{fs} = PF_t$$

$$Pf_t = \left(\frac{3\pi}{4} D_L^2 L_s\right)$$

Volumetric efficiency is given by

$$\eta_v = \frac{PF_a}{PF_t} \times 100,$$

where
PF_a = actual pump factor.
PF_t = theoretical pump factor.

For duplex pumps,

$$V_t = \frac{\pi}{4} N_c L_s \left(2D_L^2 - D_r^2\right) \eta_v$$

$$V_t = \frac{N_c L_s \left(2D_L^2 - D_r^2\right) \eta_v}{42 \times 294} \text{ bbl/stroke}$$

For triplex pumps (single-acting, three cylinders),

$$V_t = \frac{L_s \left(D^2\right) \eta_v}{42 \times 98.03} \text{ bbl/stroke}$$

60. Problem 2.10
Energy Transfer Calculations

Efficiency transfer from the diesel engines to the mud pump can be given as in Figure 2.1. Due to interrelated equipment, various efficiencies can be used:

- Engine efficiency, η_e
- Electric motor efficiency, η_{el}
- Mud pump mechanical efficiency, η_m
- Mud pump volumetric efficiency, η_v
- Overall efficiency, $\eta_o = \eta_e \times \eta_{el} \times \eta_m \times \eta_v$

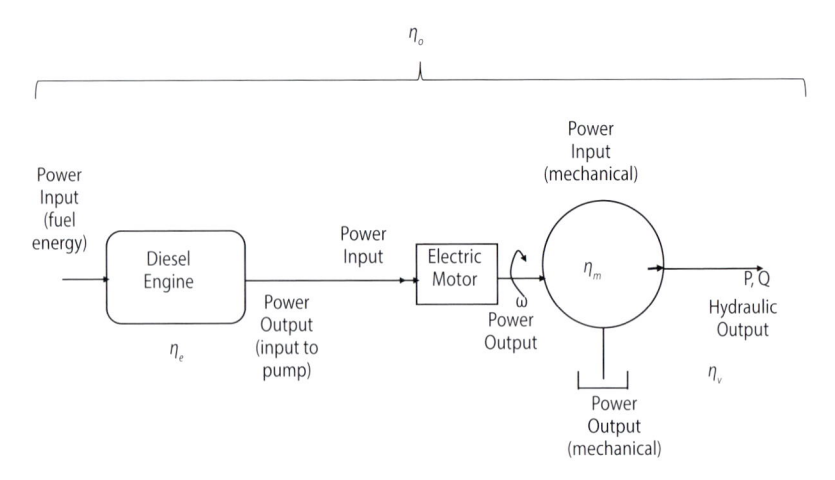

Figure 2.1 Energy transfer

Engine input (fuel energy) → engine output (input to pump) → mechanical pump output (input to hydraulic) → hydraulic output

61. Problem 2.11

Given the following pump details, estimate the liner size required for a double acting duplex pump.

Rod diameter = 2.5 in.
Stroke length = 20 in. stroke.
Pump speed = 60 strokes/min.

The maximum available pump hydraulic horsepower is 1000 hp. For optimum hydraulics the pump recommended delivery pressure is 3500 psi.

Solution:

Theoretical pump displacement for a duplex pump is

$$V_t = \frac{\pi}{4} N_c L_s \left(2D_L^2 - D_r^2\right) = \frac{2 \times 20\left(2D_L^2 - 2.5^2\right)}{294} \text{ gal/stroke}$$

The theoretical flow rate of the pump operating at 60 strokes/min is

$$V_t = \frac{2 \times 20\left(2D_L^2 - 2.5^2\right)}{294} \times 60 \text{ gal/stroke}$$

$$V_t = (16.33D_L^2 - 51.02) \text{ gpm}$$

The volumetric relationship follows:
$Q_a = Q_t \eta_v$,
$490 = (16.33D_L^2 - 51.02) \times 0.9$ gpm
$D_L = 6.03$ in.
Therefore, the liner size that can be used is 6 in.

62. Problem 2.12

Estimate the pump output for a triplex pump in gal/stroke, bbl/stroke for a 6" liner with 18 in. strokes operating at 90% efficiency.
Solution:
For triplex pumps (single-acting, three cylinders)

$$V_t = \frac{L_s(D^2)\eta_v}{42 \times 98.03} \text{ bbl/stroke}$$

Substituting the values

$$V_t = \frac{18(6^2) \times 0.90}{42 \times 98.03} = 0.1416 \text{ bbl/stroke}$$

1 US barrel = 42 gallons
Therefore, $0.1416 \times 42 = 6.00$ gal/stroke

63. Problem 2.13

Estimate the pump output for a duplex pump in gal/stroke, bbl/stoke for a 6" liner with 18 in. strokes operating at 90% efficiency. Rod diameter, 2 in.
Solution:
For duplex pumps (double-acting, two cylinders)

$$\frac{N_c L_s (2D_L^2 - D_r^2)\eta_v}{42 \times 294} = \frac{2 \times 18(2 \times 6^2 - 2^2) \times 090}{42 \times 294} = 0.1784 \text{ bbl/stroke}$$

1 US barrel = 42 gallons
Therefore, $0.1784 \times 42 = 7.5$ gal/stroke

64. Problem 2.14

A double acting duplex pump, 2.5 in. rod, 20 in. stroke, and liner size 7" is to be operated at 60 strokes/min for drilling. The pump delivery pressure is recommended to be 3423 psi. Calculate the hydraulic horsepower.

Solution:

$$Q_t = \frac{\pi}{4} \times 2 \times 20 \times \frac{\left(2 \times 7^2 - 2.5^2\right)}{294} \times 60 = 588 \ \frac{\text{gal}}{\text{min}}$$

Pump pressure = 3423 psi

$$\text{Hydraulic horsepower} = \frac{(588 \times 3423)}{1714} = 1714 \text{ hp}$$

If efficiency of 90% assumed HP = 1174 × 0.90 = 1056 hp

65. Problem 2.15

Given a single acting triplex pump, with stroke length of 12 in., rod diameter of 1.5 in., and pump factor of 6.0 at 100% volumetric efficiency. The pump was operated at 100 strokes/min and 3428 psi for 4 mm. The amount of mud collected at the flowline was 2040 gals.

a. Determine the pump input hydraulic horsepower.
b. Determine the liner diameter.

Solution:

Hydraulic horsepower is given as

$$\text{HHP}_p = \frac{P_p Q}{1714} = \frac{3428 \times 2040/4}{1714} = 1020 \text{ hp}$$

Pump factor of 6 gal/stroke at 100 efficiency is $Q_t = 6 \times 100 = 600$ gpm
The actual flowrate is

$$Q_a = \frac{2040}{4} = 510 \text{ gpm}$$

$$\text{Volumetric efficiency} = \frac{510}{600} = 0.85$$

Mud Pumps

Therefore, using the power factor equation, the diameter of the liner is

$$D_L = \sqrt{\frac{294 \times 6}{12 \times 3}} = 7 \text{ in.}$$

Alternatively, the flowrate equation can also be used.

66. Problem 2.16

A double-acting duplex pump (liner diameter = 6.5"; rod diameter = 35% of liner diameter; stroke length = 16") is being used while drilling a well at 15,000 ft. At 70 strokes/min, the pump is delivering 350 gal/min of mud at a pressure of 4000 psi. The pump mechanical efficiency is 70%.

A. What must be the input horsepower to the pump?

B. For an engine efficiency of 55%, how many gallons of diesel fuel are being consumed per day?

Use the following data:

- Heating value of diesel oil = 19,000 BTU/lbm
- Density = 7.2 ppg
- Cost = $2.16/gal

Determine the total cost of diesel oil.

Solution:

A. Volume per stroke = $\dfrac{32(2 \times 6.5^2 - 5.175625)}{294} = 8.63$ gal/stroke

Actual hydraulic horsepower is

$$H_{HP} = \frac{QP}{1714} = \frac{350 \times 4000}{1714} = 816.8 \text{ hp}$$

The theoretical flow rate is $Q_t = 8.63 \times 70 = 604$ gpm.
Therefore, volumetric efficiency is

$$\eta_v = \frac{Q_a}{Q_t} 100 = \frac{350}{604} \times 100 = 58\%$$

Input horsepower = 816.8/0.58 = 1408 hp
Considering the mechanical efficiency, the pump horsepower required = 1408/0.7 = 2012 hp

B. The amount of diesel consumed for a single engine is

$$Q_f = \frac{2010 \times 2545 \times 24}{19{,}000 \times 0.55 \times 7.2} = 1633 \text{ gal/day}$$

The cost of the diesel consumption = $3227.

67. Problem 2.17

It was decided to circulate and clean the hole before pulling out for a logging run. Calculate the number of strokes and time in minutes to displace the annulus and drillstring for the wellbore configuration.

Wellbore details:

- Total depth: 12,000 ft
- Last intermediate casing: 9 ⅝" × 8.681" set at depth 11,200 ft
- Open hole size: 8 ½"

Drillstring details:

- Drillpipe: 5" × 4.276"
- HWDP: 5" × 2 ¹³⁄₁₆"; length = 300 ft
- Drill collar: 6 ½" × 3"; length = 600 ft

Pump details:

- Pump: 7" liner
- Type: duplex, double-acting
- Stroke length: 17"
- Rod diameter: 2 ½"
- Strokes per minute: 60
- Volumetric efficiency: 90%

Mud Pumps

Solution:

Theoretical pump displacement for a duplex pump is

$$V_t = \frac{2 \times 17 \left(2 \times 7^2 - 2.5^2\right)}{294} \left(\frac{\text{gal}}{\text{stroke}}\right) \left(\frac{\text{bbl}}{42.09 \text{gal}}\right) = 0.2520 \text{ bbl/stroke}$$

1 bbl = 42.09 gal

Pipe capacities:

Capacity of drillpipe = $\dfrac{4.276^2}{1029.4}$ = 0.017762 bbl/ft

Total volume of drillpipe = 0.017762 × (12,000 – 600 – 300) = 197.158 bbl

Capacity of heavyweight drillpipe = $\dfrac{2.8125^2}{1029.4}$ = 0.0076842 bbl/ft

Total volume of heavyweight drillpipe = 0.0076842 × 300 = 2.305272 bbl

Capacity of drill colar = $\dfrac{3^2}{1029.4}$ = 0.008743 bbl/ft

Total volume of drill colar = 0.008743 × 600 = 5.245774 bbl

Total pipe volume = 195.3817 + 2.305272 + 5.245774 = 204.6871 bbl

Annular capacities:

Annular capacity in open hole against the drillpipe

$$= \frac{\left(8.681^2 - 5^2\right)}{1029.4} = 0.0489215 \text{ bbl/ft}$$

Annular volume in casing against the drillpipe = 0.0489215 × 11,100 = 543.03 bbl

Annular volume in casing against the heavyweight drillpipe = 0.0489215 × 100 = 4.89 bbl

Annular capacity in open hole against the drillpipe

$$= \frac{\left(8.5^2 - 5^2\right)}{1029.4} = 0.0459005 \text{ bbl/ft}$$

Annular volume in open hole against the heavyweight drillpipe
= 0.0459005 × 200 = 9.1801 bbl

Annular capacity in open hole against the drill collar

$$= \frac{(8.5^2 - 6.5^2)}{1029.4} = 0.0291432 \text{ bbl/ft}$$

Annular volume against the drill collar = 0.0291432 × 600 = 17.486 bbl

Total annular volume = 543.03 + 4.89 + 9.1801 + 17.486 = 574.5861 bbl

Total strokes needed to displace string volume $= \frac{204.6871}{.227} = 902$ strokes.

Total strokes needed to displace annular volume

$$= \frac{574.5865}{.227} = 2532 \text{ strokes}$$

Total number of strokes = 902 + 2532 = 3434 strokes.

Total time required $= \dfrac{3434\,(\text{strokes})}{60\left(\dfrac{\text{strokes}}{\text{min}}\right)} = 57.23$ min

68. Problem 2.18

Calculate the number of strokes and time in minutes to displace the annulus and drillstring for the wellbore configuration.

Wellbore details:
Total wellbore length = 12,000 ft
Hole size = 12 ¼"
Drillpipe = 5" × 2 ¹³⁄₁₆"; length = 11,100 ft
HWDP = 5" × 2 ¹³⁄₁₆"; length = 300 ft
Drillcollar = 6 ½" × 3" = 600 ft
Pump details:
Pump = 7" liner
Type = duplex
Stroke = 17"
Rod diameter = 2 ½"
Strokes per minute = 60
Efficiency = 90%

Mud Pumps

a) Capacity of the drilling pipe:

$$A_{dp} = \frac{d_{dp}^2}{1029.4}$$

$$A_{dp} = \frac{2.8125^2}{1029.4}$$

$A_{dp} = 0.007684$ bbl/ft
$V_{dp} = A_{dp} \times L_{dp}$
$V_{dp} = 0.007684 \times 11,100$
$V_{dp} = 85.3$ bbl

b) Capacity of the heavy weight:

$$A_{HW} = \frac{d_{HW}^2}{1029.4}$$

$$A_{HW} = \frac{2.8125^2}{1029.4}$$

$A_{HW} = 0.007684$ bbl/ft
$V_{HW} = A_{HW} \times L_{HW}$
$V_{HW} = 0.007684 \times 300$
$V_{HW} = 2.3$ bbl

c) Capacity of the drill collars:

$$A_{DC} = \frac{d_{DC}^2}{1029.4}$$

$$A_{DC} = \frac{3^2}{1029.4}$$

$A_{DC} = 0.008743$ bbl/ft
$V_{DC} = A_{DC} \times L_{DC}$
$V_{DC} = 0.008743 \times 600$
$V_{DC} = 5.2$ bbl

d) Capacity of the drill pipe annulus:

$$A_{adp} = \frac{d_2^2 - d_1^2}{1029.4}$$

$$A_{adp} = \frac{12.25^2 - 5^2}{1029.4}$$

$A_{adp} = 0.12149$ bbl/ft
$V_{adp} = A_{adp} \times L_{dp}$
$V_{adp} = 0.007684 \times 11{,}100$
$V_{adp} = 1348.5$ bbl

e) Capacity of the heavy weight annulus:

$$A_{aHW} = \frac{d_2^2 - d_1^2}{1029.4}$$

$$A_{aHW} = \frac{12.25^2 - 5^2}{1029.4}$$

$A_{aHW} = 0.12149$ bbl/ft
$V_{aHW} = A_{aHW} \times L_{HW}$
$V_{aHW} = 0.12149 \times 300$
$\mathbf{V_{adp} = 36.4}$ **bbl**

e) Capacity of the heavy weight annulus:

$$A_{aDC} = \frac{d_2^2 - d_1^2}{1029.4}$$

$$A_{aDC} = \frac{12.25^2 - 6.5^2}{1029.4}$$

$A_{aDC} = 0.10473$ bbl/ft
$V_{aDC} = A_{aDC} \times L_{DC}$
$V_{aHW} = 0.10473 \times 600$
$V_{adp} = 62.8$ bbl

e) Total capacity:

$$V_t = V_{dp} + V_{HW} + V_{DC} + V_{adp} + V_{aHW} + V_{aDC}$$
$$V_t = 85.3 + 2.3 + 5.2 + 1348.5 + 36.4 + 62.8$$
$$V_t = 1540.5 \text{ bbl}$$

f) Pump factor for a duplex pump:

$$F_p = \frac{\pi}{4} N_c L_s \left(2 \times D_L^2 - D_r^2\right)$$

$$F_p = \frac{3.1416}{4} \times 2 \cdot 17 \times \left(2 \times 7^2 - 2.5^2\right)$$

$F_p = 2450.0$ in²/stroke
As 1 bbl = 9702 in²
$F_p = 0.253$ bbl/stroke
As the efficiency of the pump is 90%, the pump factor is
$F_p = 0.253 \times 0.9$
$F_p = 0.228$ bbl/stroke

g) Number of strokes to displace the drillstring and annulus of the well:

$$N_s = \frac{V_t}{F_p}$$

$$N_s = \frac{1540.5}{0.228}$$

Number of strokes = 6756.6 strokes

h) Time to displace the drillstring and annulus of the well:

$$t = \frac{N_s}{N_s / \min}$$

$$t = \frac{6756.6}{60}$$

$$t = 112.6 \text{ min}$$

69. Problem 2.19

A duplex pump (liner diameter = 6.5"; rod diameter = 35%; stroke length = 18") is being used while drilling a well at 15,000 ft. At 50 strokes/min, the pump is delivering 350 gal/min of mud at a pressure of 4000 psi. The pump's volumetric efficiency is 70%.

a) What must be the input horsepower to the pump?

b) If the dynamic pressure in the circulating system is 2200 psi, what percent of the horsepower calculated in part (a) is wasted due to friction pressure losses?

Solution:

The theoretical flowrate is

$$\frac{Q_a}{\eta_v} = \frac{350}{0.7} = 500 \text{ gpm}$$

The theoretical hydraulic horsepower is

$$\text{HHP} = \frac{QP}{1714} = \frac{500 \times 4000}{1714} = 1167 \text{ hp}$$

Since the horsepower is directly proportional to pressure drop, the waste due to frictional pressure losses can be given as

$$\% \text{ change} = \frac{4000 - 2200}{4000} = 45\%$$

70. Problem 2.20

A 1200 hp single acting duplex pump has a volumetric efficiency of 90%. A delivery pump pressure of 3000 psi is required to drill the last 1000 ft of hole. What liner size would you specify if the pump has a 20 in stroke, 2.5 in rod diameter, and operating at 60 cycles/min.

Solution:

Actual flowrate is

$$Q_a = 1714 \times 1200/3000 = 685.6 \text{ gpm}$$

Mud Pumps

With the volumetric efficiency of 90% the theoretical flowrate can be given as

$$Q_t = 685.6/0.9 = 761.7 \text{ gpm}$$

For triplex single acting pump the theoretical flowrate can be given as

$$Q_t = \frac{nN_c LD_L^2}{294} = 761.7 = \frac{60 \times 2 \times 20 \times D_L^2}{294}$$

Solving for liner diameter

$$D_L = \sqrt{\frac{761.7 \times 294}{60 \times 2 \times 20}} = 9.66 \approx 10 \text{ in.}$$

71. Problem 2.21

A 1000 hp single-acting duplex pump has a volumetric efficiency of 80% when operated at 60 cycles/min. A delivery pump pressure of 3000 psi is recommended to drill the last 1000 ft of hole. What liner size would you specify if the pump has a 20 in. stroke, 2.5 in. rod diameter, and is operating at 60 cycles/min?

Solution:

Actual flowrate is

$$Q_a = 1714 \times 1000/3000 = 571 \text{ gpm}$$

With the volumetric efficiency of 80% the theoretical flowrate can be given as

$$Q_t = 571/0.8 = 714 \text{ gpm}$$

For triplex single acting pump the theoretical flowrate can be given as

$$Q_t = \frac{nN_c LD_L^2}{294} = 714 = \frac{60 \times 2 \times 20 \times D_L^2}{294}$$

Solving for liner diameter

$$D_L = \sqrt{\frac{714 \times 294}{60 \times 2 \times 20}} = 9.35 \approx 10 \text{ in.}$$

72. Problem 2.22

There is a single-acting triplex mud pump powered by a diesel engine. The pump data follows:

- Rod size = 2 ½"
- Liner size = 7"
- Stroke length = 12"
- Operating delivery pressure = 3400 psi at 60 strokes/minute
- Volumetric efficiency = 90%
- Mechanical efficiency = 85%

For an engine efficiency of 55%, how many gallons of diesel fuel are being consumed per day?

Given: Heating value of diesel oil = 19,000 BTU/lb
density = 7.2 ppg

Solution:

For a single-acting triplex pump, the flowrate is calculated as:

$$Q_t = \frac{\pi}{4} D_L^2 L_S N_c = \frac{\pi}{4} \times 7^2 \times 12 \times 3 \left(\frac{in^3}{stroke}\right) \times 60 \left(\frac{stroke}{min}\right) \times \left(\frac{1 ft^3}{12\ in^3}\right)$$

$$\times \left(\frac{7.48\ gal}{1\ ft^3}\right) = 360\ gpm$$

Actual flowrate is

$$Q_a = Q_t \eta_v = 360 \times 0.90 = 324\ gpm$$

Hydraulic horsepower is

$$HHP = \frac{P_p Q}{1714} = \frac{324 \times 3400}{1714} = 642\ hp$$

Output mechanical horsepower is

$$\text{Input } H_{HP} = \frac{QP}{1714} = \frac{360 \times 3400}{1714 \times 0.85} = 840\ hp$$

Mud Pumps

$$= \frac{QP}{1714} = \frac{324 \times 3400}{1714 \times 0.85 \times 0.9} = 840 \text{ hp}$$

$$HH_{HP} = \frac{840}{0.55} = 1527 \text{ hp}$$

$$Q_f = \frac{1527 \times 2545}{19{,}000} = 682 \text{ gal/day}$$

Consider the following:

Engine input (fuel energy) → engine output (input to pump) → mechanical pump output (input to hydraulic) → hydraulic output

Overall efficiency can also be used to find the engine input to the pump output horsepower.

73. Problem 2.23

An 8 ½" hole is being drilled at a depth of 8000 ft with 4 ½" OD drillpipe. Calculate the strokes and time required to displace the string and annulus with the following details:

Duplex pump details:
Liner diameter 6", stroke length 15", and rod diameter 2"
Volumetric efficiency 90% and pump strokes 50 spm
Pipe and hole details:
4 ½" DP capacity 0.01422 bbl/ft 4 ½" and 8 ½" annulus capacity 0.05 bbl/ft
Volumetric output is

$$V_t = \frac{\pi}{4} N_c L_s \left(2D_L^2 - D_r^2 \right)$$

$$V_a = \frac{2 \times 20 \left(2 \times 6^2 - 2^2 \right)}{294} \times .90 = 8.32 \text{ gal/stroke}$$

1 bbl = 42.09 gal

Total flowrate is

$$8.32\frac{\text{gal}}{\text{stroke}} \times \frac{1}{42.09}\left(\frac{\text{bbl}}{\text{gal}}\right) = 0.1976717 \text{ bbl/stroke}$$

Drillstring capacity is

$$0.01422\frac{\text{bbl}}{\text{ft}} \times 8000 \text{ ft} = 113.76 \text{ bbl}$$

Annulus capacity is

$$0.05\frac{\text{bbl}}{\text{ft}} \times 8000 \text{ ft} = 400 \text{ bbl}$$

Drillstring capacity + annulus capacity = 513.76 bbl

$$\text{Number of strokes} = \frac{513.76}{0.1976717} = 2600 \text{ strokes}$$

Total time required to displace the string and annulus

$$= \left(\frac{2600}{50}\right)\left(\frac{\text{stroke}}{\frac{\text{strokes}}{\text{min}}}\right) = 52 \text{ min}$$

74. Problem 2.24

At a certain depth, the drillstring is being pulled out of the hole. The triplex pump with 6" liner and 20" stroke length and volumetric efficiency 90% operating at 50 strokes/min is used to fill the hole. Calculate the number of strokes and time to fill the hole after pulling out 10 stands of drillpipe. Assume displacement capacity of the drillpipe as 0.00813 bbl/ft. Assume 1 stand = 90 ft.

Solution:

$$\text{Volume pumped} = \frac{20 \times 6^2 \times 0.90 \times 50}{98.03} = 330.51 \text{ gpm}$$

Volume of mud displaced = 0.00813 bbl/ft × 90 × 10 = 7.317 bbl
= 307.31 gal

Time taken to fill = 307/330 = 56 sec
Strokes required to fill = 307/(330/50) = 47 strokes

75. Problem 2.25

A double-acting duplex pump with a 6.5" liner, 2.5" rod diameter, and 18" stroke is operated at 3000 psi and 40 strokes per minute for 5 minutes with the suction tank isolated from the return mud flow. The mud level in the suction tank, which is 7' wide and 30' long, was observed to fall 1 ft during this period. Determine the volumetric efficiency and hydraulic horsepower of the pump.

Solution:

Using the data given, the theoretical flowrate can be calculated as

$$Q_t = \frac{\pi}{4} \times 2 \times 18 \times \frac{(2 \times 6.5^2 - 2.5^2)}{294} \times 40 = 301 \text{ gpm}$$

Amount of fluid collected for five minutes = 301 × 5 = 1505 gal

Total amount of fluid from the tank is the actual flow during the same amount of time

$$Q_a = 30 \times 7 \times 1 \times 7.48 = 1570.8 \text{ gal}$$

Using the theoretical flow and the actual flow, volumetric efficiency can be calculated as

$$\eta = \frac{Q_a}{Q_t} = \frac{1505}{1570.8} = 96\%$$

Hydraulic horsepower is

$$\text{HHP} = \frac{P_p Q_a}{1714} = \frac{3000 \times 3016}{1714 \times 0.96} = 550 \text{ hp}$$

76. Problem 2.26

If a pump with a piston displacement of 500 cubic inches pumps 450 cubic inches of mud per stroke, its volumetric efficiency is 90%.

a. True
b. False

Solution:
a. True
Actual volumetric displacement is equal to the theoretical displacement × volumetric efficiency.
In this case = 500 × 0.90 = 450 cubic inches

77. Problem 2.27

An 8 ½" hole is being drilled at a depth of 8000 ft with 4 ½" OD drillpipe. Calculate the strokes and time required to displace the string and annulus with the following details: Duplex pump with 6" liner, stroke 15", rod diameter 2", volumetric efficiency 90%, and pump strokes 50 strokes per minute.

4 ½" DP capacity = 0.01422 bbl/ft
4 ½" and 8 ½" annulus capacity = 0.05 bbl/ft

Volumetric output = $V_t = \dfrac{\pi}{4} N_c L_s \left(2D_L^2 - D_r^2\right)$

$$V_a = \dfrac{2 \times 20\left(2 \times 6^2 - 2^2\right)}{294} \times 0.90 = 8.32 \, \dfrac{\text{gal}}{\text{stroke}}$$

1 bbl = 42.09 gal

Total flowrate = $8.32 \dfrac{\text{gal}}{\text{stroke}} \times \dfrac{1}{42.09} \left(\dfrac{\text{bbl}}{\text{gal}}\right) = 0.1976717$ bbl/stroke

Drillstring capacity = $0.01422 \dfrac{\text{bbl}}{\text{ft}} \times 8000 \text{ ft} = 113.76$ bbl

Annulus capacity = $0.05 \dfrac{\text{bbl}}{\text{ft}} \times 8000 \text{ ft} = 400$ bbl

Drillstring capacity + annulus capacity = 513.76 bbl

Number of strokes = $\dfrac{513.76}{0.1976717} = 2600$ strokes

Mud Pumps

Total time required to displace the string and annulus

$$= \left(\frac{2600}{50}\right)\left(\frac{\text{stroke}}{\frac{\text{strokes}}{\text{min}}}\right) = 52 \text{ min.}$$

78. Problem 2.28

Given: Single acting triplex pump with stroke length of 12 in. and pump factor, F_p of 6.0 gal/stroke at η_v = 100%. The pump was operated at 100 strokes/min and 2428 psi for 3 minutes. Calculate the increase in the mud level in the trip tank (8 ft wide and 8 ft long) if the pump volumetric efficiency is 93%. Determine the pump hydraulic horsepower and the liner diameter.

Trip tank = collection or supply tank at the surface.

Solution:

Pump factor for triplex pump can be given as $Pf_t = N_c L_s D_L^2 \eta_v$
The pump factor is

$$Pf_t = \frac{3 \times 12 \times D_L^2 \times 1}{294} = 6 \text{ gal/stroke}$$

Solving results in a liner, size of D_L = 7 in.
Theoretical flowrate = Q_t = 6 × 100 = 600 gpm
Actual flowrate for a volumetric efficiency of 93% is
$Q_a = Q_t \eta_v$ = 600 × 0.93 = 558 gpm
The increase in mud level in the trip tank can be calculated as

$$H = \frac{558 \times 3}{8 \times 8 \times 7.48} = 3.5 \text{ ft}$$

Hydraulic horsepower of the pump is

$$\text{HHP} = \frac{P_p Q_a}{1714} = \frac{2428 \times 558}{1714} = 790 \text{ hp}$$

79. Problem 2.29

A drilling engineer is planning to conduct a simple test to calculate the volumetric efficiency of the mud pump by placing a sample of calcium carbide in the drillstring during the pipe connection. Calcium carbide reacts with mud to form acetylene gas and is detected at the surface after pumping 450 strokes. Drillstring consists of 800 ft. of 8" OD and 3" ID drill collars and is at bottom of the well that has an open hole diameter of 12 ¼".

Pump details:

- Type: single-acting triplex pump
- Pump factor of 6.0 gal/stroke at 100% volumetric efficiency.

Neglect gas slip and fluid mixing. Estimate the volumetric efficiency of the pump.

Assume uniform open hole size from surface to bottom.

Solution:

Pump factor = 6 × 0.0322 = 0.19358 bbl/stroke
Total inside volume = 6.99 bbl
Total outside volume = 66.88 bbl
Total strokes needed for inside volume = 36
Total strokes needed of outside volume = 345
Total strokes = 36 + 345 = 381
Volume efficiency = 85%

Drillstring capacity is

$$0.01422 \frac{bbl}{ft} \times 8000 \text{ ft} = 113.76 \text{ bbl}$$

Annulus capacity is

$$0.05 \frac{bbl}{ft} \times 8000 \text{ ft} = 400 \text{ bbl}$$

Drillstring capacity + annulus capacity = 513.76 bbl

$$\text{Number of strokes} = \frac{513.76}{0.1976717} = 2600 \text{ strokes}$$

Total time required to displace the string and annulus

$$= \left(\frac{2600}{50}\right)\left(\frac{\text{stroke}}{\frac{\text{strokes}}{\text{min}}}\right) = 52 \text{ min}$$

80. Problem 2.30

A pump is capable of a maximum pressure of 2500 psi at a circulation rate of 450 gpm when the mud weight is 12 ppg and the frictional pressure loss in the system is 1000 psi. Calculate the available bit hydraulic horsepower if the mud weight is increased to 15 ppg and the circulation rate and maximum pump pressure remaining the same at 450 gpm. The frictional pressure loss in the system is given as $C\rho^{0.8}$ psi where ρ is mud density in ppg and C is a constant.

Solution:

Pump pressure $P_p = P_f + P_b$
Using the given relation $C\rho^{0.8}$ it can be deduced

$$C = \frac{1000}{12^{0.8}} = 137$$

$$HP_b = \frac{\left(2500 - 137 \times 15^{0.8}\right) 450}{1714} = 342 \text{ hp}$$

81. Problem 2.31

A mud logging engineer is planning to calculate the length of the formation drilled after encountering a formation with gas at 9800 ft. The gas is monitored at the surface using a gas chromatograph detector at the shale shaker. If the penetration rate of the bit is 28 ft/hr and the pump is operating at 100 strokes per minute, how many feet have drilled after the formation of gas is encountered? Triplex pump with 6" liner, stroke length 18", volumetric efficiency 85%. Annulus capacity of the hole is 0.05 bbl/ft. Neglect gas slip and assume gauged hole.

Solution:

Total annular volume = 9800 × 0.05 = 490 bbl = 490 × 42 = 20,580 gal

$$\text{Volume pumped} = \frac{18 \times 6^2 \times 0.85 \times 100}{98.03} = 561.8688 \text{ gpm}$$

$$\text{Feet drilled} = \frac{28}{60} \times \frac{20{,}580}{561.86} = 17.1 \text{ ft}$$

82. Problem 2.32

Determine the daily cost of running a diesel engine to power a duplex pump under the following conditions:

Pump: single-acting duplex
Stroke length = 18 in.
Liner size = 8 in.
Delivery pressure = 1000 psi at 40 spm
Volumetric efficiency = 90%
Mechanical efficiency = 85%
Diesel engine efficiency = 50%
Diesel oil: weight = 7.2 lb/gal
Heating value = 19,000 BTU/lb
Cost = $1.15/gal

Solution:

Consider the following:

Engine input (fuel energy) → engine output (input to pump) → mechanical pump ouput (input to hydraulic) → hydraulic output.

The theoretical flowrate for a single-acting duplex pump is

$$Q_t = \frac{\pi}{4} 8^2 \times 18 \times 2\left(\frac{7.48}{12^3}\right) \times 40 = 313.3 \text{ gpm}$$

The actual flowrate with the given volumetric efficiency of 90% is

$$Q_a = 313.3 \times 0.90 = 282 \text{ gpm}$$

The hydraulic horsepower of the mud pump is

$$HP_p = \frac{P_p Q_a}{1714} = \frac{1000 \times 282}{1714} = 164 \text{ hp}$$

Mud Pumps

The input horsepower of the pump can be calculated using the mechanical efficiency of 85%:

$$HP_m = \frac{164}{0.85} = 193.56 \text{ hp}$$

The engine horsepower is

$$H_{HP} = \frac{193.56}{0.50} = 387 \text{ hp}$$

The input power is expressed in terms of the rate of fuel consumption, Q_f, and the fuel heating value, H:

$$P_i = \frac{Q_f H}{2545} = \frac{Q_f \times 19{,}000}{2545}$$

Hence,

$$Q_f = \left(\frac{387 \times 2545}{19{,}000}\right) = 51.84 \text{ lb/hr}$$

or

$$Q_f = \frac{51.84 \text{ lb}}{\text{hr}} \times \frac{24 \text{ hr}}{\text{day}} \times \frac{\text{gal}}{7.2 \text{ lb}} = 172.8 \text{ gal/day}$$

Cost = 172.8 × 1.15 = $198.7/day

83. Problem 2.33

A duplex pump (liner diameter = 6.5"; rod diameter = 35%; stroke length = 18") is being used while drilling a well at 15,000 ft. At 50 strokes/min, the pump is delivering 350 gal/min of mud at a pressure of 4000 psi. The pump volumetric efficiency is 70%.

a) What must be the input horsepower to the pump?

b) If the dynamic pressure in the circulating system is 2200 psi what percent of the horsepower calculated in part (a) is wasted due to friction pressure losses?

Solution:

$$\text{Theoretical flowrate} = \frac{Q_a}{\eta_v} = \frac{350}{0.7} = 500 \text{ gpm}$$

Theoretical hydraulic horsepower

$$H_{HP} = \frac{QP}{1714} = \frac{500 \times 4000}{1714} = 1167 \text{ hp}$$

Since the horsepower is directly proportional to pressure drop the wasted due to frictional pressure losses can be given as

$$\% \text{ change} = \frac{4000 - 2200}{4000} = 45\%$$

84. Problem 2.34

Replacing a smaller liner in a reciprocating pump with a larger liner _____ output pressure and _____ output volume.

    a. lowers; raises
    b. lowers; lowers
    c. raises; raises
    d. raises; lowers

Solution:
    a. lowers; raises

85. Problem 2.35

Replacing a smaller liner in a reciprocating pump with a larger liner will result in _____ output pressure and _____ output volume.

Solution:

Replacing a smaller liner in a reciprocating pump with a larger liner will result in <u>lower</u> output pressure and <u>increased</u> output volume.

CHAPTER 3

WELL PATH DESIGN

This chapter focuses on different basic calculations involved in designing a well path as well as in monitoring well trajectory while the well is drilled.

86. Problem 3.1

a. A well is directionally drilled for 6000 ft has a true vertical depth of 6000 ft.
   i. True
   ii. False
b. What are the three elements needed to drill a horizontal hole successfully using the conventional rotary drilling?

Solution:

a. False
b. Weight on bit, bit rotation, and flow rate or W, N, and Q.

87. Problem 3.2

A horizontal well is being planned which bypasses an obstruction. The following combinations are used below the kick-off depth:

Three build sections, three hold sections, and one drop section. The wellplaner is planning to use the third build section to bypass the obstruction. Horizontal hold section should be the last section. Using the best combination to design the well, draw the wellpath below the kick-off depth.

Solution:

The wellpath is shown in Figure 3.1.

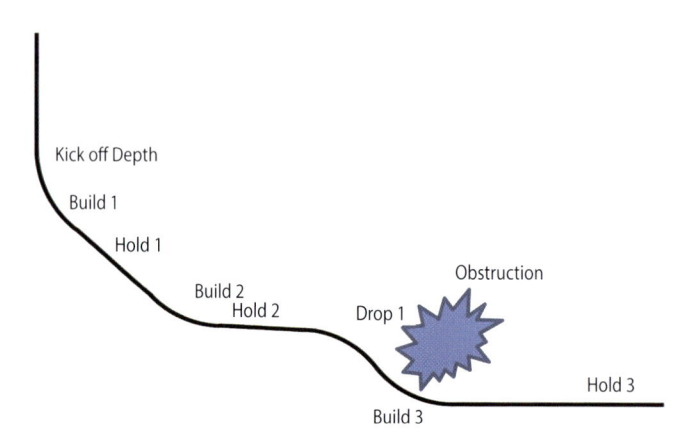

Figure 3.1   Wellpath

88. Problem 3.3

A horizontal well is being planned so that the wellpath bypasses an obstruction.

The following combinations are used:

Two build sections, three hold sections, and one drop section. Using the best combination to design the well, draw the wellpath below the kick-off depth.

Solution:

The wellpath with the different combinations is shown in Figure 3.2.

89. Problem 3.4

Average Curvature—Average Dogleg Severity (DLS) Calculations

The equations commonly used to calculate the average curvature of a survey interval are:

$$\overline{\kappa} = \sqrt{\left(\frac{\Delta\alpha}{\Delta L}\right)^2 + \left(\frac{\Delta\phi}{\Delta L}\right)^2 \sin^2 \overline{\alpha}}$$

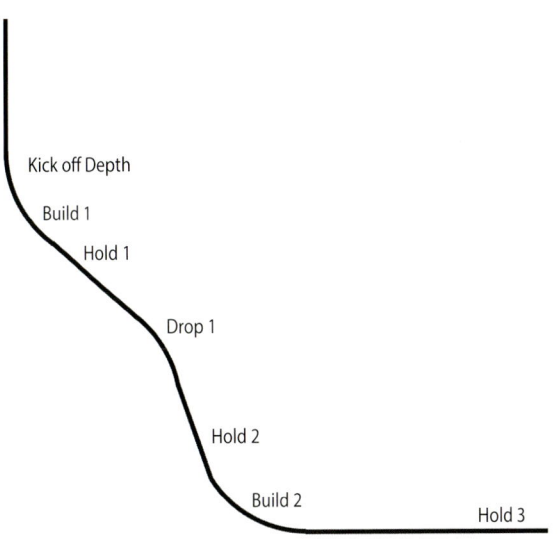

Figure 3.2 Wellpath

and

$$\bar{\kappa} = \frac{\beta}{\Delta L},$$

where

α = inclination (°).
ϕ = azimuth or direction (°).
$\bar{\alpha}$ = average inclination angle (°).
$\bar{\kappa}$ = average borehole curvature.
$\beta = a\cos(\cos\alpha_1 \cos\alpha_2 + \sin\alpha_1 \sin\alpha_2 \cos\Delta\phi)$.

90. Problem 3.5
Vertical and Horizontal Curvatures

Vertical and horizontal curvatures can be calculated, respectively, using the following equations:

$$\kappa_V = \kappa_\alpha,$$

$$\kappa_V = \frac{d\alpha}{dL} = \kappa_\alpha$$

$$\kappa_H = \frac{\kappa_\phi}{\sin\alpha},$$

$$\kappa_H = \frac{d\phi}{dS} = \frac{d\phi}{dL\sin\alpha} = \frac{\kappa_\phi}{\sin\alpha},$$

where

κ_V = curvature of wellbore trajectory in a vertical position plot, °/30 m or °/100 ft.

κ_H = curvature of wellbore trajectory in a horizontal projection plot, °/30 m or °/100 ft. $\Delta L = L_2 - L_1$.

ΔL = curved section length, m or ft.

S = arc length in the azimuthal direction, m or ft.

91. Problem 3.6
Borehole Curvature Calculations
The general formula for borehole curvature can be given as

$$\kappa = \sqrt{\kappa_\alpha^2 + \kappa_\phi^2 \sin^2 \alpha}$$

The general formula for borehole curvature with the vertical and horizontal curvatures can be given as

$$\kappa = \sqrt{\kappa_V^2 + \kappa_H^2 \sin^4 \alpha},$$

where

κ_V = curvature of wellbore trajectory in a vertical expansion plot, °/30 m or °/100 ft.

κ_H = t curvature of wellbore trajectory in a horizontal projection plot, °/30 m or °/100 ft.

92. Problem 3.7
Borehole Radius of Curvature Calculations
There are two ways of deducing the applied formula for borehole curvature and torsion. The corresponding curvature radius and torsion radii can be calculated as follows:

$$R = \frac{180 C_\kappa}{\pi \kappa}$$

$$\rho = \frac{180 C_\kappa}{\pi \tau},$$

where

C_κ = constant related to the unit of borehole curvature.
If the unit for borehole curvature is °/30 m and °/30 ft, then C_κ = 30 and 100, respectively.
κ = curvature of wellbore trajectory, °/30 m, (°)/100 ft.
τ = torsion of wellbore trajectory, °/30 m, (°)/100 ft.

93. Problem 3.8

Calculate the radius curvature for a build section with a build rate of 2°/100 ft.

Solution:

The radius of curvature is

$$R = \frac{180 C_K}{\pi \kappa},$$

where

$$C_K = 100.$$

With the given data, the radius of curvature is

$$R = \frac{180 \times 100}{\pi \times 2} = 2863.63 \text{ ft.}$$

94. Problem 3.9

Calculate the buildup rate of a build section with a radius of curvature of 2291.83 ft.

Solution:

The radius of curvature is

$$R = \frac{180 C_K}{\pi \kappa},$$

where
$C_K = 100.$

With the given data, the buildup rate is

$$R = \frac{180 C_K}{\pi \kappa} = \frac{180 \times 100}{\pi \times 2291.83} = 2.5°/100 \text{ ft.}$$

95. Problem 3.10

Bending Angle Calculations

With the given inclination angles and directions, the borehole bending angle can be given as

Well Path Design

$$\cos\beta = \cos\alpha_1 \cos\alpha_2 + \sin\alpha_1 \sin\alpha_2 \cos\Delta\phi,$$

where

$\Delta\phi = \phi_2 - \phi_1$, β = bending angle, (°).
$\Delta\phi$ = section increment of azimuth angle, (°).
α_1 = inclination angle at survey point 1, (°).
α_2 = inclination angle at survey point 2, (°).

96. Problem 3.11
Tool Face Angle Calculations

With the given inclination angles and directions, the tool face rotation angle is

$$\gamma = \arccos\left(\frac{\cos\alpha_1 \cos\beta - \cos\alpha_2}{\sin\alpha_2 \sin\beta}\right)$$

97. Problem 3.12

Calculate the dogleg severity (DLS) with the following survey data:

Measured Depth (ft)	Angle (°)	Direction (°)
15,000	16.5	150
15,045	17.5	148

Solution:
The DLS between the first two survey stations is

$$\overline{\kappa} = \sqrt{\left(\frac{17.5-16.5}{45}\right)^2 + \left(\frac{148-150}{45}\right)^2 \sin^2\overline{\frac{16.5+17.5}{2}}} = 25.4°/100 \text{ ft}$$

98. Problem 3.13

Calculate the dogleg severity (DLS) for the following data:

MD	Angle (°)	Direction (°)
5000 ft	5.5	150
5120 ft	7.5	148

Solution:

The overall angle change is given as

$\beta = \arccos(\cos 2 \sin 7.5 \sin 5.5 + \cos 5.5 \cos 7.5)$

$\beta = 2.01$

The dogleg severity (DLS) can be calculated as

$$DLS = \frac{2.01}{120} \times 100 = 1.67°/100 \text{ ft}$$

99. Problem 3.14

Using the following data calculate the hole curvature between the following survey stations:

MD	Angle (°)	Direction (°)
5100	3.5	N 15° E
5226	6.5	N 25° E

Solution:

$\Delta\varepsilon = 10°$

$\beta = \arccos(\cos \Delta\varepsilon \sin \alpha_n \sin \alpha + \cos \alpha \cos \alpha_n)$

$\beta = \arccos(\cos 10 \sin 6.5 \sin 3.5 + \cos 3.5 \cos 6.5)$

$\beta = 2.47°/100 \text{ ft}$

100. Problem 3.15

Use the survey data below to determine the dogleg severity:

First Survey

Vertical angle: 8° 15' (8¼°)

Direction: S 34° E

Depth: 6400 ft

Second Survey

Vertical angle: 2° 45'(2¾°)

Direction: S 9° E

Depth: 6475 ft

Solution:

$\Delta\varepsilon = 25°$

Dogleg angle is given as
$\beta = \arccos(\cos\Delta\varepsilon \sin\alpha_n \sin\alpha + \cos\alpha \cos\alpha_n)$
$\beta = \arccos(\cos 25 \sin 8.5 \sin 2.75 + \cos 2.75 \cos 8.5)$
This angle change is between 6475 – 6400 = 75 ft
Dog-leg severity is expressed for 100 ft, so
$\beta = 5.87° \times 100/75 = 7.82°/100$ ft

101. Problem 3.16

A whipstock was set to change the direction from N85E to S88E. After drilling a sidetrack of 90 ft, it was found that the inclination has changed from 12° to 14.5°. Compute the dogleg severity.

Solution:

Azimuth change is from 85° (N85E) to 92° (S88E). Total azimuth change is 7°.

The overall angle change is given as
$\beta = \arccos(\cos 7 \sin 12 \sin 14.5 + \cos 12 \cos 14.5)$
$\beta = 2.97$
The dogleg severity is

$$DLS = \frac{2.97}{90} \times 100 = 3.29°/100 \text{ ft}$$

102. Problem 3.17

Calculate the toolface angle for the following data:

MD	Angle (°)	Direction (°)
5000 ft	5.5	150
5120 ft	7.5	148

Solution:

The bending angle β can be calculated as shown in Problem 16 and is 2.01°/100 ft.

With other given data

$$\alpha_1 = 5.5°, \alpha_2 = 7.5° \text{ and}$$

The toolface angle is given as

$$\gamma = \arccos\left(\frac{\cos\alpha_1 \cos\beta - \cos\alpha_2}{\sin\alpha_2 \sin\beta}\right)$$

Substituting the values

$$\gamma = \arccos\left[\frac{(\cos 2.01 \cos 5.5 - \cos 7.5)}{(\sin 7.5 \sin 2.01)}\right] = 43.270°$$

103. Problem 3.18

While drilling a directional well, a survey shows the original inclination of 32° and an azimuth of 112°. The inclination and directional curvatures are to be maintained at 4°/100 ft and 7°/100 ft, respectively, for a course length of 100 ft. Calculate the new inclination angle and final direction. Also, determine the curvature, vertical curvature, and horizontal walk curvatures.

Solution:

Given data: $\alpha_1 = 32°$
$\phi_1 = 112°$
$\kappa_\alpha = 4°/100$ ft
$\kappa_\phi = 7°/100$ ft
$\Delta L = 100$ ft

The new inclination angle is

$$\alpha_2 = \alpha_1 + \kappa_\alpha(L_2 - L_1) = 32 + \frac{4}{100} \times 100 = 36°$$

The new direction is

$$\phi_2 = \phi_1 + \kappa_\phi(L_2 - L_1) = 112 + \frac{7}{100} \times 100 = 119°$$

The average inclination angle is

$$\bar{\alpha} = \frac{\alpha_1 + \alpha_2}{2} = \frac{32 + 36}{2} = 34°$$

The average curvature is

$$\kappa = \sqrt{4^2 + 7^2 \sin^2 34} = 5.596°/100 \text{ ft}$$

Using the bending or dogleg angle,

$$\cos\varepsilon = \cos\alpha_1 \cos\alpha_2 + \sin\alpha_1 \sin\alpha_2 \cos\Delta\phi$$

$$\overline{\kappa} = \frac{a\cos(\cos 32 \cos 36 + \sin 32 \sin 36 \cos 7)}{100} = 5.591°/100 \text{ ft}$$

It can be seen that a small difference is observed between the two calculation methods.

Vertical curvature is

$$\kappa_V = \kappa_\alpha = 4°/100 \text{ ft}$$

Horizontal curvature is

$$\kappa_H = \frac{7}{\sin 34} = 12.518°/100 \text{ ft}$$

104. Problem 3.19
Borehole Torsion Calculations

With the given inclination angles and directions, the borehole bending angle can be given as

$$\tau = \frac{\kappa_\alpha \dot{\kappa}_\phi - \kappa_\phi \dot{\kappa}_\alpha}{\kappa^2} \sin\alpha + \kappa_\phi \left(1 + \frac{\kappa_\alpha^2}{\kappa^2}\right) \cos\alpha,$$

where

τ = the torsion of wellbore trajectory, °/30 m or °/100 ft.

$\dot{\kappa}_\alpha$ = the first derivative of inclination change rate, viz., the second derivative of inclination angle.

$\dot{\kappa}_\phi$ = the first derivative of azimuth change rate, viz., the second derivative of azimuth angle.

105. Problem 3.20
Borehole Torsion Calculation—Cylindrical Helical Method

When the wellbore curvature equals zero, the wellpath is a straight line that will result in zero torsion. When $\kappa \neq 0$, the torsion equation for the cylindrical helical model is

$$\tau = \kappa_H \left(1 + \frac{2\kappa_V^2}{\kappa^2}\right) \sin\alpha \cos\alpha$$

106. Problem 3.21

The original hole inclination is 22°. To reach the limits of the target, it is desired to build an angle of 26° in a course length of 100 ft. The directional change is −5°. What is the resulting curvature and torsion that will achieve the desired objective?

Solution:

Given data: $\alpha_1 = 22°$; $\alpha_2 = 26°$; $\Delta\phi = -5°$; $\kappa_\alpha = 4°/100$ ft; $\kappa_\phi = 7°/100$ ft; $\Delta L = 100$ ft

The vertical curvature is

$$\kappa_\alpha = \frac{\alpha_2 - \alpha_1}{(L_2 - L_1)} = \frac{4}{100} \times 100 = 4°/100 \text{ ft}$$

The directional curvature is

$$\kappa_\phi = \frac{\phi_2 - \phi_1}{(L_2 - L_1)} = \frac{-10}{100} \times 100 = -10°/100 \text{ ft}$$

The average inclination angle is

$$\bar{\alpha} = \frac{\alpha_1 + \alpha_2}{2} = \frac{22 + 26}{2} = 24°$$

$$\kappa_\alpha = \kappa_V = 4°/100 \text{ ft}$$

$$\kappa_H = \frac{\kappa_\phi}{\sin\alpha} = \frac{-10}{\sin 24} = -12.294°/100 \text{ ft}$$

The wellbore curvature can be calculated as

$$\kappa = \sqrt{\kappa_V^2 + \kappa_H^2 \sin^4 \alpha} = 4.487(°)/100 \text{ ft}$$

The torsion is calculated as

$$\tau = \kappa_H \left(1 + \frac{2\kappa_V^2}{\kappa^2}\right) \sin\alpha \cos\alpha$$

Substituting the values the torsion is

$$\tau = -12.294 \left(1 + \frac{2 \times 4^2}{4.487^2}\right) \sin 24 \cos 24 = -11.827°/100 \text{ ft}$$

107. Problem 3.22
Wellpath Length Calculations
Circular arc:

$$\text{Arc length} = L_c = \frac{\alpha_2 - \alpha_2}{\text{BRA}},$$

where
BRA = build rate angle (°/100 ft)

Vertical distance is

$$V = R_b (\sin\alpha_2 - \sin\alpha_1),$$

where
R = radius of the build.

$$R_b = \frac{180}{\pi \times \text{BRA}}.$$

Horizontal distance, $H = R_b (\cos\alpha_1 - \cos\alpha_2)$.
Tangent section:
Length of tangent = L_t.
Vertical distance = $V = L_t \cos\alpha$.
Horizontal distance, $H = L_t \sin\alpha$.

108. Problem 3.23

Determine the build rate radius given the following data for a build-and-hold pattern type well:

Build rate angle = 30/100 ft

Solution:

The radius of the build is

$$R_b = \frac{1}{BRA}\left(\frac{180}{\pi}\right) = \frac{100}{3}\left(\frac{180}{\pi}\right) = 1910 \text{ ft}$$

109. Problem 3.24

Wellpath Trajectory Calculations from Survey Data—Minimum Curvature Method

The coordinates x_i, y_i, and z_i, at survey station i, which represent the west to east, south to north, and vertical depth, respectively, are given below:

$$\Delta N_i = \lambda_i (\sin\alpha_{i-1} \cos\phi_{i-1} + \sin\alpha_i \cos\phi_i)$$
$$\Delta E_i = \lambda_i (\sin\alpha_{i-1} \sin\phi_{i-1} + \sin\alpha_i \sin\phi_i)$$
$$\Delta H_i = \lambda_i (\cos\alpha_{i-1} + \cos\alpha_i),$$

where

$$\lambda_i = \frac{180}{\pi} \frac{\Delta L_i}{\varepsilon_i} \tan\frac{\beta_i}{2}$$

$$\cos\varepsilon_i = \cos\alpha_{i-1} \cos\alpha_i + \sin\alpha_{i-1} \sin\alpha_i \cos\Delta\phi_i.$$

β_i = angle change between stations i, and $i-1$ and is given by the following:

For the i-th measured section at the survey station i, the coordinate increment can be calculated and the coordinate of the next measured point also can be obtained. So,

$$N_i = N_{i-1} + \Delta N_i$$
$$E_i = E_{i-1} + \Delta E_i$$
$$H_i = H_{i-1} + \Delta H_i$$

110. Problem 3.25

The inclination and azimuth at two survey stations are given below:
Measured Depth = 10,023.5 ft, Inclination 22°, Azimuth S22W°
Measured Depth = 10,068.0 ft, Inclination 24°, Azimuth S19W°.

Calculate the incremental TVD, northing and easting with the minimum curvature method. Also compute the radius of curvature and dogleg severity.

Solution:

The coordinates x_i, y_i, and z_i which represent the west to east, south to north, and vertical depth, respectively, are given below:

$$\Delta N_i = \lambda_i (\sin\alpha_{i-1} \cos\phi_{i-1} + \sin\alpha_i \cos\phi_i)$$
$$\Delta E_i = \lambda_i (\sin\alpha_{i-1} \sin\phi_{i-1} + \sin\alpha_i \sin\phi_i)$$
$$\Delta H_i = \lambda_i (\cos\alpha_{i-1} + \cos\alpha_i),$$

where

$$\lambda_i = \frac{180}{\pi} \frac{\Delta L_i}{\varepsilon_i} \tan\frac{\beta_i}{2}.$$

$$\cos\varepsilon_i = \cos\alpha_{i-1} \cos\alpha_i + \sin\alpha_{i-1} \sin\alpha_i \cos\Delta\phi_i.$$

β_i = angle change between stations i, and $i-1$

For the i-th measured section, the coordinate increment can be calculated and the coordinate of the next measured point also can be obtained. So,

$$\begin{cases} N_i = N_{i-1} + \Delta N_i \\ E_i = E_{i-1} + \Delta E_i \\ H_i = H_{i-1} + \Delta H_i \end{cases}$$

ε_i = arc cos (cos24 cos22 + sin24 sin22 cos3) = 2.3° = 0.04 radians
 Since it is < 0.25 radians $\lambda_i = 1$

$$\Delta N_i = \frac{10{,}068 - 10{,}023.5}{2}(\sin 22 \cos 202 + \sin 24 \cos 199) = -16.28 \text{ ft}$$

$$\Delta E_i = \frac{44.5}{2}(\sin 22 \sin 202 + \sin 24 \sin 199) = -6.07 \text{ ft}$$

$$\Delta H_i = \frac{44.5}{2}(\cos 22 + \cos 24) = 40.96 \text{ ft}$$

Dogleg severity is

$\varepsilon_i = \arccos(\cos 24 \cos 22 + \sin 24 \sin 22 \cos 3) = 2.3$

$$DLS = \frac{2.3}{(44.5)} \times 100 = 5.17^0/100 \text{ ft}$$

Build rate angle $= BRA = \dfrac{2}{44.5} \times 100 = 4.49^0/100 \text{ ft}$

Radius of curvature $R = \dfrac{180}{(\pi)} \times \dfrac{100}{4.49} = 1276.7 \text{ ft}$

111. Problem 3.26

The inclination and azimuth at two survey stations in a horizontal well are given below:

Measured Depth = 11,200 ft, Inclination 90°, Azimuth 192°
Measured Depth = 11,291 ft, Inclination 90°, Azimuth 195°

Calculate the incremental TVD, northing and easting with the minimum curvature method. Also compute the radius of curvature and dogleg severity.

Solution:

The coordinates x_i, y_i and z_i which represent the west to east, south to north and vertical depth, respectively, are given below:

The angle change between the survey station is

$\beta_i = \arccos(\cos 90 \cos 90 + \sin 90 \sin 90 \cos 3) = 3^0 = 0.05$ radians

$$\lambda_i = \frac{180(12{,}291-12{,}200)}{\pi} \frac{\tan\dfrac{3}{2}}{0.05} = 2607$$

$\Delta N_i = 2607(\sin 90 \cos 192 + \sin 90 \cos 195) = 53$ ft
$\Delta E_i = 2607(\sin 90 \sin 192 + \sin 90 \sin 195) = -21$ ft
$\Delta H_i = 2607(\cos 90 + \cos 90) = 0$ ft

Dogleg severity is

$\varepsilon_i = \arccos(\cos 24 \cos 22 + \sin 24 \sin 22 \cos 3) = 2.3$

$$DLS = \frac{3}{(90)} \times 100 = 3.29°/100 \text{ ft}$$

Build rate angle $= BRA = \frac{0}{(90)} \times 100 = 0°/100 \text{ ft}$

Radius of curvature $R = \frac{180}{(\pi)} \times \frac{100}{0} = \infty \text{ ft}$

112. Problem 3.27
Wellpath Trajectory Calculations from Survey Data -Radius of Curvature Method

From survey data at two consecutive stations $i-1$ and i, the coordinates can be determined by the following equations:

$$\Delta N_i = r_i (\sin\phi_i - \sin\phi_{i-1})$$
$$\Delta E_i = r_i (\cos\phi_{i-1} - \cos\phi_i)$$
$$\Delta H_i = R_i (\sin\alpha_i - \sin\alpha_{i-1}),$$

where

$$R_i = \frac{180}{\pi} \frac{\Delta L_i}{\Delta \alpha_i}, \quad r_i = \frac{180}{\pi} \frac{R_i}{\Delta \phi_i} (\cos\alpha_{i-1} - \cos\alpha_i).$$

$$\Delta \alpha_i = \alpha_i - \alpha_{i-1}.$$

R = radius of curvature in the vertical plane, (°)/30 m or (°)/100 ft.
r = radius of curvature on the horizontal projection, (°)/30 m or (°)/100 ft.

Obviously, if either $\Delta\alpha_i$ or $\Delta\phi_i$ is zero, then it is not possible to calculate R_i or r_i. Following are the general formulas that are applicable for all the cases:

$$\Delta H_i = \begin{cases} \Delta L_i \cos\alpha_i, & \text{if } \Delta\alpha_i = 0 \\ R_i(\sin\alpha_i - \sin\alpha_{i-1}), & \text{if } \Delta\alpha_i \neq 0 \end{cases},$$

$$\Delta S_i = \begin{cases} \Delta L_i \sin\alpha_i, & \text{if } \Delta\alpha_i = 0 \\ R_i(\cos\alpha_{i-1} - \cos\alpha_i), & \text{if } \Delta\alpha_i \neq 0 \end{cases},$$

$$\Delta N_i = \begin{cases} \Delta S_i \cos\phi_i, & \text{if } \Delta\phi_i = 0 \\ r_i(\sin\phi_i - \sin\phi_{i-1}), & \text{if } \Delta\phi_i \neq 0 \end{cases},$$

$$\Delta E_i = \begin{cases} \Delta S_i \sin\phi_i, & \text{if } \Delta\phi_i = 0 \\ r_i(\cos\phi_{i-1} - \cos\phi_i), & \text{if } \Delta\phi_i \neq 0 \end{cases},$$

where

$$R_i = \frac{180}{\pi}\frac{\Delta L_i}{\Delta\alpha_i}$$

$$r_i = \frac{180}{\pi}\frac{\Delta S_i}{\Delta\phi_i}$$

113. Problem 3.28

Data at two survey depths, 1015 ft and 1092.6 ft, are given below.
Depth 1015 ft: inclination 12°, azimuth 123°
Depth 1092.6 ft: inclination 14°, azimuth 119°
Calculate the incremental true vertical depth (TVD), incremental northing, and incremental easting. Also, compute the dogleg severity.
Solution:

$$R_i = \frac{180}{\pi}\frac{(1092.6-1015)}{2} = 2223.07;$$

$$r_i = \frac{180}{\pi} \times \frac{2223.07}{4} = 250.03$$

$\Delta N_i = 250.03(\sin 119 - \sin 123) = 8.98$ ft

$\Delta E_i = 250.03(\cos 119 - \cos 123) = 14.96$ ft

$\Delta H_i = 2223.06(\sin 14 - \sin 12) = 75.6$ ft

Alternatively, the displacements can be calculated as follows:

$$\Delta H_i = \frac{\Delta MD_i\,(\sin\alpha_i - \sin\alpha_{i-1})}{\alpha_i - \alpha_{i-1}}\left(\frac{180}{\pi}\right)$$

$$= \frac{(1092.6-1015)\,(\sin 14 - \sin 12)}{14-12}\left(\frac{180}{\pi}\right) = 78.53 \text{ ft}$$

$$\Delta L_{(N/S)i} = \frac{(1092.6-1015)\,(\cos 14 - \cos 12)\,(\sin 119 - \sin 123)}{(14-12)(123-119)}\left(\frac{180}{\pi}\right)^2$$

$$= 9.33 \text{ ft}$$

$$\Delta L_{(E/W)i} = \frac{(1092.6-1015)\,(\cos 14 - \cos 12)\,(\cos 119 - \cos 123)}{(14-12)(123-119)}\left(\frac{180}{\pi}\right)^2$$

$$= 15.53 \text{ ft}$$

The dogleg angle is
$\beta = arc\,\cos(\cos 4 \sin 14 \sin 12 + \cos 12 \cos 12) = 2.19°$
The dogleg severity is

$$\frac{\beta}{\Delta L_i} \times 100 = \frac{2.19}{(1092.6-1015)} \times 100 = 2.83°/100 \text{ ft}$$

114. Problem 3.29

The inclination and azimuth at two survey stations are given below:
Measured Depth = 1012 ft, Inclination 12°, Azimuth 123°
Measured Depth = 1092.6 ft, Inclination 14°, Azimuth 119°.
Calculate the displacements with the radius of curvature method. Also compute the dogleg severity.
Solution:

$$\Delta TVD_i = \frac{\Delta MD_i\,(\sin\alpha_i - \sin\alpha_{i-1})}{\alpha_i - \alpha_{i-1}}\left(\frac{180}{\pi}\right)$$

$$= \frac{(1092.6-1012)\,(\sin 14 - \sin 12)}{14-12}\left(\frac{180}{\pi}\right) = 78.53$$

$$\Delta L_{(N/S)i} = \frac{(1092.6-1012)(\cos 14 - \cos 12)(\sin 119 - \sin 123)}{(14-12)(123-119)}$$

$$\left(\frac{180}{\pi}\right)^2 = 9.33 \text{ ft}$$

$$\Delta L_{(E/W)i} = \frac{(1092.6-1012)(\cos 14 - \cos 12)(\cos 119 - \cos 123)}{(14-12)(123-119)}$$

$$\left(\frac{180}{\pi}\right)^2 = 15.53$$

Dogleg angle is given as

$\beta = \arccos(\cos \Delta\varepsilon \sin \alpha_n \sin \alpha + \cos \alpha \cos \alpha_n)$
$\beta = \arccos(\cos 4 \sin 14 \sin 12 + \cos 14 \cos 12)$
$\beta = 2.19$

So the dogleg severity is $DLS = \dfrac{2.19}{(1092.6-1012)} \times 100 = 2.72°/100 \text{ ft}$

115. Problem 3.30

Wellpath Trajectory Calculations from Survey Data—Natural Curve Method

The course coordinates are given by

$$\Delta N_i = \frac{1}{2}\left[F_C\left(A_{P,i}, \kappa_{P,i}, \Delta L_i\right) + F_C\left(A_{Q,i}, \kappa_{Q,i}, \Delta L_i\right)\right],$$

$$\Delta E_i = \frac{1}{2}\left[F_S\left(A_{P,i}, \kappa_{P,i}, \Delta L_i\right) - F_S\left(A_{Q,i}, \kappa_{Q,i}, \Delta L_i\right)\right],$$

$$\Delta H_i = F_S\left(\alpha_{i-1}, \kappa_{a,i}, \Delta L_i\right),$$

where

$$\begin{cases} \kappa_{a,i} = \dfrac{\Delta \alpha_i}{\Delta L_i} \\[6pt] \kappa_{\phi,i} = \dfrac{\Delta \phi_i}{\Delta L_i} \end{cases}$$

$$\begin{cases} A_{P,i} = \alpha_{i-1} - \phi_{i-1} \\ A_{Q,i} = \alpha_{i-1} + \phi_{i-1} \end{cases}$$

$$\begin{cases} \kappa_{P,i} = \kappa_{\alpha,i} - \kappa_{\phi,i} \\ \kappa_{Q,i} = \kappa_{\alpha,i} + \kappa_{\phi,i} \end{cases}.$$

$$F_C(\theta, \kappa, \lambda) = \begin{cases} \lambda \sin \theta & \text{if } \kappa = 0 \\ \dfrac{180}{\pi \kappa}[\cos \theta - \cos(\theta + \kappa \lambda)] & \text{if } \kappa \neq 0 \end{cases}$$

$$F_S(\theta, \kappa, \lambda) = \begin{cases} \lambda \cos \theta & \text{if } \kappa = 0 \\ \dfrac{180}{\pi \kappa}[\sin(\theta + \kappa \lambda) - \sin \theta] & \text{if } \kappa \neq 0 \end{cases}$$

116. Problem 3.31

Wellpath Trajectory Calculations from Survey Data—Constant Tool Face Angle Method

Using the constant tool face angle method, course coordinates can be calculated using the following equations:

$$\Delta N_i = \begin{cases} R_i(\cos \alpha_{i-1} - \cos \alpha_i)\cos \phi_i, & \text{when } \alpha_{i-1} = 0 \text{ or } \alpha_i = 0 \\ r_i \sin \alpha_i (\sin \phi_i - \sin \phi_{i-1}), & \text{when } \Delta \alpha_i = 0 \\ \int_{L_1}^{L_2} \sin \alpha(L) \cos \phi(L) dL, & \text{for rest of the conditions} \end{cases},$$

$$\Delta E_i = \begin{cases} R_i(\cos \alpha_{i-1} - \cos \alpha_i)\sin \phi_i, & \text{if } \alpha_{i-1} = 0 \text{ or } \alpha_i = 0 \\ r_i \sin \alpha_i (\cos \phi_{i-1} - \cos \phi_i), & \text{if } \Delta \alpha_i = 0 \\ \int_{L_1}^{L_2} \sin \alpha(L) \sin \phi(L) dL, & \text{for rest of the conditions} \end{cases},$$

$$\Delta H_i = \begin{cases} \Delta L_i \cos \alpha_i, & \text{if } \Delta \alpha_i = 0 \\ R_i(\sin \alpha_i - \sin \alpha_{i-1}), & \text{if } \Delta \alpha_i \neq 0 \end{cases},$$

where

$$R_i = \frac{180}{\pi} \frac{\Delta L_i}{\Delta \alpha_i}$$

$$r_i = \frac{180}{\pi} \frac{\Delta L_i}{\Delta \phi_i}$$

$$\tan \omega_i = \frac{\pi}{180} \frac{\Delta \phi_i}{\ln \frac{\tan \frac{\alpha_i}{2}}{\tan \frac{\alpha_{i-1}}{2}}},$$

$$\alpha(L) = \alpha_{i-1} + \frac{\Delta \alpha_i}{\Delta L_i}(L - L_{i-1})$$

$$\phi(L) = \phi_{i-1} + \frac{180}{\pi} \tan \omega_i \cdot \ln \frac{\tan \frac{\alpha(L)}{2}}{\tan \frac{\alpha_{i-1}}{2}}$$

117. Problem 3.32

Original hole inclination is 22°. To reach the limits of the target, it is desired to build an angle of 26° in a course length of 100 ft. The directional change is −5°. What is the resulting curvature and torsion that will achieve the desired objective?

Solution:

Given data: $\alpha_1 = 22°$;
$\alpha_2 = 26°$;
$\Delta \phi = -5°$;
$\kappa_\alpha = 4°/100$ ft;
$\kappa_\phi = 7°/100$ ft; and $\Delta L = 100$ ft

The vertical curvature is

$$\kappa_\alpha = \frac{\alpha_2 - \alpha_1}{(L_2 - L_1)} = \frac{4}{100} \times 100 = 4°/100 \text{ ft}$$

The horizontal curvature is

$$K_\phi = \frac{\phi_2 - \phi_1}{(L_2 - L_1)} = \frac{-10}{100} \times 100 = -10°/100 \text{ ft}$$

The average inclination angle is

$$\bar{\alpha} = \frac{\alpha_1 + \alpha_2}{2} = \frac{22 + 26}{2} = 24$$

Average curvature is

$$K = \sqrt{4^2 + (-10)^2 \sin^2 24} = 4.487°/100 \text{ ft}$$

Calculating the torsion,

$$\tau = -10\left(1 + \frac{4^2}{4.487^2}\right)\cos 24 = -8.197°/100 \text{ ft}$$

118. Problem 3.33

Calculate the course coordinates with the following data:

- Initial direction and azimuth are 35° and 182°, respectively
- $K_\alpha = 4\ °/100$ ft
- $K_\phi = 0\ °/100$ ft
- The tangent length = 200 ft

Calculate the course coordinates.

Solution:

The new angle and azimuth are

$$\alpha = 35 + \frac{4}{100} \times 200 = 43°$$

and

$$\phi = 182 + \frac{0}{100} \times 200 = 182°$$

$$\kappa_V = 4°/100 \text{ ft}$$

$$\kappa_H = \frac{0}{\sin 43} = 0°/100 \text{ ft}$$

$$\kappa = \sqrt{4^2 + 0} = 4°/100 \text{ ft}$$

Calculating the intermediate variables

$$\begin{cases} A_P = 35 + 182 = 217 \\ A_Q = 35 - 182 = -147 \end{cases} \text{ and } \begin{cases} \kappa_P = 4 + 0 = 4 \text{ deg}/100 \text{ ft} \\ \kappa_Q = 4 - 0 = 4 \text{ deg}/100 \text{ ft} \end{cases}$$

$$F_S(217, 4, 200) = \frac{180 \times 100}{\pi \times 4} \left[\sin\left(217 + \frac{4}{100} \times 200\right) - \sin 217 \right] = -150.8 \text{ ft}$$

$$F_C(217, 4, 200) = \frac{180 \times 100}{\pi \times 4} \left[\cos 217 - \sin\left(217 + \frac{4}{100} \times 200\right) \right] = -131 \text{ ft}$$

$$F_S(-147, 4, 200) = \frac{180 \times 100}{\pi \times 4} \left[\sin\left(-147 + \frac{4}{100} \times 200\right) - \sin(-147) \right]$$
$$= -159.6 \text{ ft}$$

$$F_C(-147, 4, 200) = \frac{180 \times 100}{\pi \times 4} \left[\cos(-147) - \sin\left(-147 + \frac{4}{100} \times 200\right) \right]$$
$$= -120.26 \text{ ft}$$

$$F_S(35, 4, 200) = \frac{180 \times 100}{\pi \times 4} \left[\sin\left(35 + \frac{4}{100} \times 200\right) - \sin(35) \right] = 155.3 \text{ ft.}$$

$$F_C(35, 4, 200) = \frac{180 \times 100}{\pi \times 4} \left[\cos(35) - \sin\left(35 + \frac{4}{100} \times 200\right) \right] = 125.76 \text{ ft}$$

The course coordinates are

$$\Delta N = \frac{1}{2} \left[F_C(217, 4, 200) + F_C(-147, 4, 200) \right] = -125.68 \text{ ft,}$$

$$\Delta E = \frac{1}{2}\left[F_S(-147,4,200) - F_S(-147,4,200)\right] = -4.4\,\text{ft},$$

$$\Delta H = F_S(50,4,200) = 155\,\text{ft},$$

and

$$\Delta S = F_C(50,4,200) = 125.76\,\text{ft}$$

119. Problem 3.34

Using SI units, Problem 33 is worked out below:
The new angle and azimuth are

$$\alpha = 35 + \frac{4}{30.48} \times 60.96 = 43°$$

and

$$\phi = 182 + \frac{0}{100} \times 200 = 182°$$

Therefore,

$$\kappa = \sqrt{4^2 + 0} = 4°/30.48\,\text{m}$$

Calculating the intermediate variables

$$\begin{cases} A_P = 35+182 = 217° \\ A_Q = 35-182 = -147° \end{cases} \text{ and } \begin{cases} \kappa_P = 4+0 = 4\,\text{deg}/30.48\text{m} \\ \kappa_Q = 4-0 = 4\,\text{deg}/30.48\text{m} \end{cases}.$$

$$F_S(217,4,60.96) = \frac{180 \times 30.48}{\pi \times 4}\left[\sin\left(217 + \frac{4}{30.48} \times 60.96\right) - \sin 217\right]$$
$$= -45.96\,\text{m}$$

$$F_C(217,4,60.96) = \frac{180 \times 30.48}{\pi \times 4}\left[\cos 217 - \sin\left(217 + \frac{4}{30.48} \times 60.96\right)\right]$$
$$= -39.96\,\text{m}$$

$$F_S(-147,4,60.96) = \frac{180 \times 30.48}{\pi \times 4}\left[\sin\left(-147 + \frac{4}{30.48} \times 60.96\right) - \sin(-147)\right]$$
$$= -48.64\,\text{m}$$

$$F_c(-147,4,60.96) = \frac{180 \times 30.48}{\pi \times 4}\left[\cos(-147) - \sin\left(-147 + \frac{4}{30.48} \times 60.96\right)\right]$$
$$= -36.65\,\text{m}$$

$$F_S(35,4,60.96) = \frac{180 \times 30.48}{\pi \times 4}\left[\sin\left(35 + \frac{4}{30.48} \times 60.96\right) - \sin(35)\right] = 47.33\,\text{m}$$

$$F_c(35,4,60.96) = \frac{180 \times 30.48}{\pi \times 4}\left[\cos(35) - \sin\left(35 + \frac{4}{30.48} \times 60.96\right)\right]$$
$$= 38.33\,\text{m}$$

The course coordinates are

$$\Delta N = \frac{1}{2}\left[F_C(217,4,60.96) + F_C(-147,4,60.96)\right] = -38.30\,\text{m}$$

$$\Delta E = \frac{1}{2}\left[F_S(-147,4,60.96) - F_S(-147,4,200)60.96\right] = -1.33\,\text{m}$$

$$\Delta H = F_S(35,4,60.96) = 47.33\,\text{m}$$

and

$$\Delta S = F_C(35,4,60.96) = 38.33\,\text{m}$$

120. Problem 3.35

For the following survey data calculate the TVD, N/S, E/W coordinates using minimum curvature, radius of curvature, and natural curve methods.

Estimate the relative errors between the methods.

MD (ft)	INC (°)	AZ (°)
13,000	0	0
13,100	5.39	170

Solution:

$\Delta MD = 100$ ft
$K_\alpha = 8; K_\phi = 0.0°/100$ ft

$$\begin{cases} A_{P,i} = -164.61 \\ A_{Q,i} = 175.39 \end{cases} ; \begin{cases} \kappa_{P,i} = 8 \\ \kappa_{Q,i} = 8 \end{cases}$$

$$F_C(\theta, \kappa, \lambda) = \frac{18{,}000}{\pi \times 8} \left[\cos(-164.61) - \cos(-164.61 + 0.08 \times 100) \right] = -33.17$$

$$F_S(\theta, \kappa, \lambda) = \frac{18{,}000}{\pi \times 8} \left[\sin(-164.61 + 0.08 \times 100) - \sin(-164.61) \right] = 1.06$$

$$\Delta N_i = \frac{1}{2}[-33.17 + 1.06] = 16.05 \text{ ft}$$

121. Problem 3.36

Draw the following wellpath designs in a N-E-H coordinate.

- J-type
- S-type
- Horizontal
- Extended reach
- Undercut

Solution:
The different designs are shown in Figure 3.3.

122. Problem 3.37

Tool Face Angle Change Calculations

The following equations provide different combinations between α, new inclination α_n, dogleg severity β, and the toolface angle γ.

1. If the inclination α, new inclination α_n, and the dogleg severity β are known the toolface angle can be calculated using the following relationship:

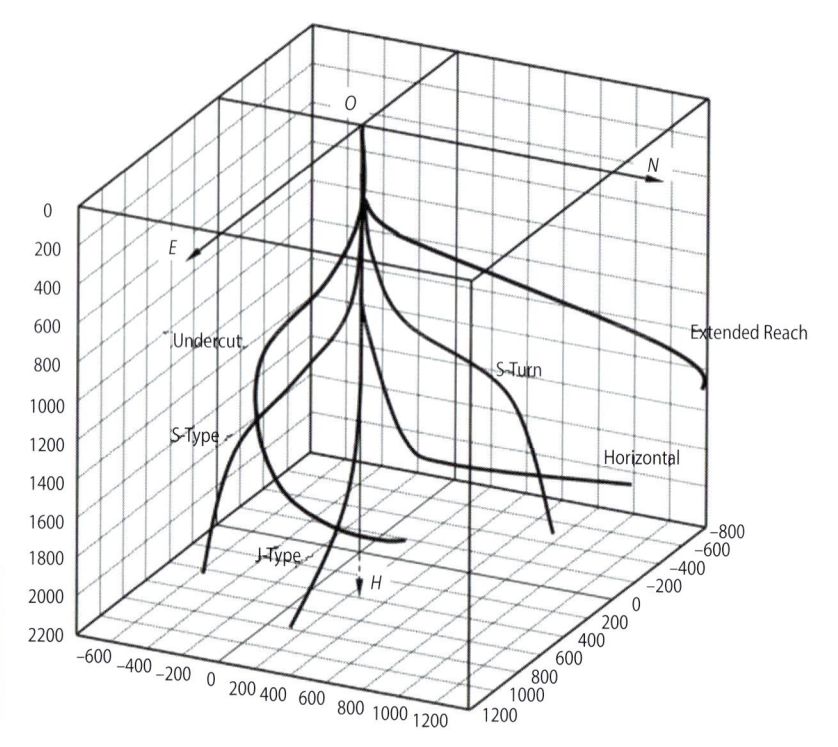

Figure 3.3   Wellpath designs

$$\gamma = arc\cos\left(\frac{\cos\alpha\cos\beta - \cos\alpha_n}{\sin\alpha\sin\beta}\right)$$

2. If the new inclination α_n, azimuth change $\Delta\phi$, and the dogleg severity β are known the toolface angle can be calculated using the following relationship:

tool face rotation angle

$$\gamma = arc\sin\left(\frac{\sin\alpha_n \sin\Delta\phi}{\sin\beta}\right)$$

3. If the inclination α, the dogleg severity β, and the toolface angle γ are known the new inclination angle can be calculated using the following relationship:

New direction

$$\alpha_n = \arccos(\cos\alpha\cos\beta - \sin\beta\sin\alpha\cos\gamma)$$

4. If the inclination α, the dogleg severity β, and the toolface angle γ are known the change in azimuth can be calculated using the following relationship:

$$\Delta\varepsilon = \arctan\left(\frac{\tan\beta\sin\gamma}{\sin\alpha + \tan\beta\cos\alpha\cos\gamma}\right)$$

123. Problem 3.38

Derive an equation to find the overall angle change in terms of $\Delta\phi$, α, and γ, the toolface rotation angle.

Solution:

Starting from the azimuth change

$$\Delta\phi = \arctan\frac{\tan\beta\sin\gamma}{\sin\alpha + \tan\beta\cos\alpha\cos\gamma}$$

$$\tan\Delta\phi = \frac{\tan\beta\sin\gamma}{\sin\alpha + \tan\beta\cos\alpha\cos\gamma}$$

Rearranging the equation

$$\tan\Delta\phi\sin\alpha + \tan\Delta\phi\tan\beta\cos\alpha\cos\gamma - \tan\beta\sin\gamma = 0$$

$$\tan\Delta\phi\sin\alpha + \tan\beta(\tan\Delta\phi\cos\alpha\cos\gamma - \sin\gamma) = 0$$

Solving the above equation to get the overall angle change β

$$\beta = \arctan\frac{\tan\Delta\phi\sin\alpha}{\sin\gamma - \tan\Delta\phi\cos\alpha\cos\gamma}$$

124. Problem 3.39

A drilling engineer plans to kick off the well from the surface with an initial holding inclination of α_1. The complete path planning parameters and the well configuration (straight inclined build-hold-drop-hold) are depicted in Figure 3.4.

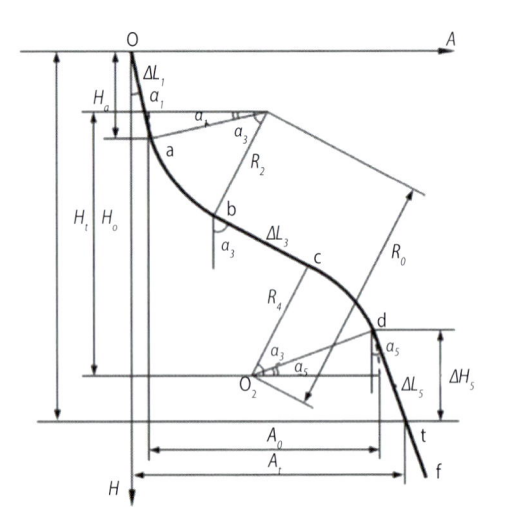

Figure 3.4 Wellpath design for problem 3.39

Prove that $\tan\dfrac{\alpha_3}{2} = \dfrac{H_o - \sqrt{H_o^2 + A_o^2 - R_o^2}}{R_o - A_o}$

Also find out the condition at which $\alpha_3 = 2\tan^{-1}\left(\dfrac{A_o}{2H_o}\right)$

Solution:

From the figure, it can be written as $R_0 = R_2 + R_4$.
The following relationship can be obtained:

$$A_0 = \Delta L_3 \sin\alpha_3 + R_0(1 - \cos\alpha_3)$$

$$H_0 = \Delta L_3 \cos\alpha_3 + R_0 \sin\alpha_3$$

Multiplying the second equation by $\cos\alpha_3$ and the third equation by $\sin\alpha_3$ and some manipulation will result in $H_0 \sin\alpha_3 + (R_0 - A_0)\cos\alpha_3 = R_0$

Using the identity $\sin\alpha_3 = \dfrac{2\tan\dfrac{\alpha_3}{2}}{1+\tan^2\dfrac{\alpha_3}{2}}, \cos\alpha_3 = \dfrac{1-\tan^2\dfrac{\alpha_3}{2}}{1+\tan^2\dfrac{\alpha_3}{2}}$

$$(2R_0 - A_0)\tan^2\dfrac{\alpha_3}{2} - 2H_0 \tan\dfrac{\alpha_3}{2} + A_0 = 0$$

Well Path Design

Solving the quadratic equation and using the positive root will result in

$$\tan\frac{\alpha_3}{2} = \frac{H_0 \pm \sqrt{H_0^2 + A_0^2 - 2R_0 A_0}}{2R_0 - A_0}$$

In the result above, if $(2R_0 - A_0) = 0$, then

$$\tan\frac{\alpha_3}{2} = \frac{A_0}{2H_0}$$

i.e., $2R_0 = A_0$

Based on the equations and condition derived, similar derivations can be done with the condition:

$$\begin{cases} \Delta L_3 \cos\alpha_3 + R_0 \sin\alpha_3 = H_0 \\ \Delta L_3 \sin\alpha_3 - R_0 \cos\alpha_3 = A_0 \end{cases}$$

$$\tan\frac{\alpha_3}{2} = \frac{H_0 - \sqrt{H_0^2 + A_0^2 - R_0^2}}{R_0 - A_0}$$

Therefore,

$$\alpha_3 = 2\tan^{-1}\left(\frac{A_0}{2H_0}\right)$$

125. Problem 3.40

A sidetracking was planned using a whipstock after a futile attempt to retrieve a fish. The survey data at the planned sidetrack depth are an inclination angle of 10° and an azimuth of N25W. The whipstock selected will produce a total angle change of 4°/100 ft. The new angle desired is 11°. Calculate the toolface setting of the whipstock with reference to the high side of the wellbore for a course length of 30 ft. Also, calculate the new direction.

Solution:

The overall angle change (dogleg) β is

$$\beta = \frac{4}{100} 30 = 1.2°$$

If α, β, $α_n$ are known tool face rotation angle

$$\gamma = \arccos\left(\frac{\cos\alpha\cos\beta - \cos\alpha_n}{\sin\alpha\sin\beta}\right)$$

Toolface angle = $\gamma = \arccos\left(\frac{\cos 10 \cos 1.2 - \cos 11}{\sin 10 \sin 1.2}\right)$ 35.4^0

Since β, α, γ are known the change of direction

$$\Delta\varepsilon = \arctan\left(\frac{\tan\beta\sin\gamma}{\sin\alpha + \tan\beta\cos\alpha\cos\gamma}\right)$$

Substituting the respective values

$$\Delta\varepsilon = 335^0 + \arctan\left(\frac{\tan 1.2 \sin 35.4}{\sin 10 + \tan 1.2 \cos 10 \cos 35.4}\right)$$
$$= 338.6^0 \, (N21.6W)$$

126. Problem 3.41

Survey was taken at a depth of 4100 ft and found to be 5° inclination and direction N15W. The directional driller desires to make a course correction using a bent sub and a motor. The bent sub is expected to result in an angle change of 3° per 100 ft. The tool face setting is 35° to the right from the high side. Calculate the new direction and the angle at 4190 ft. What would be the bent sub angle response needed if the final inclination angle is 7°.
Solution:

$$\beta = \frac{3}{100} \times 90 = 2.7°$$

The azimuth change is

$$\Delta\phi = 345^0 + \arctan\left(\frac{\tan 2.7 \sin 35}{\sin 5 + \tan 2.7 \cos 5 \cos 35}\right) = 357.2^0 \, (N2.8W)$$

$\alpha_n = \arccos(\cos 5 \cos 2.7 - \sin 2.7 \sin 5 \cos 35) = 7.38$

Well Path Design

$$\delta = \frac{arc\cos(\cos 12.2 \sin 7.38 \sin 5 + \cos 5 \cos 7.38)}{90} \times 100 = 2.6°/100 \text{ ft}$$

Also, you can find it iteratively.

127. Problem 3.42
Horizontal Displacement Calculations

On the horizontal plane, the displacement of the target point can be calculated as

$$H_t = \sqrt{(N_t - N_o)^2 + (E_t - E_o)^2},$$

where
N_t = northing of target, ft.
N_o = northing of slot, ft.
E_t = easting of target, ft.
E_o = easting of slot, ft.
Target bearing is given as

$$\varphi_t = \tan^{-1}\left(\frac{E_t - E_o}{N_t - N_o}\right)$$

128. Problem 3.43

With the slot coordinates 1200.5 ft N, 700 ft E and target coordinates 3800 ft N, 4520 ft E, calculate the horizontal displacement of the well. Also, calculate the target bearing.

Solution:
Given: N_t = 3800 ft, N_o = 1200.5 ft, E_t = 4520 ft, and 700 ft
The horizontal displacement is

$$H_t = \sqrt{(3800 - 1200.5)^2 + (4520 - 700)^2} = 4620.6 \text{ ft}$$

The target bearing is

$$\varphi_t = \tan^{-1}\left(\frac{4520 - 700}{3800 - 1200.5}\right) = 55.76° \text{ or N55.76E}$$

129. Problem 3.44

A well is to be drilled in a 500-acre lease land. The surface location is pegged at 1000 ft from the north line and 2200 ft from the east line. The target location is 500 ft from the south line and 300 ft from the east line. Estimate the horizontal departure.

Solution:

Surface and the bottom locations can be shown in Figure 3.5.

Solution:

1 acre = 43,560 sq. ft
Therefore, the total lease area = 500 (acre) × 43,560 (sq. ft/acre) = 21,780,000 sq. ft
Assuming a square lease area, the side of the lease area is

$\sqrt{21,780,000} = 4666.9$ ft

The horizontal displacement is

$H_t = \sqrt{(4666.9 - 1000 - 500)^2 + (22,200 - 300)^2} = 3693.14$ ft

130. Problem 3.45

Determine the horizontal departure of the following build-hold type well:

Target TVD = 10,000 ft
TVD at KOP depth = 2500 ft

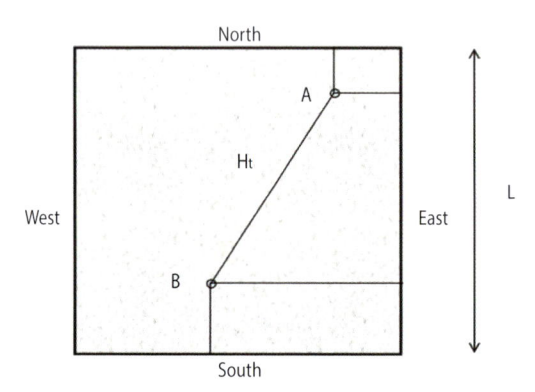

Figure 3.5 Surface location for problem 3.44

Build rate = 2°/ 100 ft
Surface coordinates: 20 ft N 5ft E
Target coordinates: 1800 ft N 5000 ft E

Solution:

Horizontal departure:

$$X_3 = \sqrt{(5000-5)^2 + (1800-2)^2} = 5302.68 \text{ ft}$$

131. Problem 3.46

Plan a build and hold directional well whose surface location on 600 acre section is 700 ft from the north line and 1000 ft from the east line as shown in Figure 3.6.

Bottom location is 1200 ft from the south line and 2000 ft from the east line.

Calculate the horizontal departure to the end of the build and at TVDs of 5450, 6000 and 8500 ft.

Solution:

Area of the field = 600 acre

Since 1 acre = 43,560 sq. ft, area of the field is 26,136,000 sq. ft

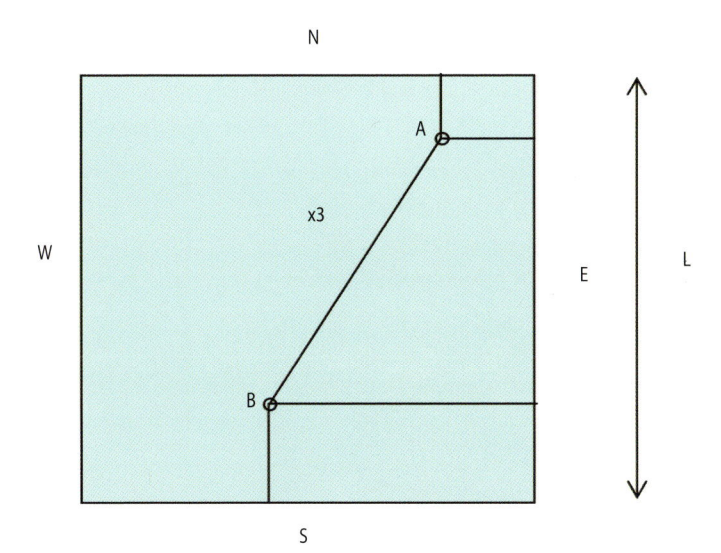

Figure 3.6 Surface location for problem 3.46

Surface coordinates:
700 ft from the north line
1000 ft from the east line
Bottom location:
1200 ft from the South line
2000 ft from the East line
Assume a square area and calculate the horizontal departure:
Side of the square field will be 5112.34 ft.

$$X_3 = \sqrt{(B_E - S_E)^2 + (L - B_S - S_N)^2}$$
$$X_3 = \sqrt{(2000 - 1000)^2 + (5112.34 - 1200 - 700)^2}$$
$$X_3 = 3364.4 \text{ ft}$$

132. Problem 3.47

A drilling engineer is planning to estimate the vertical depth of a payzone, (Fig. 3.7) with the data from the portion of a straight inclined pilot hole drilled though the formation. The formation top and bottom were calculated from the rate of penetration (ROP) and cuttings sample changes. Extrapolate from the last survey point and calculate the formation depth and thickness. Survey details at Point S_1 is 8321 ft MD, Inc 43⁰ AZM 132⁰, and TVD 6987 ft.

Top of the formation, 8.411ft (MD) and bottom of the formation, 8.561 ft (MD).

Calculate the TVD of top and bottom formation.

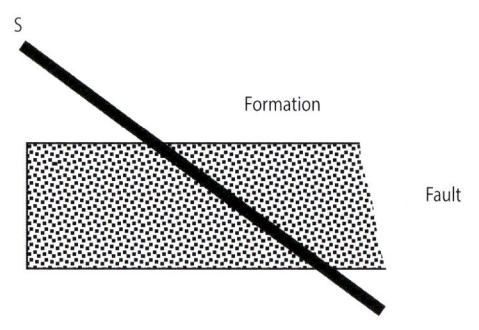

Figure 3.7 Surface location for problem 3.47

Solution:

The true vertical distance from the survey point to the top of the formation is

$X_1 = 90 \times \cos 43 = 65.82$ ft

Total measured distance to the bottom of the formation from S_1 is

$(8411 - 8321) + (8561 - 8411) = 240$ ft

The true vertical distance from the survey point to the bottom of the formation is

$$X_2 = 240 \times \cos 43 = 175.52 \text{ ft}$$

TVD of formation top is $6987 + 65.82 = 7052.82$ ft
TVD of formation bottom is $6987 + 175.52 = 7162.52$ ft

133. Problem 3.48

If the curvature is $\kappa \equiv 0$ then the condition is satisfied only if the curve is a _____

Solution:

If the curvature is $\kappa \equiv 0$ then the condition is satisfied only if the curve is a <u>straight line.</u>

134. Problem 3.49

For a course length of 300 ft and a dogleg severity of 3°/100 ft. determine the tool-face angle for a jetting run whose purpose is to obtain maximum allowable drop while changing the original direction from S10E to S30E. The inclination angle before the course change is 15°.

Solution:

Initial inclination is 15°
Using the dogleg of 3°/100 ft, the final inclination after 300 ft is 9°.
Given dogleg, initial inclination, and change in direction

$\beta = \arccos(\cos\Delta\varepsilon \sin\alpha_n \sin\alpha + \cos\alpha \cos\alpha_n)$
$9 = \arccos(\cos 20 \sin\alpha_n \sin 15 + \cos 15 \cos\alpha_n)$
$9 = \arccos(0.2432 \sin\alpha_n + 0.9660 \cos\alpha_n)$
$0.9876 = (0.2432 \sin\alpha_n + 0.9660 \cos\alpha_n)$
Solving will yield 6.65°

$$\gamma = \arcsin\left(\frac{\sin\alpha_n \sin\Delta\varepsilon}{\sin\beta}\right)$$

$$\gamma = \arcsin\left(\frac{\sin 6.65 \times \sin 20}{\sin 9}\right) = 14.5°$$

1. **Tortuosity**

Wellbore tortuosity is given as

$$T = \frac{\sum_{i=1}^{m} \alpha_{n-1} + \Delta D \times \beta_i}{D_i - D_{i-1}},$$

where
β = dogleg angle
ΔD = calculation course length
D = depth
i = index for survey stations

135. Problem 3.50
Absolute and Relative Tortuosity Calculations

$$\Gamma_{(abs)_n} = \left(\frac{\sum_{i=1}^{i=n} \alpha_{adj}}{D_n + \Delta D_n}\right),$$

where
$\alpha_{adj} = \alpha_i + \Delta D_i \times \beta_i$ and is the dogleg adjusted, summed total inclination angle.

$$\Gamma_{(rel)_n} = \Gamma_{(abs)_n}^{tor} - \Gamma_{(abs)_n}^{notor}\ °/100\ \text{ft}$$

The period of the sine wave should not be of $\frac{2}{n}(n = 1, 2, 3...)$

Different methods include (1) the sine wave method, (2) the helical method, (3) random inclination and the azimuth method, and (4) the random inclination dependent azimuth method.

136. Problem 3.51
Tortuosity—Sine Wave Method

$$\Delta\alpha = \sin\left(\frac{D}{P} \times 2\pi\right) \times M,$$

where
D = measured depth (ft).
P = period.
M = magnitude.
If the measured depth of the survey point is an exact integer multiple of the period, then $\Delta\alpha = \sin\left(\frac{MD}{P} 2\pi\right) = 0$.

In addition, the inclination angle is modified so that it does not become less than zero, since negative inclination angles are not allowed.
If the $\Delta\alpha$ is negative and $\phi_n = \phi + \Delta\alpha$ is negative, then $\phi_n = 360 + \Delta\alpha$:
$\alpha_n = \alpha + \Delta\alpha$
$\phi_n = \phi + \Delta\alpha + \psi_{xvc}$,
where
ψ_{xvc} = cross vertical correction.
If $\alpha_n < 0$, then cross vertical correction = $180°$.
$\alpha_n = |\alpha_n|$.
If $\alpha_n \geq 0$, then cross vertical correction = $0°$.

137. Problem 3.52
Using the survey data given below and calculate the tortuosity using the sine wave method.
Measured depth = 3900 ft, Inclination = 5.15°, and Azimuth = 166°.
Measured depth = 3725 ft, Inclination = 3.25°, and Azimuth = 165°.
Solution:

$$\Delta\alpha = \sin\left(\frac{3725}{1000} 2\pi\right) \times 1 = -0.99°$$

$\phi_n = 165 - 0.99 + 0 = 164.01°$
$\alpha_n = 3.25 - 0.99 = 2.26°$

If $\alpha_n < 0$, then cross vertical correction $= 180°$
$\alpha_n = |\alpha_n|$.
If $\alpha_n \geq 0$, then cross vertical correction $= 0°$

138. Problem 3.53
Tortuosity—Helical Method

The helical method modifies the inclination and azimuth of the survey points by superimposing a helix along the wellbore path using the magnitude (radius of the cylinder in the parametric equation) and period (pitch) specified. This method uses the circular helix defined as
$$f(u) = a\cos(u) + a\sin(u) + bu.$$
The generalized parametric set of equations for the helix used to superimpose the wellbore path is given by

$$\begin{cases} x(u) = M\cos(u) \\ y(u) = M\sin(u) \\ z(u) = \dfrac{P}{2\pi} u \end{cases}$$

139. Problem 3.54
Tortuosity—Random Inclination Azimuth Method Calculations

The following calculation is used for random inclination and azimuth method:
$$\alpha_n = \alpha + \Delta\alpha$$
$$\phi_n = \phi + \Delta\alpha + \psi_{cvc}$$
$$\psi_{xvc} = \text{cross vertical correction}$$

140. Problem 3.55

Using the survey data and calculate the tortuosity using the random inclination azimuth method.

Measured depth = 3725 ft, Inclination = 3.25°, and Azimuth = 165°.
Measured depth = 3900 ft, Inclination = 5.15°, and Azimuth = 166°.
ζ (Rand number) = 0.375.

Solution:

$$\Delta\alpha = \frac{0.375 \times (3900 - 3725)}{100} \times 1 = 0.66°$$

$$\alpha_n = 5.15 + 0.66 = 5.81°$$

$$\phi_n = 166 + 0.66 + 0 = 166.66°$$

141. Problem 3.56

Tortuosity Calculations—Random Inclination Dependent Azimuth Method

The following calculation is used for Random Inclination and azimuth method:

$$\Delta\alpha = \zeta \times \delta$$

$$\delta = \frac{\Delta MD}{P} M,$$

where ζ = random number.

$$\alpha_n = \alpha + \Delta\alpha.$$

$$\phi_n = \phi + \frac{\Delta\alpha}{2\sin\alpha_n} + \psi_{cvc}.$$

142. Problem 3.57

Using the following survey data, calculate the tortuosity using the random inclination dependent azimuth method.

Measured depth = 3725 ft, Inclination = 3.25°, and Azimuth = 165°.
Measured depth = 3900 ft, Inclination = 5.15°, and Azimuth = 166°.
Random number = 0.375.

Solution:

$$\Delta\alpha = \frac{0.375 \times (3900 - 3725)}{100} \times 1 = 0.66°$$

$$\alpha_n = 5.15 + 0.66 = 5.81°$$

$$\phi_n = 166 + \frac{0.66}{2\sin 5.81} + 0 = 169.26°$$

143. Problem 3.58
Well Profile Energy Calculations

Wellpath profile energy between survey stations can be given as

$$E_{(abs)_n} = \left(\frac{\sum_{i=1}^{n}\left(\kappa_i^2 + \tau_i^2\right)\Delta D_i}{D_n + \Delta D_n} \right)$$

To quantify the change of the well trajectory after applying the artificial tortuosity, a relative energy is defined, which is the energy of the wellbore relative to the absolute energy and is given below:

$$E_{s(rel)_n} = E^{tor}_{s(abs)_n} - E^{notor}_{s(abs)_n}$$

144. Problem 3.59

With the data given in Table 3.1, calculate the absolute and relative tortuosity:

Tortuosity magnitude = 0.25°

Solution:

Table 3.2 shows the calculated results of the absolute and relative tortuosity at various survey points.

Table 3.1 Data for problem 3.59

L, ft	a, deg.	f, deg.
13,200	59.52	224.36
14,400	59.71	224.55
15,600	60.29	225.13
16,200	59.52	224.36
17,400	59.71	224.55
18,600	60.29	225.13
19,800	60.48	225.32
20,000	60	224.84

Table 3.2 Survey results with absolute and relative tortusosities

L, ft	a, deg.	f, deg.	H, ft	DLS, (°/100ft)	Abs.Tort, (°/100ft)	Rel. Tort, (°/100ft)
13,200	59.52	224.36	9842.2	0.16	0.47	0.01
14,400	59.71	224.55	10443.5	0.31	0.45	0.03
15,600	60.29	225.13	11043.5	0.36	0.44	0.05
16,200	59.52	224.36	11342.2	0.16	0.43	0.06
17,400	59.71	224.55	11943.5	0.31	0.42	0.07
18,600	60.29	225.13	12543.5	0.36	0.41	0.09
19,800	60.48	225.32	13142.1	0.09	0.4	0.1
20,000	60	224.84	13241.3	0.41	0.4	0.1

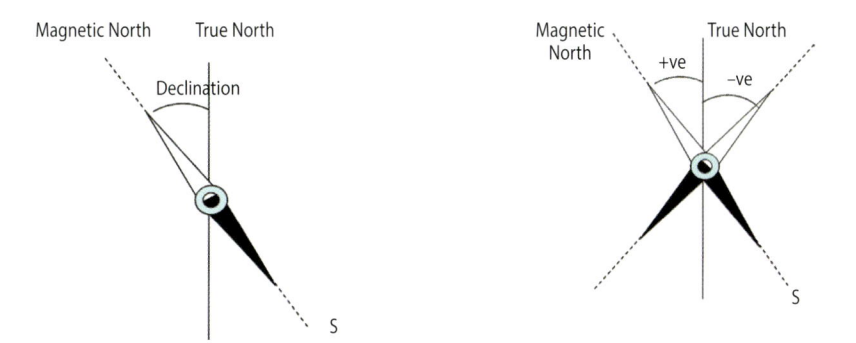

Figure 3.8 Declination/east-west declination

145. Problem 3.60

Magnetic Reference and Interference

Declination is the angular difference in azimuth readings between magnetic north and true north as shown in Figure 3.8. Magnetic declination is positive when magnetic north lies east of true north, and it is negative when magnetic north lies west of true north. It is actually the error between the true north and magnetic north for a specific location.

Azimuth correction can be given as:

Azimuth (true) = Azimuth (magnetic) + Magnetic declination

Westerly Declination

Azimuth correction:

Azimuth (true) = Azimuth (magnetic) + (-Magnetic declination)

For true North and grid North, true north uses latitude and longitude coordinates of the curved earth as the reference. Longitudes converge upon the rotational pole. In the grid system, the Y-axis does not converge to single point.

Magnetic Declination − Grid Convergence = Total Correction

Magnetic Azimuth + Total Correction = Corrected Azimuth

146. Problem 3.61

Determine the azimuth with respect to the true north of the following wells:

- N45E, declination 5° west
- N45E, declination 5° east

Solution:

Azimuth = 45 − 5 = 40° = N40 E

Azimuth = 45 + 5 = 50 = N50 E

They are shown in Figure 3.9

147. Problem 3.62

Calculate the total correction if the magnetic declination is −4° and grid convergence is − 6°.

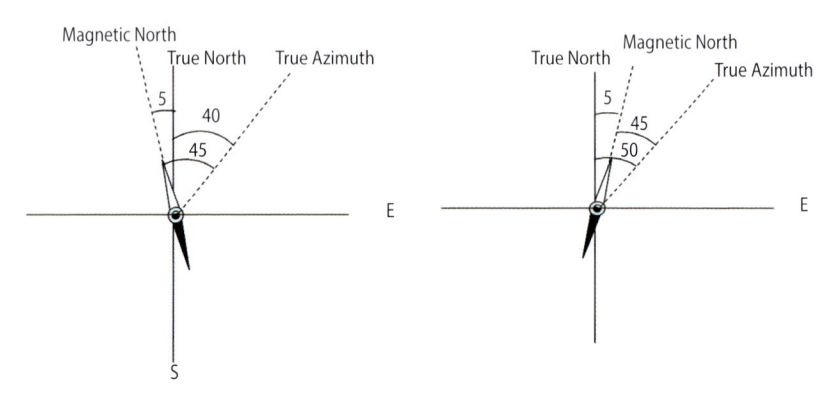

Figure 3.9   Declinations for problem 3.61

Solution:

The total correction = magnetic declination – grid convergence:
Total correction = –4 – (–6) = 2°

148. Problem 3.63

Wellbore Trajectory Uncertainty

Nominal distance between two points, i.e., from the point concerned in a reference wellbore (the reference point) to the point observed in the object wellbore (the object point), is

$$S_{or} = \sqrt{(x_o - x_r)^2 + (y_o - y_r)^2 + (z_o - z_r)^2},$$

where

x_o, y_o, z_o = coordinates of the object point in the OXYZ system.
x_r, y_r, z_r = coordinates of the object point in the OXYZ system, S_{or} (Figure 3.10).

If r_S represents the sum of radii, then

$$r_S = r_o + r_r,$$

where
r_o - radius of an object well.
r_r - radius of a reference well.

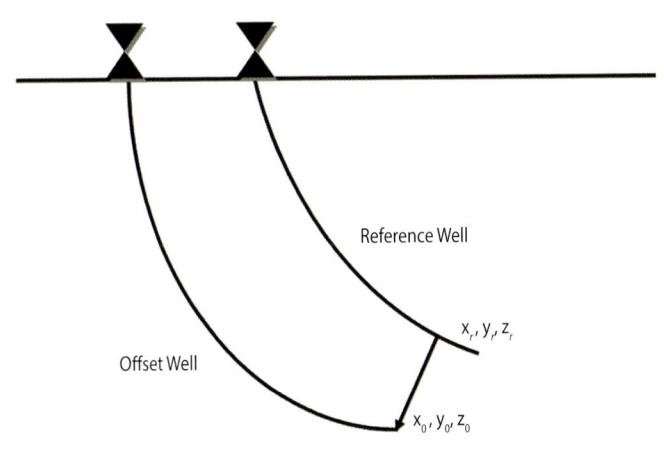

Figure 3.10 Reference and offset well

Relative covariance matrixes are

$$\Sigma_{RR} = \Sigma_{P_oP_o} + \Sigma_{P_rP_r},$$

where

$\Sigma_{P_oP_o}$ and $\Sigma_{P_rP_r}$ are covariance matrixes of the points P_o and P_r, respectively.

Azimuth and the dip angle of the signal point can be determined by

$$\phi = \arcsin\frac{\sqrt{U_r^2 + V_r^2}}{S_{or}},$$

and

$$\theta = \arctan\frac{V_r}{U_r}.$$

The amplifying factor corresponding to the ellipsoid is given as

$$k = \frac{S}{\sqrt{\sigma_1^2 \cos^2\theta \sin^2\varphi + \sigma_2^2 \sin^2\theta \sin^2\varphi + \sigma_3^2 \cos^2\varphi}},$$

where

ϕ = the azimuth angle.
θ = the dip angle.

Minimum probability of intersection is given as

$$P_P = 1 - \frac{4}{\sqrt{\pi}} \int_0^{\frac{K_P}{\sqrt{2}}} \exp[-r^2] r^2 dr$$

or

$$P_P = \frac{4}{\sqrt{\pi}} \int_{\frac{K_P}{\sqrt{2}}}^{\infty} \exp[-r^2] r^2 dr$$

149. Problem 3.64

A pair of parallel horizontal wells is shown in Figure 3.10. Through survey, calculation, and error analysis, a group of data is obtained in Table 3.3 for analyzing the probability of intersection of the two points given.

Well Path Design

Table 3.3 Basic parameters for the calculation

	Coordinate (m)	Diameter (mm)	Covariance Matrix (m2)		
Object well (point)	X = 1000	215	2	0	0
	Y = 1000		0	1	$\frac{\sqrt{2}}{2}$
	Z = 1000		0	$\frac{\sqrt{2}}{2}$	1
Reference well (point)	X = 1001	215	1	0	0
	Y = 1001		0	1	$\frac{\sqrt{2}}{2}$
	Z = 1005		0	$\frac{\sqrt{2}}{2}$	2

Solution:

Nominal distance between the two points is

$$S_{or} = \sqrt{1+1+25} = 5.196 \text{ (m)}$$

The sum of radii is

$$r_S = (215 + 215)/2 = 215 \text{ (mm)}$$

Relative covariance matrixes are

$$\Sigma_{RR} = \begin{bmatrix} 2 & 0 & 0 \\ 0 & 1 & \frac{\sqrt{2}}{2} \\ 0 & \frac{\sqrt{2}}{2} & 1 \end{bmatrix} + \begin{bmatrix} 1 & 0 & 0 \\ 0 & 1 & \frac{\sqrt{2}}{2} \\ 0 & \frac{\sqrt{2}}{2} & 2 \end{bmatrix} = \begin{bmatrix} 3 & 0 & 0 \\ 0 & 2 & \sqrt{2} \\ 0 & \sqrt{2} & 3 \end{bmatrix}.$$

(1) Determining the orthogonal matrix (T)

(a) Calculating the eigenvalue and eigenvectors of Σ_{RR}:

From

$$\begin{vmatrix} \lambda-3 & 0 & 0 \\ 0 & \lambda-2 & \sqrt{2} \\ 0 & \sqrt{2} & \lambda-3 \end{vmatrix} = 0$$

we get $\lambda_1 = 4, \lambda_2 = 3, \lambda_3 = 1$

When $\lambda_1 = 4$, from

$$\begin{bmatrix} 1 & 0 & 0 \\ 0 & 2 & \sqrt{2} \\ 0 & \sqrt{2} & 1 \end{bmatrix} \begin{pmatrix} x_1 \\ x_2 \\ x_3 \end{pmatrix} = 0$$

we get the eigenvector: $a_1 = [0 \quad 1 \quad -\sqrt{2}]^T$

When $\lambda_1 = 3$, from

$$\begin{bmatrix} 0 & 0 & 0 \\ 0 & 1 & \sqrt{2} \\ 0 & \sqrt{2} & 0 \end{bmatrix} \begin{pmatrix} x_1 \\ x_2 \\ x_3 \end{pmatrix} = 0$$

we get $a_2 = [1 \quad 0 \quad 0]^T$

When $\lambda_1 = 1$, from

$$\begin{bmatrix} -2 & 0 & 0 \\ 0 & -1 & \sqrt{2} \\ 0 & \sqrt{2} & -2 \end{bmatrix} \begin{pmatrix} x_1 \\ x_2 \\ x_3 \end{pmatrix} = 0$$

we get $a_3 = [0 \quad \sqrt{2} \quad 1]^T$

(b) Making the unit vectors from above three eigenvectors:

$e_1 = [0 \quad \frac{\sqrt{3}}{3} \quad -\sqrt{\frac{2}{3}}]^T$; $e_2 = [1 \quad 0 \quad 0]^T$; $e_3 = [0 \quad \sqrt{\frac{2}{3}} \quad \frac{\sqrt{3}}{3}]^T$

Thus, the orthogonal matrix T can be written as

$$T = \begin{bmatrix} 0 & 1 & 0 \\ \frac{\sqrt{3}}{3} & 0 & \sqrt{\frac{2}{3}} \\ -\sqrt{\frac{2}{3}} & 0 & \frac{\sqrt{3}}{3} \end{bmatrix}$$

(2) Make an orthogonal transformation using the matrix T. The relative uncertainty ellipsoid cluster can be obtained as

$$\frac{U^2}{4} + \frac{V^2}{3} + \frac{W^2}{1} = k^2$$

(3) Determining amplifying factor
   (a) Distance from the signal point to the center point: $S = 5.196 - 0.215 = 4.98$ m
   (b) Relative coordinates of the reference point in the OUVW system:

$$\begin{pmatrix} U \\ V \\ W \end{pmatrix} = \begin{bmatrix} 0 & 1 & 0 \\ \frac{\sqrt{3}}{3} & 0 & \sqrt{\frac{2}{3}} \\ -\sqrt{\frac{2}{3}} & 0 & \frac{\sqrt{3}}{3} \end{bmatrix} \begin{pmatrix} 1 \\ 1 \\ 5 \end{pmatrix} = \begin{pmatrix} 1 \\ 4.695 \\ 1.255 \end{pmatrix}$$

   (c) Azimuth and dip angle of the signal point:

$$\varphi = \arcsin \frac{\sqrt{1+4.695^2}}{5.196} = 67.486° \ ; \ \theta = \arctan \frac{4.695}{1} = 77.976°$$

   (d) Amplifying factor:

$$k_P = \frac{4.98}{\sqrt{4 \times \cos^2 77.976 \times \sin^2 67.486 + 3 \times \sin^2 77.976 \times \sin^2 67.486 + \cos^2 67.486}}$$
$$= 3.009$$

Thus, taking $k_E = 3$.

(4) Probability of intersection:

$$P_E = 1 - \frac{4}{\sqrt{2\pi}}\left(\frac{3^3}{6} - \frac{3^5}{20} + \frac{3^7}{112} - \frac{3^9}{864} + \cdots\right) < 5\%$$

150. Problem 3.65

Define an ERD well.

Solution:

ERD—extended reach well.

Measured depth to true vertical depth ration should be greater than 2.

151. Problem 3.66

Derive an equation to find the overall angle change in terms of $\Delta\varepsilon, \alpha$, and γ the toolface rotation angle.

Solution:

Starting from the azimuth change

$$\Delta\varepsilon = \arctan\frac{\tan\beta\sin\gamma}{\sin\alpha + \tan\beta\cos\alpha\cos\gamma}$$

$$\tan\Delta\varepsilon = \frac{\tan\beta\sin\gamma}{\sin\alpha + \tan\beta\cos\alpha\cos\gamma}$$

$\tan\Delta\varepsilon \sin\alpha + \tan\Delta\varepsilon \tan\beta \cos\alpha\cos\gamma - \tan\beta \sin\gamma = 0$
$\tan\Delta\varepsilon \sin\alpha + \tan\beta (\tan\Delta\varepsilon \cos\alpha\cos\gamma - \sin\gamma) = 0$
Overall angle change β

$$\beta = \arctan\frac{\tan\Delta\varepsilon \sin\alpha}{\sin\gamma - \tan\Delta\varepsilon \cos\alpha\cos\gamma}$$

152. Problem 3.67

Determine the trajectory of a build-hold-drop type well:

Horizontal displacement: 6000 ft
Target TVD = 12,000 ft
TVD at KOP depth = 1500 ft

Well Path Design

Build rate = 2 °/100 ft
Drop rate = 1.5 °/100 ft

Solution:

Horizontal displacement:	6000	ft	X_4
Target TVD =	12,000	ft	D_4
TVD at KOP depth =	1500	ft	D_1
Build rate =	2	°/100 ft	q
Drop rate =	1.5	°/100 ft	q

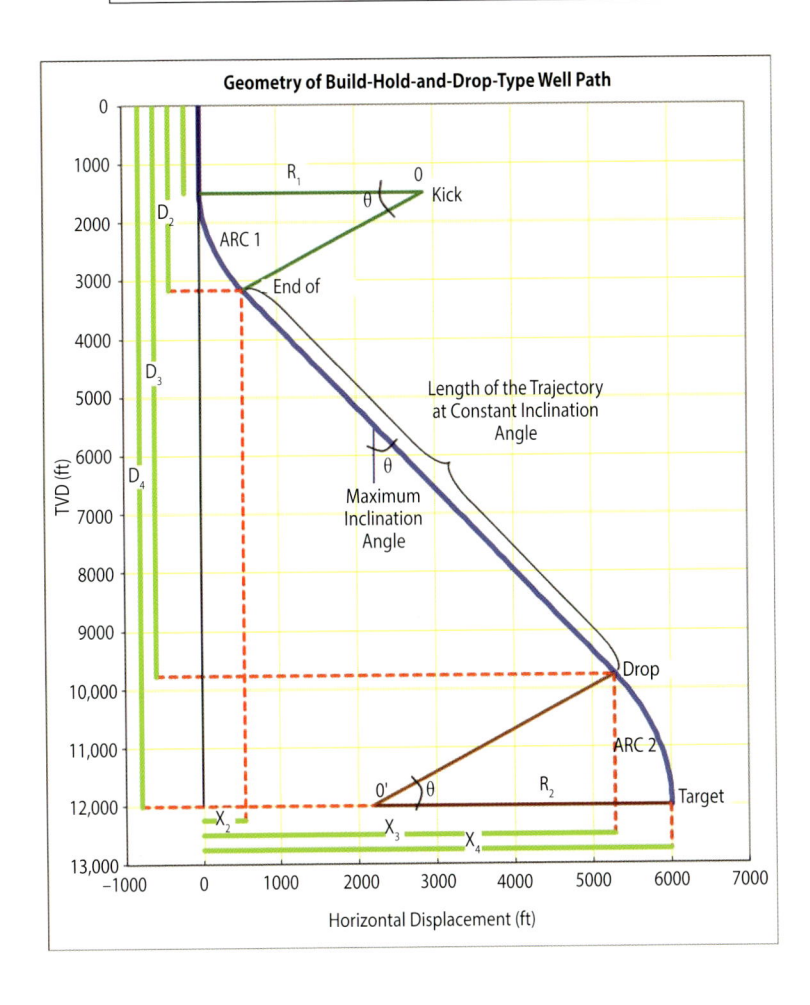

Geometry of Build-Hold-and-Drop-Type Well Path

CALCULATIONS		
$R_1 =$	2865	ft
$R_2 =$	3820	ft
$R_1 + R_2 =$	6685	ft
$q =$	35.7	°
$X_2 =$	539	ft
$D_2 =$	3172	ft
$L_{arc1} =$	1786	ft
$X_3 =$	5282	ft
$D_3 =$	9770	ft
$L_{arc2} =$	2381	ft
$L_{constant\ q} =$	8126	ft
Total MD =	13,792	ft

(a) Radius of curvature ; R_1 (build) R_2 (drop)

(i) $R_1 = \dfrac{18,000}{2\pi} = 2864.78$ ft

and

$$R_2 = \dfrac{18,000}{1.5\pi} = 3819.72 \text{ ft}$$

Since $R_1 + R_2 = 6684.5 > X_4$;

$$\theta = \arctan\left(\dfrac{a}{b+R_2}\right) - \arccos\left\{\left(\dfrac{R_1+R_2}{a}\right) \times \sin\left[\arctan\left(\dfrac{a}{b+R_2}\right)\right]\right\}$$

a = (12,000 − 1500) = 10,500
b = $R_1 - X_4$ = (2864.78 − 6000) = − 3135.22
= 86.27° − arc cos [0.636 × 0.9978] = 86.27 − 50.56 = 35.71°

(b) Length of build-up arc section of DC :

$L_{DC} = \pi/180 \times R_1 \times \theta = \pi/180 \times 2864.78 \times 35.71 = 1785.49$ ft

Well Path Design

(c) Length of drop angle arc section

Drop angle is the same as build angle which can be derived from the above build angle equations which is 35.71°

$L_{drop} = \pi/180 \times R_2 \times \theta_2 = \pi/180 \times 3819.72 \times 35.71 = 2380.66$ ft

(d) Length of hold section; CB:

Vertical section of build angle; $DC = R_1 \sin \theta$

Vertical section = $2864.78 \sin (35.71°) = 1672.12$ ft

Vertical section of drop angle = $R_2 \sin \theta = 3819.72 \sin (35.71°) = 2229.50$ ft

Therefore the vertical section of hold section:

$D_4 - D_1 - R_1 \sin \theta - R_2 \sin \theta = 12,000 - 1500 - 1672.12 - 2229.50 = 6598.38$ ft

MD for hold section, $CB = 6598.38 / \cos \theta = 6598.38 / \cos (35.71) = 8126.26$ ft

(e) TMD well trajectory :

$MD_{total} = KOP + L_{build} + L_{hold} + L_{drop} = 1500 + 1785.49 + 8126.26 + 2380.66 = 13{,}792.4$ ft

CALCULATIONS

TVD	INC	MD	$\Delta L_{(N-S)}$
0	0	0	0
1500	0.0	1500.0	0
1600	2.0	1600.1	3
1700	4.0	1700.3	10
1800	6.0	1800.9	21
1900	8.0	1901.8	35
2000	10.0	2003.4	53
2100	12.0	2105.6	74
2200	14.0	2208.7	99
2300	16.0	2312.7	128
2400	18.0	2417.9	160
2500	20.0	2524.3	196
2600	22.0	2632.1	237

TVD	INC	MD	$\Delta L_{(N-S)}$
2700	24.0	2741.6	281
2800	26.0	2852.8	330
2900	28.0	2966.1	383
3000	30.0	3081.6	441
3100	32.0	3199.5	504
3172	35.7	3288.3	555
3200	35.7	3322.6	575
3300	35.7	3445.8	647
3400	35.7	3569.0	719
3500	35.7	3692.1	791
3600	35.7	3815.3	863
3700	35.7	3938.4	935
3800	35.7	4061.6	1007
3900	35.7	4184.7	1079
4000	35.7	4307.9	1150
4100	35.7	4431.1	1222
4200	35.7	4554.2	1294
4300	35.7	4677.4	1366
4400	35.7	4800.5	1438
4500	35.7	4923.7	1510
4600	35.7	5046.8	1582
4700	35.7	5170.0	1654
4800	35.7	5293.1	1726
4900	35.7	5416.3	1797
5000	35.7	5539.5	1869
5100	35.7	5662.6	1941
5200	35.7	5785.8	2013
5300	35.7	5908.9	2085
5400	35.7	6032.1	2157
5500	35.7	6155.2	2229

TVD	INC	MD	$\Delta L_{(N-S)}$
5600	35.7	6278.4	2301
5700	35.7	6401.5	2373
5800	35.7	6524.7	2444
5900	35.7	6647.9	2516
6000	35.7	6771.0	2588
6100	35.7	6894.2	2660
6200	35.7	7017.3	2732
6300	35.7	7140.5	2804
6400	35.7	7263.6	2876
6500	35.7	7386.8	2948
6600	35.7	7509.9	3019
6700	35.7	7633.1	3091
6800	35.7	7756.3	3163
6900	35.7	7879.4	3235
7000	35.7	8002.6	3307
7100	35.7	8125.7	3379
7200	35.7	8248.9	3451
7300	35.7	8372.0	3523
7400	35.7	8495.2	3595
7500	35.7	8618.3	3666
7600	35.7	8741.5	3738
7700	35.7	8864.7	3810
7800	35.7	8987.8	3882
7900	35.7	9111.0	3954
8000	35.7	9234.1	4026
8100	35.7	9357.3	4098
8200	35.7	9480.4	4170
8300	35.7	9603.6	4242
8400	35.7	9726.8	4313
8500	35.7	9849.9	4385

TVD	INC	MD	$\Delta L_{(N-S)}$
8600	35.7	9973.1	4457
8700	35.7	10,096.2	4529
8800	35.7	10,219.4	4601
8900	35.7	10,342.5	4673
9000	35.7	10,465.7	4745
9100	35.7	10,588.8	4817
9200	35.7	10,712.0	4888
9300	35.7	10,835.2	4960
9400	35.7	10,958.3	5032
9500	35.7	11,081.5	5104
9600	35.7	11,204.6	5176
9700	35.7	11,327.8	5248
9770	35.7	11,414.0	5298
9800	34.2	11,448.7	5316
9900	32.7	11,567.5	5380
10,000	31.2	11,684.5	5441
10,100	29.7	11,799.6	5498
10,200	28.2	11,913.1	5551
10,300	26.7	12,025.0	5602
10,400	25.2	12,135.6	5649
10,500	23.7	12,244.8	5693
10,600	22.2	12,352.8	5734
10,700	20.7	12,459.7	5771
10,800	19.2	12,565.6	5806
10,900	17.7	12,670.6	5838
11,000	16.2	12,774.7	5867
11,100	14.7	12,878.1	5893
11,200	13.2	12,980.8	5917

Well Path Design

TVD	INC	MD	$\Delta L_{(N-S)}$
11,300	11.7	13,082.9	5938
11,400	10.2	1,3184.6	5956
11,500	8.7	13,285.7	5971
11,600	7.2	13,386.5	5984
11,700	5.7	13,487.0	5994
11,800	4.2	13,587.3	6001
11,900	2.7	13,687.4	6006
12,000	1.2	13,787.4	6008

CHAPTER 4

PRESSURE CALCULATIONS

This chapter focuses on different basic calculations associated with wellbore pressure and equivalent mud weight calculations.

153. Problem 4.1

Gauge and Absolute Pressure Calculations

Pressure, expressed as the difference between the fluid pressure and that of the surrounding atmosphere, is relative to ambient or atmosphere.

Absolute pressure of a fluid is expressed relative to that of a vacuum and is given as

$$P_{abs} = P_{atm/ambient} + P_{gauge}$$

The standard value of atmospheric pressure $=14.696$ psi $= 101.3$ kPa $= 29.92$ inHg $= 760$ mmHg.

The relationship between gauge and absolute pressures is shown in Figure 4.1.

154. Problem 4.2

Hydrostatic Pressure

The hydrostatic pressure at a depth, D, column of mud having a density of γ_m can be easily derived and is given by

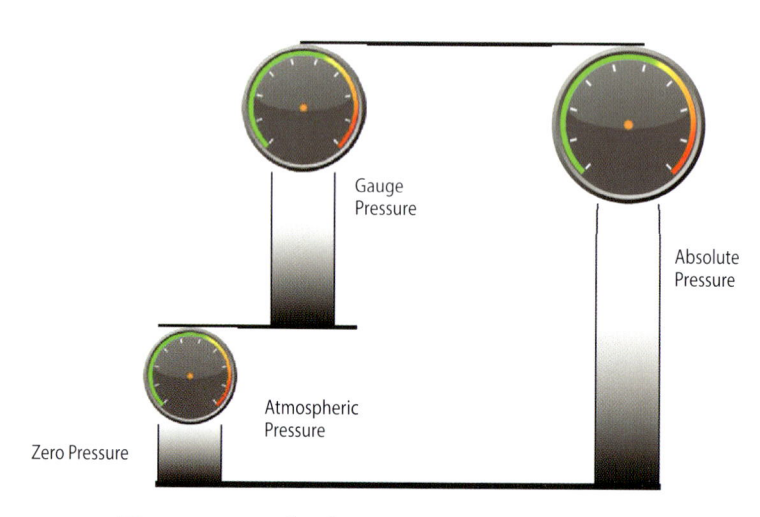

$$P_h = k\rho_m L_v,$$

Figure 4.1 Absolute-gauge-atmospheric pressures

where
D_v = vertical depth of mud column.
ρ_m = mud weight.
k = conversion unit factor.
If P_h is in psi, ρ_m in ppg, and D in ft, then the value of k = 0.052.

155. Problem 4.3
Mud Gradient Calculations

Pressure inside the wellbore is expressed in terms of gradient and is expressed in psi/ft of depth. In oil field units,

$$\rho_g = 0.052 \times \rho_m \text{ psi/ft,}$$

where
ρ_m = mud weight in ppg.

$$0.052 = \frac{748 \text{ gal}}{144 \text{ in}^2}.$$

Hydrostatic pressure, P_h, at any measured depth in the wellbore can be calculated as

$$P_h = \rho_g \times L_v \text{ psi,}$$

where
L_v = the corresponding true vertical depth at the measure depth in ft.

156. Problem 4.4

Prove that when density of drilling fluid is expressed in ρ ppg (lbm/gal) and depth in h ft, the pressure at the bottom of the wellbore can be expressed as $P = 0.052\,\rho h$ psi.

Solution:

$$P = \rho g h = \rho \left[\frac{\text{lbm}}{\text{gal}} \times 7.48 \frac{\text{gal}}{\text{ft}^3} \right] gh(\text{ft}),$$

where
g = acceleration due to gravity.

$$P = 7.48\rho gh\left[\frac{lbm}{ft^2}\right] = 7.48\rho h\left[\frac{ft^2}{144\,in^2}\right] = 0.052\,\rho h\,\text{psi}.$$

157. Problem 4.5

Prove that when density of drilling fluid is expressed in ρ (kg/m³) and depth in h m, the pressure at the bottom of the wellbore can be expressed as $P = 9.81\,\rho h$ Pa

In SI units

$$P = \rho gh = \rho\left[\frac{kg}{m^3}\right]gh(m),$$

where
g = acceleration due to gravity 9.81 m/s².

$$P = \rho gh\left[\frac{kgm}{m^2}\right] = 9.81\rho h\left[\frac{kg}{m^2} \times \frac{m}{s^2}\right] = 9.81\rho h\,\text{N/m}^2 = 9.81\rho h\,\text{Pa}$$

158. Problem 4.6

What is the hydrostatic pressure at target depth of the well with the drill string on bottom of the mud weight used is 12 ppg?

Solution:
Target depth = 13,697 ft
Mud density = 12 ppg
$P = 0.052 \times \rho \times D_v = 0.052 \times 12 \times 13{,}697 = 8547$ psi

159. Problem 4.7

What would be the static mud density required to balance a formation pressure of 6000 psi at a depth of 8000 ft (vertical)?

Solution:

$$\rho_m = \frac{P_h}{0.052 \times L_v}\,\text{psi}.$$

Substituting the appropriate values,

$$\rho_m = \frac{6000}{0.052 \times 8000} = 14.42 \text{ ppg}$$

160. Problem 4.8

Convert 10 ppg to various units and vice versa.

Solution:

Converting 10 ppg to psi/ft = 0.0519 × 10 = 0.519 or 0.52 psi/ft

Converting 10 ppg to psf/ft = 7.48 × 10 = 74.8 psf/ft

Converting ppg to specific gravity (SG) = 8.3472 × 10 = 1.198

Converting 0.519 psi/ft to mud weight lb/ft³ = $\dfrac{0.519}{0.006944}$ = 74.8 lb/ft³

Converting 0.519 psi/ft to mud weight in SG = $\dfrac{0.519}{0.433}$ = 1.198

Converting 0.519 psi/ft to mud weight in ppg = $\dfrac{0.519}{0.0519}$ = 10 ppg

161. Problem 4.9

The pressure can be measured in the U-tube by applying the static fluid equations to both legs of the manometer. For example, for the manometer with different fluids it can be written as

$P_2 = P_1 + 0.052 \times \rho_1 \times (h_2 + h_3)$
$P_3 = P_5 + 0.052 \times \rho_2 \times h_2 + 0.052 \times \rho_1 \times h_3$
Since $P_3 = P_2$,
$P_5 = P_1 + 0.052 \times h_2 (\rho_1 - \rho_2)$

162. Problem 4.10

Find the pressure, P_6, shown in Figure 4.2.

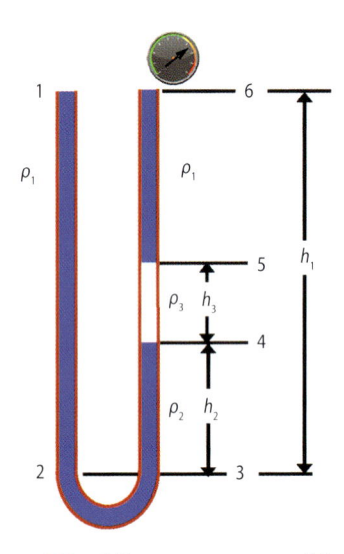

Figure 4.2 Manometer—problem 4.10

Solution:

Applying pressure balance in both the legs of the manometer at the points 2 and 3,

$P_2 = P_1 + 0.052 \times \rho_1 \times h_1$
$P_3 = P_6 + 0.052 \times \rho_1 \times (h_1 - h_2 - h_3) + 0.052 \times \rho_3 \times h_3 + 0.052 \times \rho_2 \times h_2$
Since $P_3 = P_2$,
$P_6 = P_1 + 0.052 \times h_2\,(\rho_1 - \rho_2) + 0.052 \times h_3\,(\rho_1 - \rho_3)$

163. Problem 4.11

A reservoir at 13,568 ft has a formation pressure gradient of 0.65 psi/ft. Mud weight planned to drill this formation is 12 ppg. Check whether it is safer to drill this formation.

Solution:

Reservoir depth = 13,568 ft
Reservoir pressure = 13,568 × 0.65 = 8819 psi
Mud weight of 12 ppg
$P = 0.052 \times \rho \times D_v = 0.052 \times 12 \times 13{,}568 = 8466.43$ psi

Mud weight if used will result in formation influx.

Mud weight required to drill should be greater than the pressure gradient of 0.65 psi/ft

Pressure gradient (psi/ft)

Mud density (ppg) × 0.052 which is 0.65/0.052 = 12.50 ppg

Alternatively, 8819/(0.052 × 13,568) = 12.50 ppg

164. Problem 4.12

1. **Equivalent Mud Weight Calculation**

The pressure (P – psi) in the wellbore or formation can be expressed in terms of equivalent mud weight (EMW) and is given as

$$\text{EMW} = \frac{P}{0.052 \times L_{tvd}} \text{ ppg},$$

where

L_{tvd} = true vertical depth (ft).

If the well is deviated at $\alpha°$ from the vertical, the EMW is given as

$$\text{EMW} = \frac{P_h}{0.052 \times D_h \cos\alpha} \text{ ppg}$$

165. Problem 4.13

Calculate the equivalent mud weight at the bottom of well depth and casing shoe depth as shown in Figure 4.3.

Mud weight 10 ppg

Total water volume required at the top of annulus is 22 bbl where the annulus volume is 0.051 bbl/ft.

Solution:

Feet of water in the annulus: 20/0.051 = 431.37 ft of water

Pressure at the bottom of water = 0.052 × 8.33 × 431.37 = 186.85 psi

Remaining depth for mud = 10,000 – 431.37 = 9568.63 ft

Pressure due to mud weigt = 0.052 × 10 × 9568.63 = 4975.68 psi

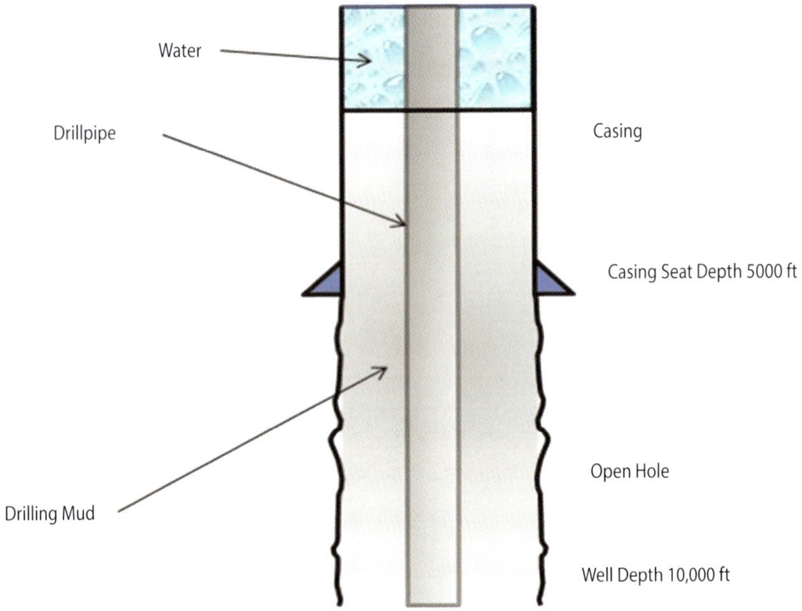
Water
Drillpipe
Casing
Casing Seat Depth 5000 ft
Open Hole
Drilling Mud
Well Depth 10,000 ft

Figure 4.3 **Well schematic**

Total bottomhole pressure = 4975.69 + 186.85 = 5162.54 psi
Equivalent mud weight = 5162.53/(0.052 × 10,000) = 9.92 ppg

166. Problem 4.14

Classification of Formation Pressures Calculation

The pressures that may be encountered during drilling are classified as the following:

- Normal pressure – the pressure gradient is approximately 0.433 psi/ft of depth
- Abnormal pressure – the pressure gradient is greater than 0.433 psi/ft of depth
- Subnormal pressure – the pressure gradient is less than 0.433 psi/ft of depth

167. Problem 4.15

Calculate the hydrostatic pressure at a well depth of 10,000 ft (TVD) with a mud weight of 10 ppg.

Solution:

The hydrostatic pressure is

$$P_h = 0.052 \times \text{mud weight (ppg)} \times \text{true vertical depth (ft) psi}$$

$$P_h = 0.052 \times \rho_m \times L_{tvd} = 0.052 \times 10 \times 10{,}000 = 5200 \text{ psi}$$

168. Problem 4.16

Calculate the equivalent mud weight at a well depth of 9000 ft (TVD) with a bottomhole pressure of 6555 psi.

Solution:

Equivalent mud weight is given as

$$\text{EMW} = \frac{P}{0.052 \times L_{tvd}}$$

Substituting the values,

$$\text{EMW} = \frac{6555}{0.052 \times 9000} = 14 \text{ ppg}$$

169. Problem 4.17

Casing is set on east Texas well at 7800 ft (TVD). It was estimated the fracture pressure was 6815 psi. Calculate the equivalent fracture mud weight and fracture gradient.

Solution:

Equivalent mud weight for fracture is given as

$$\text{EMW} = \frac{P}{0.052 \times L_{tvd}} = \frac{6815}{0.052 \times 7800} = 16.8 \text{ ppg}$$

$$\text{Fracture gradient} = \frac{6815}{7800} = 0.8737 \text{ psi/ft}$$

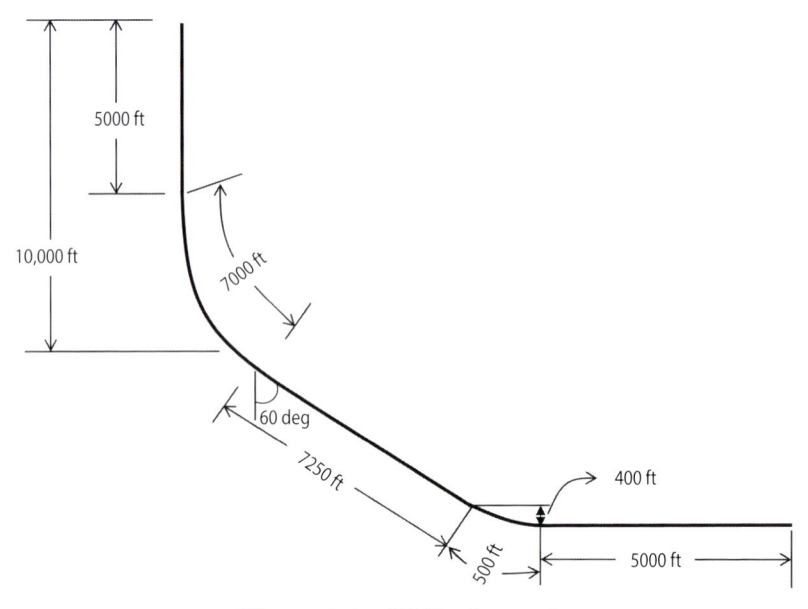

Figure 4.4 Well schematic

170. Problem 4.18

Calculate the bottomhole pressure for the well profile shown in Figure 4.4 with a drilling fluid of density 15 ppg. Assume the density of the fluid to be same inside the drillstring as well as in the annulus.

    a. when the measured depth of the well is 12,500 ft
    b. when the measured depth of the well is 20,000 ft
    c. and at the bottom of the well

Solution:

    a. when the measured depth of the well is 12,500 ft
        In order to calculate the TVD the position of the 12,500 ft in the wellpath has to be found.
        Measured depth:
        5000 ft (KOP) + 7000 ft (curved section) + 500 ft (slant section)
        True vertical depth:
        5000 ft (vertical) + 5000 ft (end of curved section) + 500 $\cos 60^0$ (slant section) = 10,250 ft

Pressure = 0.052 × 10,250 × 15 = 7995 psi

b. when the measured depth of the well 20,000 ft
Measured depth:
5000 ft (KOP) + 7000 ft (curved section) + 7250 ft (slant section) + 500 ft (curved section) + 250 (horizontal section)
True vertical depth:
5000 ft (vertical) + 5000 ft(end of curved section) + 7250 cos 60°(slant section) + 400 ft (curved section) + 0 (horizontal section) = 14,025 ft
Pressure = 0.052 × 15 × 14,025 = 10,940 psi

171. Problem 4.19

A kick was taken at 10,000 ft, while drilling with a 14 ppg mud. Well was shut-in and the observed stand pipe pressure was 600 psi. Mud weight is increased by adding barite to 800 bbls of original mud to contain formation pressure such that a differential wellbore pressure of 500 psi is achieved after killing the well. Determine the cost due to mud weight increase, assuming barite is $101/sack and 100 lbm = 1 sack.

Solution:

Equivalent mud weight to control the kick with the excess differential pressure

$$\frac{600 + 0.052 \times 14 \times 10,000 + 500}{0.052 \times 10,000} = 16.12 \text{ ppg}$$

Cost of the barite needed

$$= 42 \times 35 \times \frac{16.12 - 14}{35 - 16.11} \times 800 \times \frac{101}{100} = \$133300$$

172. Problem 4.20

The formation pressure and fracture pressure expected at location P and Q (Fig. 4.5) are 10 ppg and 11.5 ppg, respectively. The drilling engineer plans to maintain a differential pressure of 400 psi under static condition

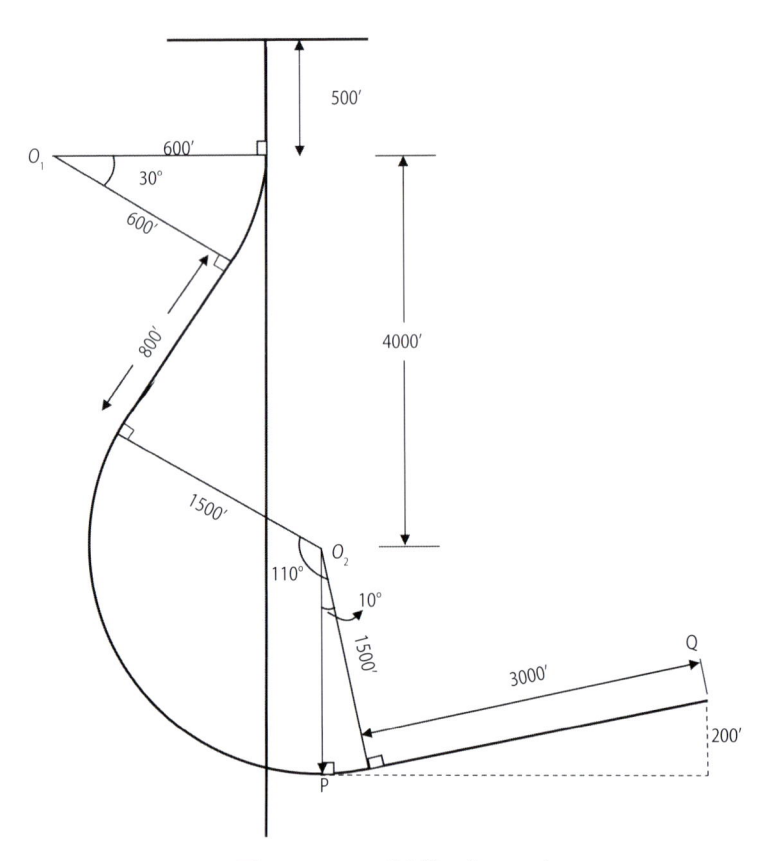

Figure 4.5 Well schematic

while drilling at the above locations. Check whether it is okay to drill while circulating. The total frictional pressure loss gradient in the annulus is 0.00769 psi/ft.

Total wellbore measured length at location

$$P = 500 + \frac{2\pi}{180} \times 600 \times 30 + 800 + \frac{2\pi}{180} \times 1500 \times 100 = 7164 \text{ ft}$$

TVD at P
500 + 4000 + 1500 = 6000 ft
Static pressure = 0.052 × 10 × 6000 + 400 = 3520 psi

Pressure Calculations

Circulating pressure = 3520 + 0.00769 × 7688 = 3575 psi
ECD = 3575/(0.052 × 6000) = 11.46 ppg
This is less than the fracture pressure.
Total wellbore measured length at location

$$P = 500 + \frac{2\pi}{180} \times 600 \times 30 + 800 + \frac{2\pi}{180} \times 1200 \times 110 + 3000 = 10{,}688 \text{ ft}$$

TVD at P
500 + 4000 + 1500 = 6000 ft
Static pressure = 0.052 × 10 × 6000 + 400 = 3416 psi
Circulating pressure = 3520 + 0.00769 × 7688 = 3598 psi
ECD = 3579/(0.052 × 6000) = 11.60 ppg

This is greater than the fracture pressure and so the formation may fracture.

173. Problem 4.21

Calculate the bottomhole pressure for the well profile (Fig. 4.6) with a drilling fluid of density 15 ppg. Assume the density of the fluid to be same inside the drillstring as well as in the annulus.

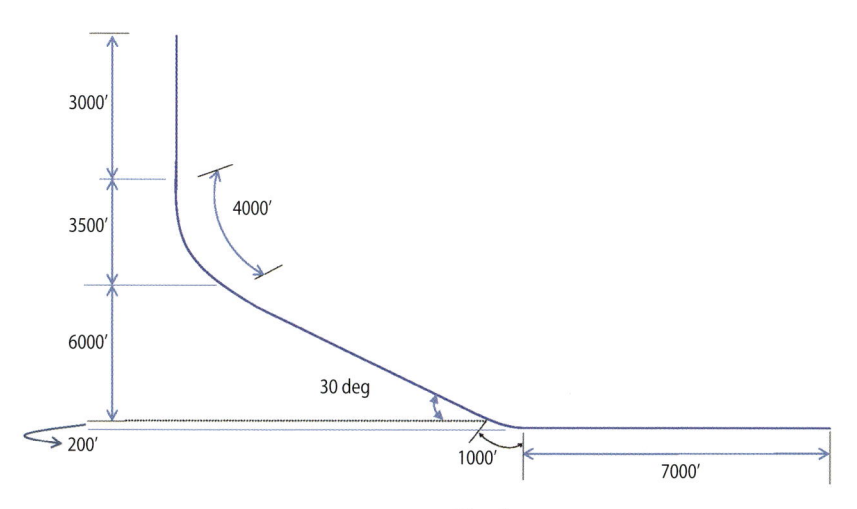

Figure 4.6   Well schematic

a. when the measured depth of the well is 9500 ft
b. and at the bottom of the well.

Solution:

a. when the measured depth of the well is 9500 ft
TVD at the measured depth of 9500 ft is = 3000 + 3500 + 2500x sin30 = 7750 ft
Pressure at 9500 ft = 0.052 × 15 × 7750 = 6045 psi

b. and at the bottom of the well
TVD at the bottom of the well = 12,700 ft
Pressure = 0.052 × 15 × 12,700 = 9906 psi

174. Problem 4.22

Calculate the bottomhole pressure for the well profile shown in Figure 4.7 with a drilling fluid of density 15 ppg. Assume the density of the fluid to be same inside the drillstring as well as in the annulus.

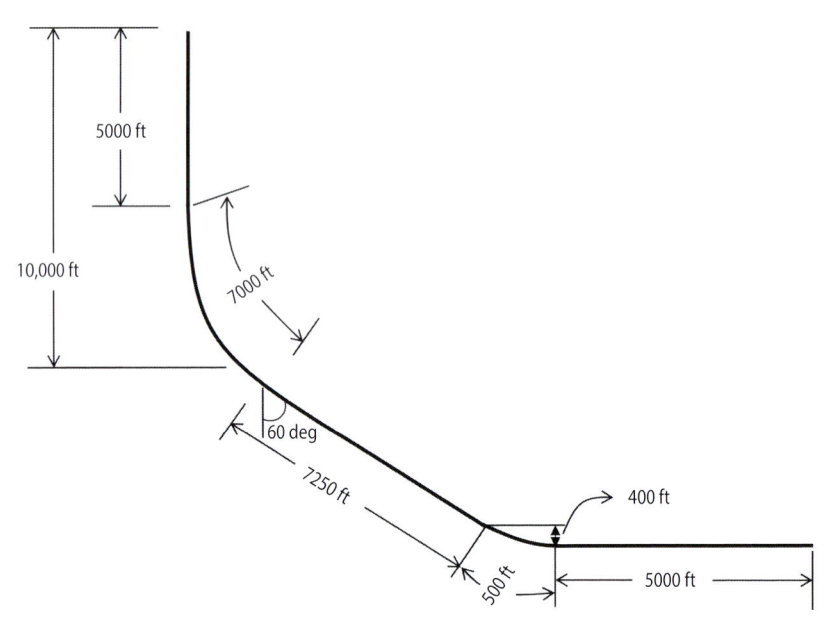

Figure 4.7 Well schematic

a. when the measured depth of the well is 12,500 ft,
b. when the measured depth of the well 20,000 ft,
c. and at the bottom of the well.

Solution:

c. when the measured depth of the well is 12,500 ft
In order to calculate the TVD the position of the 12,500 ft in the wellpath has to be found.
Measured depth:
5000 ft (KOP) + 7000 ft (curved section) + 500 ft (slant section)
True vertical depth:
5000 ft (vertical) + 5000 ft (end of curved section) + 500 cos 60⁰ (slant section) = 10,250 ft.
Pressure = 0.052 × 10,250 × 15 = 7995 psi

d. when the measured depth of the well 20,000 ft
Measured depth:
5000 ft (KOP) + 7000 ft (curved section) + 7250 ft (slant section) + 500 ft (curved section) + 250 (horizontal section)
True vertical depth:
5000 ft (vertical) + 5000 ft(end of curved section) + 7250 cos 60⁰ (slant section) + 400 ft (curved section) + 0 (horizontal section) = 14,025 ft
Pressure = 0.052 × 15 × 14,025 = 10,940 psi

c. and at the bottom of the well
Since there is no change in true vertical depth at the end of the well as it is in the horizontal section the TVD and the pressure remains the same as 10,940 psi.

175. Problem 4.23

Calculate the bottomhole pressure for the following fish hook well profile (Fig. 4.8) with a drilling fluid of density 15 ppg. Assume the density of the fluid to be same inside the drillstring as well as in the annulus.

a. when the measured depth of the well is 9500 ft
b. and at the bottom of the well.

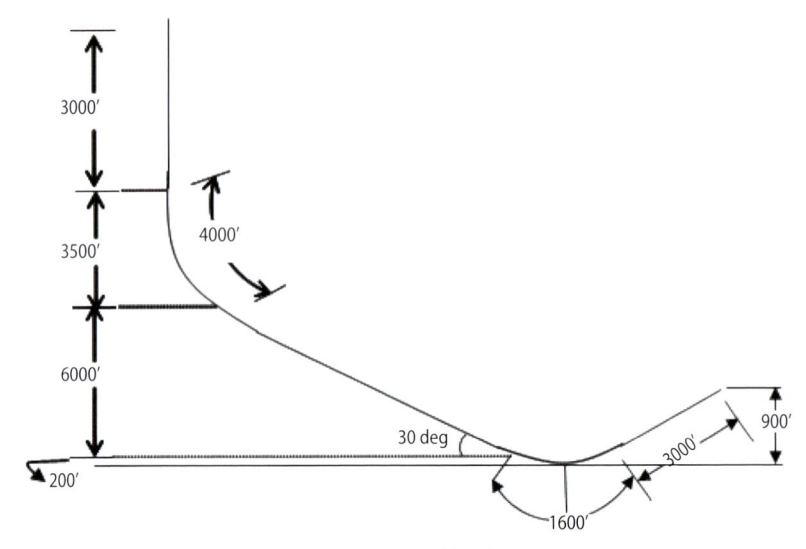

Figure 4.8 Well schematic

Solution:
    c. when the measured depth of the well is 9500 ft
        TVD at the measured depth of 9500 ft is = 3000 + 3500 + 2500x
        sin30 = 7750 ft
        Pressure at 9500 ft = 0.052 × 15 × 7750 = 6045 psi
    d. and at the bottom of the well
        TVD at the bottom of the well = 11,800 ft
        Pressure = 0.052 × 15 × 11,800 = 9204 psi

CHAPTER 5

MUD WEIGHTING

Calculation of Density, mud weight, volume of different fluids and ratio of different materials added are important when drilling fluid is prepared. This chapter focuses on different basic calculations associated with mud weight, volume and density using the basic material balance, volume and density, volume and mass relationships.

176. Problem 5.1
Mud Weight, Volume, and Density Calculation

The four fundamental equations used in developing mud weighting mathematical relations are

(1) the material balance equation,
(2) the volume balance equation,
(3) the relationship between weight and volume equation, and
(4) the volume balance in low specific gravity solids equation.

$$W_f = W_o + W_a,$$

where
W_f = final mixture weight.
W_o = original liquid weight.
W_a = added material weight.

$$V_f = V_o + V_a,$$

where
V_f, = final mixture volume.
V_o = original liquid volume.
V_a = added material weight.

$$\rho = \frac{W}{V},$$

where
ρ = weight density.
W = weight.
V = volume.

$$V_f f_{vf} = V_o f_{vo},$$

where
f_{vf} = volume fraction of low specific gravity solids in final mixture.
f_{vo} = volume fraction of low specific gravity solids in original liquid.

Mud Weighting

These equations may be combined and rearranged algebraically to suit the particular calculations desired. From these equations, the following can be obtained as a general relationship equation:

$$V_1\rho_1 + V_2\rho_2 + V_3\rho_3 + V_4\rho_4 + \ldots = V_f\rho_f,$$

$$V_1 + V_2 + V_3 + V_4 + \ldots = V_f,$$

where

$V_1, V_2, V_3,$ and V_4 = volumes of materials 1, 2, 3, and 4, respectively.
$\rho_1, \rho_2, \rho_3,$ and ρ_4 = densities of materials 1, 2, 3, and 4, respectively.

A simultaneous solution of the two equations above results in any two sought unknown and the rest of the parameters are known.

177. Problem 5.2

Average Weight, Volume, and Density Calculation

The following are example formulations for commonly encountered field problems:

(a) The amount of weighting materials required to increase original mud density, ρ_o, to a final density, ρ_f, is

$$W_{wm} = \frac{42(\rho_f - \rho_o)}{1 - \left(\dfrac{\rho_f}{\rho_{wm}}\right)},$$

where

W_{wm} = the required amount of weighting material, lbs/bbl, or original mud.

(b) The average weight density, ρ_{av}, of two added materials i and j of weights ρ_i and ρ_j, respectively, is

$$\rho_{av} = \frac{\rho_i \rho_j}{f_w f_i + (1 + f_w)\rho_j},$$

where

$$f_w = \frac{w_j}{w_i + w_j} = \text{the weight fraction of material } j \text{ with respect to added}$$

weights of materials i and j.

(c) The amount of liquid volume, V_l, required to make a mud of total volume V_f having a weight of ρ_f is

$$V_l = V_f \left(\frac{1 - \left(\frac{\rho_f}{\rho_a}\right)}{1 - \left(\frac{\rho_L}{\rho_a}\right)_j} \right),$$

where

ρ_a = the weight density of added material or average weight density, ρ_{av}, of material added to the liquid.

(d) The amount of solids i and j, such as clay and barite for example, required to make-up a specified final mud volume, V_f, and density, ρ_f, is

$$w_i = \frac{42 f_w (\rho_f - \rho_o)}{1 - f_w \left(\frac{\rho_o}{\rho_j}\right) - (1 - f_w)\left(\frac{\rho_o}{\rho_i}\right)}$$

$$w_i = \frac{(1 - f_w)(\rho_f - \rho_o)}{\left[1 - f_w \left(\frac{\rho_o}{\rho_j}\right) - (1 - f_w)\left(\frac{\rho_o}{\rho_i}\right)\right]},$$

where

w_i and w_j = solid materials i and j in lbs/bbl of final mud.

(e) The final mud density, ρ_f, when a certain liquid volume, V_l, of density ρ_1 is added to a mud system of original density, ρ_o, and volume, V_o, is

$$\rho_f = \frac{\rho_o + \alpha \rho_f}{1 + \alpha},$$

where
$\alpha = V_l/V_o$.

178. Problem 5.3

In mixing up a mud system, a basic bentonite-water suspension containing **x**% bentonite by weight was mixed before adding the barite weight material. Derive an equation to calculate the density of the basic bentonite-water suspension.

Solution:

Assume 100 lbs of total materials

$$\text{Final density} = \frac{\text{total weight}}{\text{total volume}} = \frac{\text{mass of bent} + \text{mass of water}}{\text{vol. of bent} + \text{vol. of water}}$$

Simplifying will result in the final density

$$= \frac{\text{mass of bent} + \text{mass of water}}{\dfrac{\text{mass of bent}}{\text{density of bent}} + \dfrac{\text{mass of water}}{\text{density of water}}}$$

$$= \frac{x + (100-x)}{\dfrac{x}{20} + \dfrac{100-x}{8.33}} \text{ ppg}$$

where
bent = bentonite
If you assume 6% bentonite by weight, the final density will be 8.63 ppg.

179. Problem 5.4

Particle sizes
Micron sizes:

Clay	<1
Bentonite	<1
Barite	2–60
Silt	2–74
API Sand	>74

Micron Cut Points for Solid Removal System:

Centrifuge	3–5 micron

Desilter 3" - 4" cones   12–60 micron
Desander 5" - 12" cones   30–60 micron

180. Problem 5.5
Common Weighting Materials

What are the common weighting materials?

Solution:

Average weight of the commonly used weighing materials are given in Table 5.1:

Table 5.1   Density of common weighting materials

Weighting materials	Specific gravity (gm/cm³)	Density (lbm/gal – ppg)	Density (lbm/bbl – ppb)
Barite Pure grade	4.5	37.5	
Barite API – drilling grade	4.2	35	1470
Bentonite	2.6	21.7	910
Calcium carbonate	2.7	22.5	945
Calcium chloride	1.96	16.3	686
Sodium chloride	2.16	18	756
Water	1	8.33	1001
Diesel	0.86	7.2	300
Galena	6.6	55	6007

181. Problem 5.6

Determine the percentage of clay based upon a ton (2000 lb) of clay of 10 ppg clay-water drilling mud. Final mud volume is 100 barrels.

Solution:

If x is the percentage of clay in the final mud it can be written as

$$\frac{x}{100} \times \rho_f \times V_f = 2000$$

$$\frac{x}{100} \times 10 \times 500 = 2000, \text{ and solving for } x$$

$$x = \frac{2000}{10 \times 100 \times 42} \times 100 = 4.76\%$$

182. Problem 5.7

Determine the final mud density of mud if 100 barrels of diesel of specific gravity 0.82 is mixed with 500 barrels 9 ppg mud.

Solution:

Final volume of mud after adding diesel = 500 + 100 = 600 bbl
Using the density relationship, the final density of the mud can be calculated

$$\rho_f = 9 + \left(\frac{100}{600}\right)(9 - 0.82 \times 8.33) = 8.57 \text{ ppg}$$

183. Problem 5.8

What will the final density of mud after adding 5 tons of barite to 500 barrels of 9 ppg drilling mud. Assume the barite density to be 35 ppg.

Solution:

Weighing material added is

$$w_{wm} = \frac{5 \times 2000}{500} = 20 \text{ lb/bbl}$$

The final density of the mud is:

$$\rho_f = \frac{9 + \dfrac{20}{42}}{1 + \left(\dfrac{20}{35 \times 42}\right)} = 9.35 \text{ ppg}$$

184. Problem 5.9

How many sacks of weighing material of density 35 ppg will be required to make 500 barrels of mud weighing 89.76 pound per cubic foot? Also estimate how many barrels will be needed? (1 sack = 100 lb).

Solution:

$$\text{Density of the mud} = \frac{89.76}{\text{ft}^2} \times \frac{\text{ft}^2}{7.48 \text{ gal}} = 12 \text{ ppg}$$

Using the mass balance and volume balance equation it can be written as

$V_{mf} \gamma_{mf} = V_o \gamma_o + V_{mw} \rho_{mw}$ or $500 \times 12 = 8.33 \times (500 - V_{mw}) + V_{mw} \times 35$

Solving for $V_{mw} = 68.80$ bbl

Number of sacks of weighing material required

$$= 35 \frac{\text{lbs}}{\text{gal}} \times 68.80 \text{ bbl} \times \frac{42 \text{ gal}}{\text{lbs}} \times \frac{1 \text{ sack}}{100 \text{ bbl}} = 1011 \text{ sacks}$$

185. Problem 5.10

A mud engineer is planning to calculate the volume of mud that can be prepared with 20% of clay by weight from one ton of clay. Assume the specific gravity of the clay to be 2.5.

Solution:

1 ton of clay = 2000 lbs.
The weight of clay = $2.5 \times 8.33 = 20.82$ ppg.
The weight of mud = $2000/0.20 = 10,000$ lbs.
Using the mass balance, weight of water = $10,000 - 2000 = 8000$ lbs.
Using the volume balance, volume of mud = volume of clay + volume of water:

$$\text{Volume of mud} = \frac{2000}{20.82} + \frac{8000}{8.33} = 1056.44 \text{ gal}$$

$$= 1056.44 \text{ gal} \left(\frac{1 \text{ bbl}}{42 \text{ gal}} \right) = 25.15 \text{ bbl}$$

186. Problem 5.11

What will be the final density of an emulsion mud composed of 25% (by volume) diesel oil and a 10 ppg water-base mud? Assume the specific gravity of oil to be 0.75.

Mud Weighting

Solution:

Starting from the mass balance equation,

$$V_{mf}\rho_{mf} = V_o\rho_o + V_{mw}\rho_{mw},$$

where

V_{mf} = final mud volume.
ρ_{mf} = final density mud.
V_o = emulsion oil volume.
V_{mw} = volume of water-based mud.

Also, it is given as

$$V_{mf}\rho_{mf} = 0.25 V_{mw}\rho_o + (1 - 0.25) V_{mf}\rho_{mw}$$

since $V_o = 0.25 V_{mw}$ and $V_{mw} = (1 - 0.25) V_{mf}$

Therefore,

$$\rho_{mf} = 0.25\rho_o + (1 - 0.25)\rho_{mw}$$

Also,

$$\rho_o = \rho_w \times \rho_d$$

$$V_f\rho_f = \rho_w \times 0.25 \times \rho_d + (1 - 0.25) V_f\rho_w$$

Substituting the values, the final density can be calculated as

$$\rho_f = 8.33 \times 0.25 \times 0.75 + (1 - 0.25) \times 10 = 9.06 \text{ ppg}$$

187. Problem 5.12
Mud Dilution Calculation

When diluting mud with a liquid, the resulting density of the diluted mud can be given as

$$\rho_f = \rho_o + \left(\frac{V_a}{V_o}\right)(\rho_o - \rho_a),$$

where
ρ_o = original mud weight, ppg.

V_a = original volume in barrels.
V_o = final volume in barrels.
ρ_a = density of material added, ppg.

188. Problem 5.13

It is desired to decrease the mud weight by diluting with fluid without any restriction on the final volume handled. Prove that volume of the dilution fluid can be calculated as

$$V_a = V_i \left[\frac{\rho_i - \rho_f}{\rho_f - 8.33\rho_a} \right] \text{ bbl},$$

where
V_i = initial volume of mud, bbl.
ρ_i = initial mud weight, ppg.
ρ_f = final mud weight, ppg.
ρ_a = specific gravity of the dilution fluid.

It is desired to decrease the mud weight by dilution. When the final volume is specified, prove that initial volume of mud can be calculated as

$$V_i = \left[\frac{8.33\rho_a - \rho_f}{8.33\rho_a - \rho_i} \right] \times V_f,$$

where
V_f = final volume required, bbl.

Solution:
Final volume
$V_f = V_i + V_a$
Mass balance

$$\rho_f V_f = \rho_i V_i + 8.33\rho_a V_a$$

Substituting V_f mass balance equation and algebraically adjusting will result in

Mud Weighting

$$V_a = V_i \left[\frac{\rho_i - \rho_f}{\rho_f - 8.33\rho_a} \right]$$

Final volume
$V_f = V_i + V_a$
Mass balance

$$\rho_f V_f = \rho_i V_i + 8.33 \rho_a V_a$$

Substituting V_a mass balance equation and algebraically adjusting will result in

$$V_i = \left[\frac{8.33 \rho_a - \rho_f}{8.33 \rho_a - \rho_i} \right] \times V_f$$

189. Problem 5.14

Determine the final mud density of mud if 100 barrels of diesel with a specific gravity of 0.82 mixed with 500 barrels 9 ppg mud.
Solution:
The final volume of mud after adding diesel = 500 + 100 = 600 bbl.

Using the density relationship, the final density of the mud can be calculated as

$$\rho_f = 9 + \left(\frac{100}{600} \right) (9 - 0.82 \times 8.33) = 8.57 \text{ ppg}$$

190. Problem 5.15

While drilling it was desired to increase the mud weight to 13 ppg. The volume of existing mud in the mud system was 1000 bbl and mud weight was 12.5 ppg. Calculate the amount of barite required to increase the mud weight. Also, calculate the number of sacks required. Assume the barite density to be 35 ppg and 1 sack to be 100 lb.
Solution:
The barite required in lb per bbl of original mud is

$$W_{wm} = \frac{42(13-12.5)}{1-\left(\dfrac{13}{35}\right)} = 33.4 \text{ lb/bbl}$$

The amount of barite sacks required is

$$W_{wm} = \frac{1000 \times 33.4}{100} = 334 \text{ sacks}$$

191. Problem 5.16

A 1000 bbl unweighted fresh water clay mud has a density of 9 ppg. What mud treatment would be required to reduce the solids content to 5% by volume? The total volume must be maintained at 1000 bbl and the minimum allowable mud density is 8.8 ppg.

Solution:

Volume balance: $1000 = V_c + V_w$

Assuming the weight of the clay to be 20.83

Weight balance: $1000 \times 9 = 20.83 \times V_c + 8.33 \times V_w$

Substituting the volume equation and solving for V_c

$1000 \times 9 = 20.83 \times V_c + 8.33 \times (1000 - V_c)$

$V_c = 9464$ bbl

In order to contain 5% by volume of solid content, the volume should be $5\% \times 1000 = 50$ bbl

192. Problem 5.17

It is desired to increase a mud weight from 12 ppg to 0.75 psi/ft using barite having a 4.4 specific gravity. Determine the cost due to mud weight increase, assuming barite is $101/100 lbs.

Solution:

$$\rho_f = \frac{0.75}{0.052} = 15 \text{ ppg.}$$

The weight of the barite required per barrel of original mud can be estimated as

$$W_{wm} = \frac{42\left(\rho_f - \rho_o\right)}{1 - \left(\dfrac{\rho_f}{\rho_{wm}}\right)} = 213.34 \text{ lbs/bbl of original mud}$$

The total cost of the barite required is calculated as
Cost = 213.34 × 1.01 = $215.47/bbl of original mud

193. Problem 5.18

Show that the percent of weight of clay can be estimated from the formula

$$P = \frac{1 - \dfrac{\rho_w}{\rho_m}}{1 - \dfrac{\rho_w}{\rho_c}} \times 100,$$

where

ρ_w = density of water.
ρ_c = density of clay.
ρ_m = density of mud.

A 9.5 ppg mud contains clay and fresh water. Estimate the volume percentage and weight percentage of clay in this mud. Assume specific gravity of clay to be 2.5.

Solution:

$V_m \rho_m = V_c \rho_c + V_w \rho_w$
But → $V_m = V_w + V_c$
$V_w = V_m - V_c$
$V_m \rho_m = V_c \rho_c + (V_m - V_c)\rho_w$
$V_m \rho_m - (V_m - V_c)\rho_w = V_c \rho_c$
$V_m (\rho_m - \rho_w) = V_c (\rho_c - \rho_w)$

$$\frac{V_c}{V_m} = \frac{\rho_m - \rho_w}{\rho_c - \rho_w}$$

$$\frac{\rho_c V_c}{\rho_m V_m} = \frac{\rho_c (\rho_m - \rho_w)}{\rho_m (\rho_c - \rho_w)}$$

$$\left(\frac{W_c}{W_m}\right) = \frac{1 - \frac{\rho_w}{\rho_m}}{1 - \frac{\rho_w}{\rho_c}}$$

So, the percent increase is $P = \dfrac{1 - \frac{\rho_w}{\rho_m}}{1 - \frac{\rho_w}{\rho_c}} \times 100$

Volume percent of solids $= P = \dfrac{9.5 - 8.33}{2.5 \times 8.3 - 8.33} \times 100 = 9.4\%$

Weight percent of solids $= P = \dfrac{20.8\,(9.5 - 8.33)}{9.5 \times (20.8 - 8.3)3} \times 100 = 20.6\%$

194. Problem 5.19

How many sacks of weighing material with a specific gravity of 4.3 will be required to increase the mud weight of 100 barrels of mud from 11 ppg to 12.0 ppg, and what will be the increase in volume? Assume 1 sack to be 100 lb.

Solution:

The barite required in lb per bbl of original mud is

$$W_{wm} = \frac{42(12-11)}{1 - \left(\dfrac{12.0}{4.3 \times 8.33}\right)} = 63.15 \text{ lb/bbl}$$

The amount of barite sacks required is

$$W_{wm} = \frac{100 \times 63.15}{100} \approx 62 \text{ sacks}$$

The final volume is

$$V_{mf} P_{mf} = V_o P_o + V_{mw} P_{mw},$$

where

V_{mf} = final mud volume.
P_{mf} = final mud density.

V_o = original volume.
V_{mw} = the volume of weighting material.
The final mu density is given as

$$V_{mf} \times 12 = 100 \times 11 + (V_{mf} - 100) \times 4.33 \times 8.33$$

and solving yields 104.16 bbl.
The increase in volume is 104.16 − 100 = 4.16 bbl

195. Problem 5.20

A particular drilling phase is planned to be drilled with a mud weight of 14 ppg. Present mud weight in the tank and hole is 13 ppg. The total volume of the hole and mud tanks including reserve mud tanks are 1200 bbl. Estimate the number of sacks of barite needed. Assume 1 sack to be 100 lbs.

Solution:

The weight of the barite required per barrel of mud is calculated as

$$W_{wm} = \frac{42(14-13)}{1-\left(\frac{13}{35}\right)} = 69.47 \text{ lb/bbl}$$

The total amount of barite required = 69.47 × 1200 = 83,364 lbs of barite.

Sacks required is

$$\frac{83,364 \text{ (sack)}}{100 \left(\text{lb}/\text{sack}\right)} = 834 \text{ lbs } (\sim 42 \text{ tons})$$

196. Problem 5.21

How many sacks of barite are necessary to increase the density of 500 bbl of mud from 10 to 12 ppg. What will be final volume of mud? Assume 1 sack to be 100 lbs and the original volume of 500 bbl is to be maintained.

Solution:

Barite required in lb per bbl of original mud

$$w_{wm} = \frac{42(12-10)}{1-\left(\frac{12.0}{4.3 \times 8.33}\right)} = 126.31 \text{ lb/bbl}$$

Amount of barite sacks required

$$w_{wm} = \frac{500 \times 126.31}{100} \approx 631 \text{ sacks}$$

Final volume if the original mud is not discarded
$$V_{mf} \rho_{mf} = V_o \rho_o + V_{mw} \rho_{mw}$$
V_{mf} = final mud volume; ρ_{mf} = final density mud
V_o = original volume; V_{mw} = volume of weighting material
$V_{mf} \times 12 = 500 \times 10 + (V_{mf} - 500) \times 4.33 \times 8.33$
Solving yields, 541.52 bbl
Increase in volume is 541.52 − 500 = 41.52 bbl
Final volume if the original mud to be maintained
Volume of additives added

$$w_{wm} = \frac{500(10-12)}{(10-4.3 \times 8.33)} = 38.73 \text{ bbl}$$

Volume of original mud dumped is = 38.73 bbl
Amount of barite sacks required

$$w_{wm} = \frac{(500-38.73) \times 126.31}{100} \approx 583 \text{ sacks}$$

To check the final volume for 500 bbl using the following equation
$V_{mf} \times 12 = (500 - 38.73) \times 10 + (V_{mf} - (500 - 38.73)) \times 4.33 \times 8.33$

197. Problem 5.22

How many sacks of weighing material of density 35 ppg will be required to make 500 barrels of mud weighing 89.76 pound per cubic foot? Also, estimate how many barrels will be needed. Assume 1 sack to be 100 lb.

Mud Weighting

Solution:

The density of the mud = $\dfrac{89.76}{ft^2} \times \dfrac{ft^2}{7.48\ gal} = 12$ ppg

Using the mass balance and volume balance equation, it can be written as

$$V_{mf} \rho_{mf} = V_o \rho_o + V_{mw} \rho_{mw}$$

or

$$500 \times 12 = 8.33 \times (500 - V_{mw}) + V_{mw} \times 35$$

Solving for $V_{mw} = 68.80$ bbl

The number of sacks of weighing material required is

$$35\ \dfrac{lbs}{gal} \times 68.80\ bbl \times \dfrac{42\ gal}{lbs} \times \dfrac{1\ sack}{100\ bbl} = 1011\ sacks$$

198. Problem 5.23

What will be the volume of diesel oil that needs to be added when a final density of an emulsion mud is 9.5 ppg composed of diesel oil and a 10 ppg, 800 barrels of water-based mud? Assume the specific gravity of oil to be 0.75. Also, calculate the percentage of diesel oil by volume in the total mud.

Solution:

Density of the oil = $0.75 \times 8.33 = 6.25$ ppg

The ratio of the mud to the diesel oil added can be given as

$$\dfrac{V_m}{V_l} = \dfrac{(\rho_a - \rho_o)}{\rho_f - \rho_o} - 1$$

Substituting the values,

$$\dfrac{V_m}{V_l} = \dfrac{(\rho_a - \rho_o)}{\rho_f - \rho_o} - 1 = \dfrac{(6.25 - 10)}{9.5 - 10} - 1 = 6.505$$

Volume of diesel oil added = $800/6.505 = 123$ bbl

Percentage of diesel in the mud = $\dfrac{123}{123+800} \times 100 = 13.3\%$

199. Problem 5.24

An emulsion mud is being prepared by adding 100 barrels of oil to 800 barrels of 11.5 ppg mud. If the density of the mud is to be maintained as its original value, how much barite will also have to be added to the mud? Specific gravity of oil is 0.82; specific gravity of barite is 4.4. Determine the cost owing to mud weight increase assuming barite cost to be $18/100 lbs.

Solution:

Density of oil $\rho_f = 0.82 \times 8.33 = 6.83$ ppg

Amount of barite to be added

$$W_{wm} = \dfrac{42\,(11.5-6.8)}{1-\left(\dfrac{11.5}{36.65}\right)} = 287.68 \text{ lbs/bbl}$$

Total cost = 287.68 (lbs/bbl) × 18 ($/lbs) = $5178 per bbl of mud

200. Problem 5.25

It is desired to increase the mud weight from 12 ppg to 13 ppg without increasing the volume using weighing material with a density of 35 ppg. Calculate the amount of mud discarded if the original volume is 800 bbl.

Solution:

The volume of additives added is

$$W_{wm} = \dfrac{800\,(12-13)}{(12-35)} = 34.8 \text{ bbl}$$

The volume of original mud dumped = 34.8 bbl

201. Problem 5.26

Before drilling the next phase, it was desired to increase the 900 bbl of mud with a mud density of 16 ppg mud to 17 ppg. The volume fraction of

Mud Weighting

low specific gravity solids must be reduced from 0.055 to 0.030 by dilution with water. Compute the original mud that must be discarded while maintaining the final mud volume of 900 bbl and the amount of water and barite that should be added. Assume barite density to be 35 ppg.

Solution:

Volumetric balance is

$$V_i = V_f \frac{f_f}{f_{1i}},$$

where

$$V_f = \frac{(11 \times 5 \times 36) \times 2 \times 0.90}{5.6} = 636 \text{ bbl.}$$

Discarded volume is

$$636 - V_i = 636 - 636 \frac{0.030}{0.055} = 289 \text{ bbl}$$

For ideal mixing and performing mass balance, the volume of water added is

$$V_w = \frac{(\rho_b - \rho_f) V_f - (\rho_b - \rho_i) V_i}{(\rho_b - \rho_w)}$$

and the mass of barite needed is $m_b = (V_f - V_i - V_w) \rho_b$

Substituting respective values yields
$V_w = 204$ bbls; $m_B = 125,486$ lbs

202. Problem 5.27

Calculate the percent weight increase for the following data:

- Density of water = 8.34 ppg
- Density of solids = 22 lbs/gal
- Mud weight = 86 lb/cu. ft
- 1 bbl = 42 gal
- 7.48 lb/cu.ft = ppg

Solution:

Percent weight increase is given as

$$P = \frac{1 - \frac{\rho_w}{\rho_m}}{1 - \frac{\rho_w}{\rho_c}} \times 100$$

Substituting the values,

$$P = \frac{1 - \frac{8.34}{30.43}}{\frac{8.34}{13.45}} = \frac{0.274}{0.62} \times 100 = 44\%$$

203. Problem 5.28

Calculate the mud weight increase due to cuttings in the mud for the following data:

- Cuttings generated = 2 bbl/min
- Density of cuttings = 800 lbs/bbl
- Flowrate of mud = 400 gpm
- Mud weight = 10 ppg

Solution:

Weight of cuttings generated = 2 × 800 = 1600 lbs/min
Weight of mud circulated = 400 × 10 = 4000 lbs/min
Volume of cuttings generated = 1600 × 42/800 = 84 gpm
Volume rate of mud = 400 gpm

Mud weight in the annulus is

$$\rho_{ann} = \frac{(4000 + 600)}{(400 + 84)} = 11.57 \text{ ppg}$$

204. Problem 5.29

Calculate the volume of cuttings generated while drilling 12 ¼" hole with the rate of penetration of 50 ft/hr. Assume formation porosity to be 30%

Mud Weighting

Solution:

Using the equation
Volume of cuttings generated in barrels per hour
Volume of cuttings entering the mud system

$$V_c = \frac{(1-\phi) D_b^2 \times ROP}{1029} \text{ bbl/hr},$$

where

ϕ = average formation porosity.
D_b = diameter of the bit, in.
ROP = rate of penetration, ft/hr.

$$V_c = \frac{(1-\phi) D_b^2 \times ROP}{24.49} \text{ gal/hr or } V_c = \frac{(1-\phi) D_b^2 \times ROP}{1469.4} \text{ gpm}$$

Substituting the values

$$V_c = \frac{(1-\phi) D_b^2 \times ROP}{1029} = \frac{\left(1-\frac{30}{100}\right) \times 12.5^2 \times 50}{1029} = 5.1 \text{ bbl/hr}$$

Volume of cuttings generated in barrels per hour

$$V_c = \frac{(1-\phi) D_b^2 \times ROP}{24.49} = \frac{\left(1-\frac{30}{100}\right) \times 12.5^2 \times 50}{24.49} = 214.5 \text{ gal/hr}$$

Volume of cuttings generated in barrels per hour

$$V_c = \frac{(1-\phi) D_b^2 \times ROP}{1469.4} = \frac{\left(1-\frac{30}{100}\right) \times 12.5^2 \times 50}{1469.4} = 3.57 \text{ gpm}$$

205. Problem 5.30

A well is being drilled with a mud density of 14 ppg, and it is desired to increase the mud density to 15 ppg to prevent any fluid influx formation. The volume fraction of low specific gravity solids must be reduced from

0.055 to 0.030 by dilution with water. Compute the original mud that must be discarded and the amount of water and barite that should be added. Calculate the unnecessary volume of mud discarded/added if an error of ±0.02 was made in determining the original and final volume fractions of low specific gravity solids in the mud.

If the barite and mixing water added are 175,000 lbm and 250 bbls, respectively, calculate the present volume fraction of low specific gravity solids. Assume the barite density to be 35 ppg.

Mud tank size = 11' width × 5' height × 36' length
Number of tanks = 2 and both 90% full

Solution:

Volumetric balance is

$$V_1 = V_2 \frac{f_{s2}}{f_{s1}},$$

where

$$V_2 = \frac{(11 \times 5 \times 36) \times 2 \times 0.90}{5.6} = 636 \text{ bbl.}$$

Discarded volume is

$$636 - V_1 = 636 - 636 \frac{0.030}{0.055} = 289 \text{ bbl}$$

For ideal mixing and performing mass balance, the volume of water added is

$$V_w = \frac{(\rho_B - \rho_2)V_2 - (\rho_B - \rho_1)V_1}{(\rho_B - \rho_w)}$$

The mass of barite needed is

$$m_B = (V_2 - V_1 - V_w)\rho_B$$

Substituting respective values yields

$$V_w = 204 \text{ bbl}$$

$$m_B = 125{,}486 \text{ lbs}$$

Mud Weighting

For an error of + 0.02, mud discarded

$$= 636 - V_1 = 636 - 636 \frac{0.05}{0.075} = 212 \text{ bbl}$$

Unnecessary mud discarded in error = 77 bbl

For an error of –0.02, mud discarded

$$= 636 - V_1 = 636 - 636 \frac{0.01}{0.035} = 455 \text{ bbl}$$

Unnecessary mud discarded in error = 166 bbl

If the barite and mixing water added are 175,000 lbm and 250 bbl, respectively, the present volume fraction of low specific gravity solids is

$$f_2 = f_1 \frac{V_1}{V_2} = (0.055) \frac{267}{636} = 0.0231$$

206. Problem 5.31

Calculate the average mud density when the rate of penetration is 20.0 ft/min. Assume the hole is 17-½ in. diameter, the flow rate is 800 gpm, and the (γ_{ps}) is 9.0 ppg

Solution:

An estimate of the average annulus mud weight ($\gamma_{m,av}$) is given by

$$\gamma_m = \frac{\gamma_{ps} Q + 0.85 D_h^2 R}{Q + 0.0408 D_h^2 R},$$

where,

$\gamma_{m, av.}$ = average annular mud weight (lb/gal).
Q = flow rate (gpm).
γ_{ps} = measured mud weight at pump suction (lb/gal).
D_h = diameter of hole (in).
R = penetration rate, based on time the pump is on before, during and after joint is drilled down (fpm).

$$\gamma_{m,av} = \frac{800 \times 9 + 0.85 \times 20 \times 17.5 \times 17.5}{800 + 0.0408 \times 17.5 \times 17.5 \times 20} = 11.8 \text{ ppg}$$

207. Problem 5.32

Derive expressions for determining the amounts of barite and water that should be added to increase the density of 100 bbl of mud from ρ_1 to ρ_2. Also, derive an expression for the increase in mud volume expected upon adding barite and water. Assume a water requirement of 1 gal per sack of barite, 1 sack = 100 lb.

Solution:

Using the volume and mass balance, it can be written as

$$V_2 = V_1 + V_w + V_B = V_1 + \frac{m_B}{\rho_B} + m_B V_{wB},$$

where

V_{wB} = barite-water requirement.

Including the mass of water,

$$\rho_2 V_2 = \rho_1 V_1 + m_B + \rho_w V_{wB} m_B.$$

Solving the above equations yields

$$V_1 = V_2 \left[\frac{\rho_B \left(\frac{1 + \rho_w V_{wB}}{1 + \rho_B V_{wB}} \right) - \rho_1}{\rho_B \left(\frac{1 + \rho_w V_{wB}}{1 + \rho_B V_{wB}} \right) - \rho_2} \right],$$

$$100 = V_2 \left[\frac{35 \left(\frac{1 + 8.33 \times 0.01}{1 + 35 \times 0.01} \right) - \rho_2}{100 \left(\frac{1 + 8.33 \times 0.01}{1 + 35 \times 0.01} \right) - \rho_1} \right].$$

Given values are ρ_w = 8.33 ppg, ρ_B = 35 ppg, V_1 = 100 bbl, V_{wB} = 0.01 gal/lb.

$$m_B = \frac{\rho_B}{1 + \rho_B V_{wB}} (V_2 - V_1) = \frac{35 \times 42}{1 + 35 \times 0.01} (V_2 - 100)$$

Mud Weighting

Substituting V_2 from previous equation

$$m_B = 109,00 \frac{(\rho_2 - \rho_1)}{(28.02 - \rho_2)} \text{ lbm}$$

$$= 109 \frac{(\rho_2 - \rho_1)}{(28.02 - \rho_2)} \text{ sacks}$$

$$V_w = V_{wB} m_B = \frac{m_B}{100} \text{ gal} = \frac{m_B}{4200} \text{ bbl}$$

$$V = V_w + \frac{m_B}{\rho_s} = \frac{m_B}{4200} + \frac{m_B}{1470} = 0.0091 \, m_B$$

208. Problem 5.33

Consider the mud consisting of oil, water, and solids:

- f_o = volume fraction of oil
- f_w = volume fraction of water
- f_s = volume fraction of solid
- f_f = volume fraction of fluid including oil and water.

Derive an expression to calculate f_f in terms of oil-water ratio (OWR $= \frac{f_o}{f_w}$), and the following densities:

- Mud density = ρ_m
- ρ_o = oil density
- ρ_w = water density
- ρ_s = solid density.

Solution:

Consider the mud consisting of oil, water, and solids. The density of the mud can be written from mass balance as

$$\rho_m = \rho_o f_o + \rho_w f_w + \rho_s f_s$$

In terms of the volume fraction of oil, water, and solids, it can be written as

$$f_o + f_w + f_s = 1$$

The oil-water ratio, OWR, is

$$\frac{f_o}{f_w}$$

Also,

$$f_f + f_s = 1, f_o + f_w = f_f,$$

where
f_f = the volume fraction of fluid including oil and water.

The above equation can be re-written as

$$f_f \left(\rho_o \frac{f_o}{f_f} + \rho_w \frac{f_w}{f_f} \right) + \rho_s (1 - f_f) = \rho_m$$

Therefore, the fluid fraction can be given as

$$f_f = \frac{\rho_m - \rho_s}{\left(\rho_o \dfrac{f_o}{f_f} + \rho_w \dfrac{f_w}{f_f} - \rho_s \right)}$$

It is known that $f_o + f_w = f_f$ and can be written as

$$\frac{f_f}{f_w} = \frac{f_o}{f_w} + 1$$

From the above equation, $\dfrac{f_w}{f_f}$ can be calculated, and $\dfrac{f_o}{f_f}$ can be found out from the following equation:

$$\frac{f_o}{f_f} = 1 - \frac{f_w}{f_f}.$$

In terms of OWR, the volume fraction of the fluid can be written as

$$f_f = \frac{\rho_m - \rho_s}{\rho_o - \rho_s + (\rho_w - \rho_o)\left(\dfrac{1}{1 + OWR}\right)}$$

Mud Weighting

209. Problem 5.34

It is desired to mix X barrels of ρ_m ppg of water-based mud. Bentonite (weight = ρ_c ppg) and barite (weight = ρ_b ppg) are to be added in the ratio of α_c lb of bentonite to P_b lbs of barite. Determine the amount of water, V_w, bentonite, W_c, and barite, W_b, needed to make-up the mud system.

Solution:

Mass balance is

$$V_f \rho_m = \rho_w V_w + \rho_b V_b + \rho_c V_c,$$

where

$$V_f \rho_m = 42 X \rho_m.$$

Volume balance, using in $42X$ gal of mud, is

$$42X = V_w + V_b + V_c \text{ or } V_f = V_w + V_b + V_c$$

Another relation given is

$$\frac{\alpha_c}{P_b} = \frac{\rho_c V_c}{\rho_b V_b}$$

Solving will yield

$$V_b = \frac{V_f (\rho_m - \rho_w)}{(\rho_c - \rho_w) \frac{\alpha_c \rho_b}{P_b \rho_c} + \rho_b - \rho_w}$$

and

$$V_c = \frac{V_f (\rho_m - \rho_w)}{(\rho_c - \rho_w) \frac{\alpha_c \rho_b}{P_b \rho_c} + \rho_b - \rho_w} \cdot \frac{\alpha_c \rho_b}{P_b \rho_c}$$

$$W_c = \frac{V_f (\rho_m - \rho_w)}{(\rho_c - \rho_w) \frac{\alpha_c \rho_b}{P_b \rho_c} + \rho_b - \rho_w} \cdot \frac{\alpha_c \rho_b}{P_b}$$

$$W_b = \frac{V_f (\rho_m - \rho_w)}{(\rho_c - \rho_w) \frac{\alpha_c}{P_b \rho_c} + 1 - \left(\frac{\rho_w}{\rho_b}\right)}$$

$$V_w = V_f - V_b - V_c$$

Substituting V_b and V_c from above equation yields

$$V_w = V_f \left(1 - \frac{(P_b\rho_c - \alpha_c\rho_b)(\rho_m - \rho_w)}{\alpha_c\rho_b(\rho_c - \rho_w) + P_b\rho_c(\rho_b - \rho_w)}\right)$$

210. Problem 5.35

It is desired to mix V_f of ρ_m mud weight. Mud composition is as follows: liquid (V_w), barite (V_b, ρ_b), and bentonite (V_c, ρ_c). Weight fraction of barite to bentonite is α. Develop (derive) the equations that will allow you to make-up the mud system. V = volume, ρ = weight density.

Solution:

Using the relationship as $\alpha = \dfrac{\rho_c V_c}{\rho_b V_b}$

Solving will yield

$$V_b = \frac{V_f(\rho_m - \rho_w)}{(\rho_c - \rho_w)\dfrac{\alpha\rho_b}{\rho_c} + \rho_b - \rho_w}$$

$$V_c = \frac{V_f(\rho_m - \rho_w)}{(\rho_c - \rho_w)\dfrac{\alpha\rho_b}{\rho_c} + \rho_b - \rho_w} \cdot \frac{\alpha\rho_b}{\rho_c}$$

$$W_c = \frac{V_f(\rho_m - \rho_w)}{(\rho_c - \rho_w)\dfrac{\alpha\rho_b}{\rho_c} + \rho_b - \rho_w} \cdot \alpha\rho_b$$

$$W_b = \frac{V_f(\rho_m - \rho_w)}{(\rho_c - \rho_w)\dfrac{\alpha}{\rho_c} + 1 - \left(\dfrac{\rho_w}{\rho_b}\right)}$$

$$V_w = V_f - V_b - V_c$$

Substituting V_b and V_c from above yields

$$V_w = V_f \left(1 - \frac{(\rho_c - \alpha\rho_b)(\rho_m - \rho_w)}{\alpha\rho_b(\rho_c - \rho_w) + \rho_c(\rho_b - \rho_w)}\right)$$

Mud Weighting

211. Problem 5.36

How many sacks of Aquagel of 2.5 specific gravity are required to prepare 300 barrels of an Aquagel-water mud containing 5.5% Aquagel by weight? Assume 1 sack to be 100 lb.

Solution:

$$P = \frac{1 - \dfrac{\rho_w}{\rho_{mf}}}{1 - \dfrac{\rho_w}{\rho_{aq}}} \times 100$$

Substituting the values, $\rho_{mf} = 8.61$ ppg

It can be written as

$$V_{mf}\rho_{mf} = V_{aq}\rho_{aq} + V_w\rho_w$$

or

$$V_{aq}\rho_{aq} = V_{mf}\rho_{mf} - V_w\rho_w$$

Since $V_w = (300 - V_{aq})$

$$V_{aq}\gamma_{aq} = 300 \times 8.61 - 8.33(300 - V_{aq})$$

Solving,

$$V_{aq} = 6.825 \text{ bbl}$$

$$\text{Number of sacks} = \frac{6.825 \times 2.5 \times 8.33 \times 42}{100} = 60$$

212. Problem 5.37

The density of 900 bbl of 16 ppg mud must be increased to 17 ppg. The volume fraction of low-specific gravity solids must be reduced from 0.055 to 0.030 by dilution with water. A final mud volume of 900 bbl is desired. Compute the original mud that must be discarded and the amount of water and barite that should be added. Calculate the mud discarded if an error of ±0.02 was made in determining the original volume fraction of low specific gravity solids in the mud. Assume the barite density to be 35 ppg.

Solution:

Volumetric balance is

$$V_1 = V_2 \frac{f_{s2}}{f_{s1}}$$

The initial volume of mud needed = $900 \times \frac{0.03}{0.055} = 490.9$ bbl

Thus, (900–490.9) 409 bbl of initial 900 bbl should be discarded. New volume and volume balance can be given as below

Volume balance

$$V_2 = V_1 + V_w + \frac{m_B}{\rho_B}$$

Mass balance is

$$\rho_2 V_2 = \rho_1 V_1 + m_B + \rho_w V_{wB}$$

Solving will give

$$V_w = \frac{(\rho_B - \rho_2)V_2 - (\rho_B - \rho_1)V_1}{(\rho_B - \rho_w)}$$

Substituting,

$$V_w = \frac{(35-17)900 - (35-16)490}{(35-8.33)} = 258.34 \text{ bbl}$$

Solving, you can get the equation for the mass of barite needed:

$m_B = (V_2 - V_1 - V_w)\rho_b = (900 - 490.9 - 258.34)42 \times 35 = 221{,}620$ lbm

For an error of + 0.02, mud discarded is

$$490.9 - V_1 = 490.9 - 490.9 \frac{0.05}{0.075} = 163.63 \text{ bbl}$$

Unnecessary mud discarded in error = 245.37 bbl

For an error of -0.02, mud discarded is

$$490.9 - V_1 = 490.9 - 490.9 \times \frac{0.01}{0.035} = 350.64 \text{ bbl}$$

Unnecessary mud discarded in error = 58.36 bbl

Mud Weighting

213. Problem 5.38

A retort analysis of a mud with a density of 13 ppg shows the following percentage of components:

- Water = 57% (fresh water)
- Diesel = 8%
- Solids = 35%

Calculate the average density of the solids.

Solution:
The mass balance for the mud is

$$V_m \rho_m = \rho_w V_w + \rho_o V_o + \rho_s V_s,$$

where
V = volume and ρ = density. Subscripts m = mud, w = water, o = oil, and s = solids.
The density of water = 8.33 ppg.
The density of diesel = 7.16 ppg.
The density of mud = 13 ppg.
Substituting the values in the equation,

$$1 \times 13 = 0.57 \times 8.33 + 0.08 \times 7.16 + \rho_s \times 0.35$$

Solving for the density of the solids,

$$\rho_s = \frac{1 \times 13 - 0.57 \times 8.33 - 0.08 \times 7.16}{0.35} = 21.9 \text{ ppg}$$

214. Problem 5.39

A 1000 bbl unweighted fresh water clay mud has a density of 9 ppg. What mud treatment would be required to reduce the solids content to 5% by volume? The total volume must be maintained at 1000 bbl, and the minimum allowable mud density is 8.8 ppg.

Solution:
The volume balance is

$$1000 = V_c + V_w.$$

Assuming the weight of the clay to be 20.83, the weight balance is

$$1000 \times 9 = 20.83 \times V_c + 8.33 \times V_w$$

Substituting the volume equation and solving for V_c,

$$1000 \times 9 = 20.83 \times V_c + 8.33 \times (1000 - V_c)$$

$$V_c = 9464 \text{ bbl}$$

In order to contain 5% by volume of solid content, the volume should be 5% × 1000 = 50 bbl

215. Problem 5.40
Base Fluid—Water-Oil Ratios Calculations

The base fluid/water ratio can be calculated as the percentage by volume of base fluid in a liquid phase:

$$P_b = \left(\frac{VR_b}{VR_b + VR_w} \right) \times 100,$$

where
VR_b = the volume percentage of base fluid.
VR_w = the volume percentage of water.
The percentage of water is
$P_w = 100 - P_b$
The base fluid water ratio is

$$\frac{P_b}{P_w} = \frac{VR_b}{VR_w}$$

216. Problem 5.41

Using the following retort analysis data, calculate the base fluid water ratio:

- Base fluid = 62%
- Water = 12%
- Solids = 26%

Solution:

The percentage by volume of base fluid in the liquid phase is

$$P_b = \left(\frac{VR_b}{VR_b + VR_w}\right) \times 100 = \left(\frac{62}{62+12}\right) \times 100 = 83.78$$

The percentage of water is given as

$$P_w = 100 - P_b = 100 - 83.78 = 16.22$$

The base fluid water ratio is given as

$$\frac{P_b}{P_w} = \frac{VR_b}{VR_w} = \frac{83.78}{16.22} = \frac{62}{12} = 5.17$$

217. Problem 5.42

It was desired to reduce the base fluid water ratio to 85/100. Determine the amount of base fluid to be added to a 1200 bbl mud system. Use the data from Problem 4.27.

Solution:

The percentage of water is

$$P_w = \left(\frac{VR_w}{VR_b + VR_w + V_b}\right) \times 100,$$

where
V_b = the volume of the base fluid.

Substituting the values,

$$P_w = \left(\frac{VR_w}{VR_b + VR_w + V_b}\right) \times 100 = \left(\frac{.12}{.62 + .12 + V_b}\right) \times 100 = 0.15$$

Solving for V_b yields 0.06 bbl/bbl of mud.

The total amount of base fluid required = 0.06 (bbl/bbl of mud) × 1200 (bbl of mud) = 72 bbl

218. Problem 5.43

It was desired to reduce the base fluid water ratio to 80/100. Determine the amount of water to be added to a 1200 bbl mud system. Use the data from Problem 4.27.

Solution:

The percentage of water is

$$P_b = \left(\frac{VR_b}{VR_b + VR_w + V_w} \right) \times 100,$$

where
V_b = the volume of the base fluid.

Substituting the values,

$$P_b = \left(\frac{VR_b}{VR_b + VR_w + V_w} \right) \times 100 = \left(\frac{.80}{.62 + .80 + V_w} \right) \times 100 = 0.035$$

Solving for V_w yields 0.035 bbl of water /bbl of mud.

The total amount of water required = 0.035 (bbl /bbl of mud) × 1200 (bbl of mud) = 35 bbl

219. Problem 5.44

Determine the amount (lb/bbl) of a 5.2 specific gravity weighing material that must be added to a mud system to increase its pressure gradient from 0.52 psi/ft to 0.624 psi/ft.

Solution:

The initial pressure gradient = 0.52 psi/ft = 10 ppg
The final pressure gradient = 0.624 psi/ft = 12 ppg
Weighting material = 5.2 = 5.2 × 8.33 = 43.316 ppg:

$$w_{wm} = \frac{42\,(12-10)}{1 - \left(\dfrac{12}{43.32} \right)} = 116 \text{ lbs/bbl}$$

CHAPTER 6

FLUIDS

This chapter focuses on different basic calculations associated with drilling fluids such as rheology, fluid loss.

220. Problem 6.1
Define Viscosity with Related Units
Mud viscosity is defined as the internal resistance of a fluid to flow.

Only Newtonian fluids, such as water, have a true viscosity that can be defined by a single term. Non-Newtonian fluids, on the other hand, require more than one component of viscosity to describe their flow behavior.

It is generally expressed as plastic viscosity (PV) with the unit cP.

221. Problem 6.2
Provide Viscosity Dimensions
The dimension of the viscosity is
$$[FL^{-2}T]$$
or
$$[ML^{-1}T^{-1}]$$

Units are

Ns/m²

or

Poise (P)

or

centipoise (cP)

and

lbf-s/ft²

222. Problem 6.3
Viscosity Unit Conversion
$$1P = 1\frac{g}{cm-s}$$

$$1\frac{N-s}{m^2} = 10P$$

$$1P = 100 \text{ cP}$$

$$1 \text{cP} = \frac{1}{47,880.26} \frac{\text{lbf}-\text{s}}{\text{ft}^2}$$

223. Problem 6.4
Mud Rheology Calculations

Shear stress (τ) and shear rate (γ) for Newtonian fluid (Fig. 6.1) is given as

$$\tau = \mu\gamma,$$

where

μ = Newtonian viscosity.

In engineering units, τ = dynes/cm^2 = 4.79 lb/100 ft^2, γ = sec^{-1}, and μ = poise = dyne × sec/cm^2.

The field unit of viscosity is the centipoise (1 poise = 100 centipoise). The field unit of shear stress is lb/100 ft^2.

For Bingham plastic model the relationship is given by (Fig. 6.2)

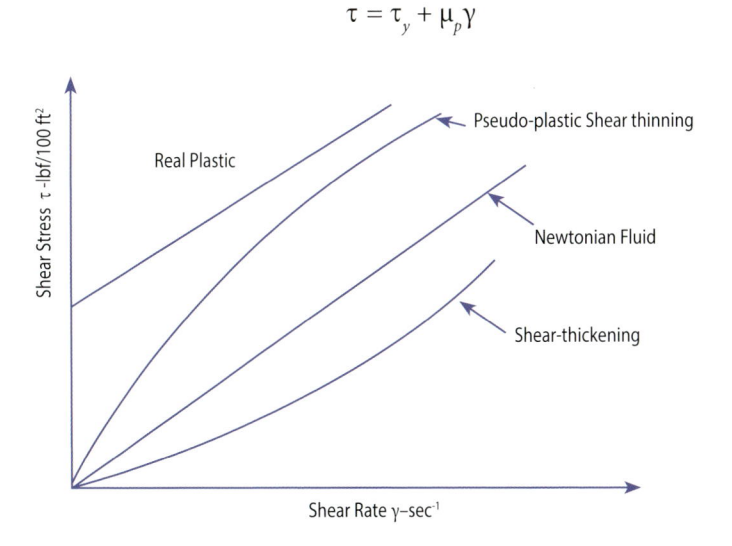

$$\tau = \tau_y + \mu_p \gamma$$

Figure 6.1 Shear rate—shear stress relationship of time independent non-Newtonian fluids

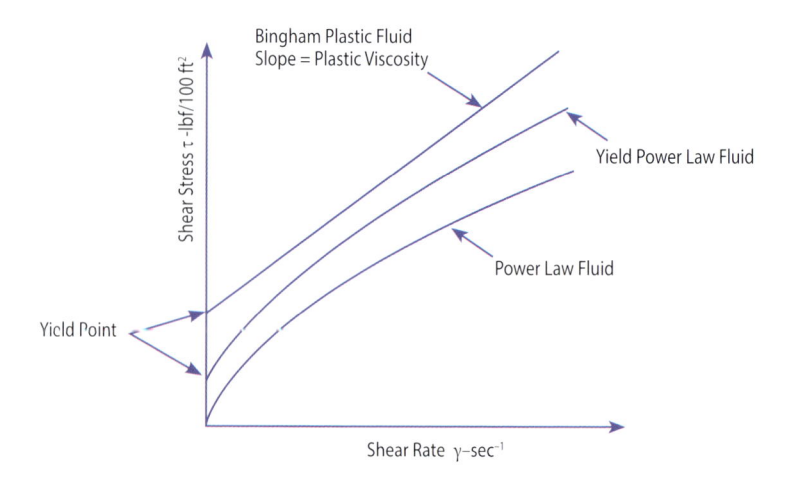

Figure 6.2 Shear rate—shear stress relationship of non-Newtonian fluids

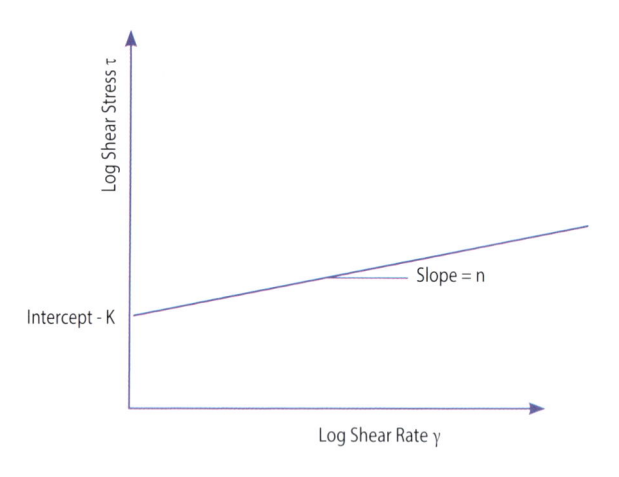

Figure 6.3 Shear rate—shear stress relationship of a power-law fluid on logarithmic scale

For the power law model shear stress (τ) and shear rate (γ) are given by (Fig. 6.3)

$$\tau = K\gamma^n$$

and for the yield power law (Herschel–Bulkley) shear stress (τ) and shear rate (γ) are given by

$$\tau = \tau_y + K\gamma^n,$$

where
τ_y = yield value or yield stress.
μ_p = Bingham plastic viscosity.
K = consistency index.
n = power law index.

224. Problem 6.5
Rheology
Explain a three parameter rheological model.
Solution:
Yield Power Law or Herschel–Bulkley Model

$$\tau = \tau_y + K\gamma^n,$$

where
τ_y = yield value or yield stress.
K = consistency index.
n = power law index.

225. Problem 6.6
Identify the following non-Newtonian rheological models and say whether they are two, three, or four parameter models.

Collins-Graves $\tau = (\tau_y + K\gamma)(1 - e^{-\beta\gamma})$

Robertson-Stiff $\tau = K(C + \gamma)^n$

Casson $\tau = (\sqrt{\tau_y} + \sqrt{\mu\gamma})^2$

Power law $\tau = K\gamma^n$

Herschel–Bulkley $\tau = \tau_y + K\gamma^n$

Generalized Herschel–Bulkley $\tau^m = \tau_y^m + (\mu\gamma)^n$

Bingham plastic $\tau = \tau_y + K\gamma$

Ellis model $\tau = \mu_0 / \left\{ 1 + \left(\tau / \tau_{\frac{1}{2}} \right) \right\}^{\alpha - 1} \gamma$

Solution:

Collins–Graves $\tau = (\tau_y + K\gamma)(1 - e^{-\beta\gamma})$ - 3 parameter model

Robertson–Stiff $\tau = K(C + \gamma)^n$ - 3 parameter model

Casson $\tau = (\sqrt{\tau_y} + \sqrt{\mu\gamma})^2$ - 2 parameter model

Power Law $\tau = K\gamma^n$ - 2 parameter model

Herschel–Bulkley $\tau = \tau_y + K\gamma^n$ - 3 parameter model

Generalized Herschel–Bulkley $\tau^m = \tau_y^m + (\mu\gamma)^n$ - 4 parameter model

Bingham plastic $\tau = \tau_y + K\gamma$ - 2 parameter model

Sisko model $\tau = a\gamma + b\gamma^n$ - 3 parameter model

Ellis model $\tau = \mu_0 / \left\{ 1 + \left(\tau/\tau_{\frac{1}{2}}\right)^n \right\}^{\alpha-1} \gamma$ - 3 parameter model

226. Problem 6.7

Identify the following non-Newtonian rheological models and say whether they are two, three, four, more parameter models.

Quemada $\tau = \eta\left(1 - \frac{1}{2}K_Q \times H \times c \times t\right)^{-2} \gamma$

Collins–Graves $\tau = (1 - e^{-\beta\gamma})(\tau_0 + \mu\gamma)$

Robertson–Stiff $\tau = A\left(\left(\frac{\tau_0}{A}\right)^{\frac{1}{n}} + \gamma\right)^n$

Casson $\tau = (\sqrt{\tau_0} + \sqrt{\mu\gamma})^2$

Ellis model $\tau = \dfrac{\tau_0}{1 + \left(\dfrac{\tau_0}{\tau_{\frac{1}{2}}}\right)^{n-1}} \gamma$

Generalized Herschel–Bulkley $\tau^m = \tau_0^m + (\mu\gamma)^n$

Sisko model $\tau = a\gamma + b\gamma^n$

Solution:

Quemada $\tau = \eta\left(1 - \frac{1}{2}K_Q \times H \times c \times t\right)^{-2} \gamma$ - 5 parameter model

Collins-Graves $\tau = (1 - e^{-\beta\gamma})(\tau_0 + \mu\gamma)$ - 3 parameter model

Robertson-Stiff $\tau = A\left(\left(\frac{\tau_0}{A}\right)^{\frac{1}{n}} + \gamma\right)^n$ - 3 parameter model

Ellis model $\tau = \dfrac{\tau_0}{1 + \left(\dfrac{\tau_0}{\tau_{\frac{1}{2}}}\right)^{n-1}} \gamma$ - 3 parameter model

227. Problem 6.8
Casson Model
Standard Model
It is given as

$$\sqrt{\tau} = \sqrt{\tau_0} + \sqrt{\mu}\sqrt{\gamma}$$

Fluids in motion

$$\sqrt{\tau} = \sqrt{\tau_0} + \sqrt{\mu}\left|\frac{\partial u}{\partial r}\right|^{\frac{1}{2}} \text{ for } |\tau| \geq \tau_0$$

Generalized Casson Model

$$\tau^{\frac{1}{p}} = \tau_0^{\frac{1}{p}} + \mu^{\frac{1}{p}}\gamma^{\frac{1}{p}}$$

This is a fundamental equation

$$\frac{d\tau}{(\tau + \theta)^\alpha} = k\frac{d\gamma}{(\gamma + \phi)^\alpha},$$

where
ϕ, θ, k = positive constants.
α = a constant with a value ≤ 1.
This relationship yields power law, Bingham and Herchel–Bulkley equations as special cases.

Casson/Steiner Model

The change is the yield point value

$$\sqrt{\tau} = \frac{2}{1+a}\sqrt{\tau_0} + \sqrt{\mu\gamma},$$

where

Form factor $a = \dfrac{R_t}{R_c} = \dfrac{1}{\delta_{cc}} \delta_{cc} = 1.0847$.

Then

$$\sqrt{\tau} = 1.04\sqrt{\tau_0} + \sqrt{\eta\gamma}.$$

228. Problem 6.9

Gel Strength Calculations

Explain gel strength with examples.

Gel Strength

Definition: The time-dependent forces in the drilling mud cause an increase in viscosity as the fluid remains quiescent for a certain period of time. The *gel strength* is a measurement of the electrochemical forces within the fluid under static conditions. Its field unit is the same as that of the yield strength. The strength is a function of suspended solids, solid contents, temperature, chemical content, and time. Usually it is caused by the high concentration of clay.

τ_g = gel strength, lbf/100 ft²

229. Problem 6.10

Gel Strength Measurement

The *gel strengths* are also measured with the rotating viscometer. The sample should be stirred at high speed for 10 sec (or constant reading), then allowed to stand undisturbed for 10 sec. With the gears in neutral, a slow (about 3 rpm), steady motion on the hand wheel is applied. The maximum reading is the initial gel in pounds per 100 ft^2. The mud is re-stirred for 10 sec and allowed to stand for 10 min. The measurement is repeated as before and the maximum reading is recorded as the 10-min gel strength in pounds per 100 ft^2. When an electrically driven instrument is used, the gel readings are the maximum value obtained at the low-speed 3 rpm setting.

The dial readings from Fann VG viscometer are shown below and the last two readings are taken at 10 sec, 10 min, and 30 min.

Speed (rpm)	Dial (°)
600	44.00
300	32.00
200	26.00
100	20.00
6	7.00
3	6.00
3$_{10\ sec}$	13.00
3$_{10\ min}$	28.00
3$_{30\ min}$	70.00

The mud tends to build gel structure when in idle state and so it needs additional force to break which can be either through the movement of the pipe or pumping fluid or both. When pumping fluid gel structure of the fluid has to be broken inside the pipe as well as in the annulus side.

The time for measurement such as 10 sec, 10 min, and 30 min are used as a basis to measure the rate at which the mud is gelled.

If the gel strength increases slowly then it is weak whereas when it increases quickly and steadily it is classified as strong (Fig. 6.4).

230. Problem 6.11
Explain How to Control Gel Strength

The gel strength is the measurement of attractive (electrochemical) forces within the mud system under no-flow conditions. It differs from the

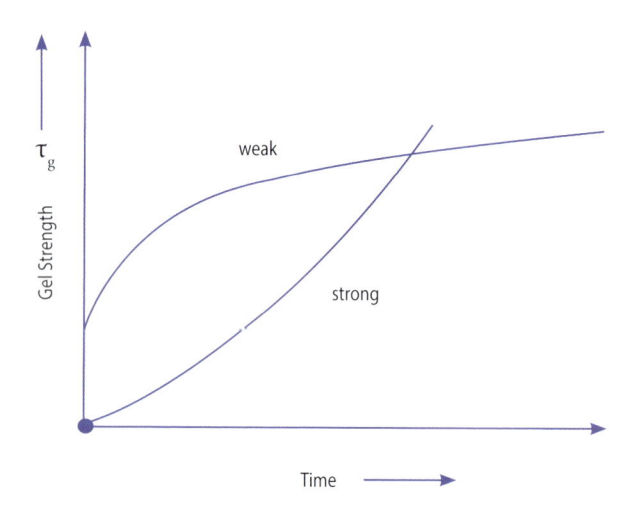

Figure 6.4 Gel strength

yield value because of its time dependence and is broken up after flow is limited. As with other properties, the ability to maintain the proper value of gel strength depends on effective solids control.

231. Problem 6.12
Filtration

The filtration test is used to measure the following…

Solution:

The filtration test is used to measure mud cake thickness, filtration loss, and spurt loss.

232. Problem 6.13
Spurt Loss

Explain spurt loss.

Solution:

The term *spurt loss* is used to characterize the initial loss of filtrate to formation at time practically equal to zero.

233. Problem 6.14

Define solids control efficiency and establish a relationship to calculate it.

Solution:

Solids control efficiency can be estimated using the following relationship:

$$\eta_{sce} = \frac{V_r \times V_d}{V_h},$$

where
V_h = hole volume in bbl or gal or bbl/hr or gal/hr.
V_r = volume or fraction of the solids removed.
V_d = volume of the solids discarded.

234. Problem 6.15

Calculate the volume dilution for the following condition:
Volume of the hole drilled: 500 bbl
Solid controlled efficiency: 75%
Allowable drilled solids in the drilling fluid: 25%

Solution:

Using the relationship $\eta_{sce} = \frac{V_r \times V_d}{V_h}$,

Volume Dilution = Volume of formation drilled × fraction of the solids left out/ fraction of allowable drilled solids
Using the above relationships, volume dilution = 500 (bbl) × 0.25 / 0.25
= 500 bbl

235. Problem 6.16

A 500 bbl of formation is being drilled and the solids control efficiency is estimated to be 75%. Calculate the solids buildup in the circulating system.

Solution:

Solids buildup in the system $V_{sb} = V_h(1 - \eta_{sce})$
The fraction of solids not removed = (1 − 0.75) = 0.25
Using the above relationships, volume dilution = 500 × (1 − 0.25)
= 375 bbl

236. Problem 6.17

Calculate the optimal solids removal efficiency with the following given data:

Expected drilled solids concentration = 5%
Drilled solids concentration in the discard = 25%

Solution:

Optimal solids removal efficiency can be estimated using the following relationship:

$$\eta_{sre} = \frac{(1-V_s)}{1-V_s+\dfrac{V_c}{V_s}},$$

where
V_s = expected drilled solids in drilling fluid.
V_c = drilled solids in discard.

Using this relationship and with the given data the optimal solids removal efficiency is

$$\eta_{sre} = \frac{(1-0.05)}{1-0.05+\dfrac{.25}{.05}} = 86\%$$

237. Problem 6.18

Plastic Viscosity, Yield Point, and Zero-Sec-Gel for Bingham Plastic Model

Plastic viscosity, yield point, and zero-sec-gel can be calculated from the Fann reading using the following relationships:

$$PV = \theta_{600} - \theta_{300},$$
$$YP = 2\theta_{300} - \theta_{600},$$
$$\tau_0 = \theta_3.$$

Alternatively, the dial readings can be reverse calculated by using PV, YP, and zero-gel as shown below:

$$\theta_{300} = PV + YP$$
$$\theta_{600} = 2PV + YP$$
$$\theta_3 = \tau_0,$$

where
θ_{600} = Fann dial reading at 600 rpm.
θ_{300} = Fann dial reading at 300 rpm.
θ_{3} = dial reading at 3 rpm.

If the Fann RPMs are anything other than 600 and 300, the following relationship can be used:

$$\mu_p = \frac{300}{N_2 - N_1}\left(\theta_{N_2} - \theta_{N_1}\right),$$

$$\tau_y = \theta_{N_1} - \mu_p \frac{N_1}{300}.$$

238. Problem 6.19

Shear Stress and Shear Rate Calculations

Shear stress and shear rate can be calculated using the following relationships:

$$\tau_1 = (0.01065)\theta \left(\frac{\text{lbf}}{\text{ft}^2}\right)$$

$$\tau_2 = (1.065)\theta \left(\frac{\text{lbf}}{100\ \text{ft}^2}\right)$$

$$\gamma = (1.703)N\left(\frac{1}{\sec}\right),$$

where
N = dial speed in $\left(\frac{1}{\sec}\right)$.

239. Problem 6.20

Plastic Viscosity, Yield Point, and Zero-Sec-Gel—Power Law

The rheological equation for the power law model can be given as

$$\tau = K\gamma^n,$$

where
γ = shear rate (1/sec).
τ = shear stress (lb/ft²).

The flow behavior index can be given as

$$n = 3.322 \log\left(\frac{\theta_{600}}{\theta_{300}}\right)$$

and for the modified power law it can be given as

$$n = 3.322 \log\left(\frac{YP + 2PV}{YP + PV}\right),$$

where
PV = plastic viscosity.
YP = yield point.
The consistency index, K, is given as

$$K = \frac{5100 \theta_{300}}{(511^n)} \text{ eq. cP}$$

K can be expressed in $\left(\frac{lb \times \sec^n}{ft^2}\right)$ using the conversion factor:

$$\left(\frac{lb \times \sec^n}{ft^2}\right) = 0.002088543 \times eq.cP$$

The consistency index, K, for the modified power law is given as

$$K = \frac{YP + 2PV}{(100)(1022^n)}$$

If the Fann RPMs are anything other than 600 and 300, the following relationship can be used:

$$n = \frac{\log\left(\frac{\theta_{N_2}}{\theta_{N_1}}\right)}{\log\left(\frac{N_2}{N_1}\right)},$$

$$K = \frac{5100 \theta_N}{(1.703 \times N)^n}.$$

240. Problem 6.21

Using the following Fann data, calculate the PV and YP for the Bingham plastic model and n and K for the power law model:

- $\theta_{600} = 48$ and $\theta_{300} = 28$
- $\theta_{400} = 44$ and $\theta_{200} = 29$

Solution:

Bingham plastic model:

Plastic and yield point are calculated as

$$PV = \theta_{600} - \theta_{300} = 48 - 28 = 20 \text{ cP}$$

and

$$YP = 2\theta_{300} - \theta_{600} = 2 \times 28 - 48 = 8 \text{ lbf/100 ft}^2$$

For the second set of readings,

$$\mu_p = \frac{300}{N_2 - N_1}(\theta_{N_2} - \theta_{N_1}) = \frac{300}{400 - 200}(44 - 29) = 22.5 \text{ cP}$$

$$\tau_y = \theta_{N_1} - \mu_p \frac{N_1}{300} = 29 - 22.5 \times \frac{200}{300} = 14 \text{ lbf/100 ft}^2$$

Power law model:

Flow behavior and consistency indices are calculated as

$$n = 3.322 \log\left(\frac{\theta_{600}}{\theta_{300}}\right) = 3.322 \log\left(\frac{48}{28}\right) = 0.777$$

and

$$K = \frac{510\theta_{300}}{(511^n)} = \frac{510 \times 28}{(511^{0.777})} = 111.84 \text{ eq.cP}$$

Using the conversion $\left(\dfrac{\text{lb} \times \sec^n}{\text{ft}^2}\right) = 0.002088543 \times \text{eq.cP}$

$$K = 111.84 \times 0.002088543 = 0.0020 \left(\frac{\text{lb} \times \sec^{0.77}}{\text{ft}^2}\right)$$

For the second set of readings,

$$n = \frac{\log\left(\dfrac{\theta_{N_2}}{\theta_{N_1}}\right)}{\log\left(\dfrac{N_2}{N_1}\right)} = \frac{\log\left(\dfrac{44}{29}\right)}{\log\left(\dfrac{400}{200}\right)} = 0.601$$

$$K = \frac{510 \times 29}{(1.703 \times 300)^{0.601}} = 443.55 \text{ .eq.cP}$$

Using the conversion $\left(\dfrac{\text{lb} \times sec^n}{\text{ft}^2}\right) = 0.002088543 \times \text{eq.cP}$

$$K = 443.55 \times 0.002088543 = 0.00926 \left(\frac{\text{lb} \times sec^{0.77}}{\text{ft}^2}\right)$$

241. Problem 6.22

Using the following Fann data, calculate the PV and YP for the Bingham plastic model and n and K for the power law model:
- $\theta_{600} = 25$ and $\theta_{300} = 14$

Solution:

Power Law Model

$$n = 3.32 \log \frac{\theta_{600}}{\theta_{300}}$$

$$n = 3.32 \log \frac{25}{14} = 0.84$$

$$K \approx \frac{510 \times 14}{(511)^n} \text{lbf} \frac{s^n}{100\,\text{ft}^2} = 0.076 \text{ lbf} \frac{s^n}{100\,\text{ft}^2}$$

Bingham Plastic Model

Plastic Viscosity
$\mu_p = \theta_{600} - \theta_{300}$
$\mu_p = 25 - 14 = 11$

Yield stress
$\tau_y = 2\theta_{300} - \theta_{600}$
$\tau_y = 2 \times 14 - 25 = 3 \text{ lbf}/100\,\text{ft}^2$

242. Problem 6.23

The following Fann data are obtained while weighing up the mud. Estimate the plastic viscosity and yield point using the Bingham plastic model.

- $\theta_{600} = 63$ and $\theta_{300} = 41$
- $\theta_{200} = 26$ and $\theta_{100} = 21$
- $\theta_{6} = 9$ and $\theta_{3} = 8$

Solution:

The Fann data given are converted to shear stress and stress as shown in Table 6.1.

Table 6.1   Shear rate and shear stress—problem 6.23

RPM	Shear rate (1/sec)	Dial reading	Shear stress lbf/100 ft²
600	1.703 × 600 = 1021.8	63	1.065 × 63 = 67.095
300	1.703 × 300 = 510.9	41	1.065 × 41 = 43.665
200	1.703 × 200 = 340.6	26	1.065 × 26 = 27.69
100	1.703 × 100 = 170.3	21	1.065 × 21 = 22.365
6	1.703 × 6 = 10.218	9	1.065 × 9 = 9.585
3	1.703 × 3 = 5.109	8	1.065 × 8 = 8.52

It is easy to use an excel spreadsheet and, using a linear fit, the slope and intercept can be found:

- Slope = 0.0576 lbf-s/ft
- Intercept = 10.048 lbf-s/100 ft²

The slope and intercept are the viscosity and yield point, respectively.

Using the conversion $\left(\dfrac{lb \times s}{ft^2} \right) = 4.79 \times 10^4 \, cP$.

Equivalent viscosity is

$$0.0576 \times \dfrac{4.79 \times 10^4}{100} \, cP$$

Intercept = yield point = 10.048 lbf/100 ft²

243. Problem 6.24

Using the following Fann data, estimate the flow behavior index and consistency index using the power law model:

- $\theta_{600} = 52$ and $\theta_{300} = 40$
- $\theta_{200} = 33.6$ and $\theta_{100} = 26$
- $\theta_{6} = 9$ and $\theta_{3} = 6.5$

Solution:

The above Fann data are converted to shear stress and stress as shown in Table 6.2.

Table 6.2   Shear rate and shear stress—problem 6.24

RPM	Shear rate (1/sec)	Dial reading	Shear stress lbf/100 ft²
600	1.703 × 600 = 1021.8	52	1.065 × 52 = 55.38
300	1.703 × 300 = 510.9	40	1.065 × 40 = 42.6
200	1.703 × 200 = 340.6	33.6	1.065 × 33.6 = 35.784
100	1.703 × 100 = 170.3	26	1.065 × 26 = 27.69
6	1.703 × 6 = 10.218	9	1.065 × 9 = 9.585
3	1.703 × 3 = 5.109	6.5	1.065 × 6.5 = 6.922

Using an Excel spreadsheet and a power law fit, $n = 0.3878$ and $K = 0.0377$ lbf-s/ft².

244. Problem 6.25

Volume Fraction of Solids Calculations

The volume fraction of solids in mud, f_{vm}, can be written as

$$f_{vm} = \frac{V_s}{V_m} = f_{vc} + \frac{hA}{V_f + hA},$$

where
V_f = amount of filtrate volume.
h = cake thickness.
A = filtration area.

V_m = total volume of mud filtered.
V_s = volume of solids deposited in mud cake.

245. Problem 6.26
API Fluid Loss Calculations
API water loss is given as

$$V_{30} = 2V_{7.5} - V_{sp} \text{ cm}^3,$$

where
$V_{7.5}$ = water loss in 7.5 min, cm³.
V_{sp} = spurt loss, cm³.

Filtrate loss is always estimated with reference to the square root of time.

Filtrate volume is given as

$$V_f = A \left[\frac{2k\left(\frac{f_{vc}}{f_{vm}} - 1\right) \Delta P t}{\mu} \right]^{\frac{1}{2}} + V_s,$$

where
ΔP = differential pressure.
μ = filtrate viscosity.
t = filtration time.
k = cake permeability.

246. Problem 6.27

During the filtration loss test it is obesrved that the total volume of fitrate collected is 6 cm³ in 10 min. The initial spurt loss is 1 cm³. Calculate the API water loss.

The API water loss is given as

$$V_{30} = 2(V_{7.5} - V_{sp}) + V_{sp},$$

where
V_{30} = filtrate volume at 30 min.
$V_{7.5}$ = filtrate volume at 7.5 min.
V_{sp} = spurt loss.

Solution:

We know that

$V_t = c\sqrt{t} + V_{sp}$, $V_{sp} = 1$ and $V_{10} = 6$ hence, $c = 5/\sqrt{10}$ that gives

$$V_{7.5} = 5\sqrt{7.5}/\sqrt{10} + 1$$

and

$$V_{30} = 2\left(5\sqrt{7.5}/\sqrt{10} + 1 - 1\right) + 1 = \sqrt{75} + 1 = 9.66\,\text{cm}^3$$

247. Problem 6.28

The following data were obtained from a filtration loss test:

Time (min)	Filtrate Volume (cm³)
1.0	5
8	12

Calculate the spurt loss.

Solution:

Using the equation

$$\frac{V_{t1} - V_{t2}}{V_{t2} - V_{sp}} = \frac{\sqrt{t_1}}{\sqrt{t_2}},$$

where
V_{t1} = volume of filtrate at time t_1.
V_{t2} = volume of filtrate at time t_2.
V_{sp} = spurt loss at time zero.

$$2\frac{5 - V_{t2}}{V_{t2} - V_{sp}} = \frac{\sqrt{1}}{\sqrt{8}}$$

$$V_{sp} = \frac{5\sqrt{8} - 12}{\sqrt{8} - 1} = 1.17\,\text{cm}^3$$

248. Problem 6.29

The spurt loss of a mud is known to be 2 cc. If a filtrate of 10 cc is collected in 15 min using a filter press, what is the standard API filtrate for this mud?

Solution:

The filter loss at 7.5 min is found by extrapolating the data given at time 0 and 15 min.

Let x be the filter loss at 7.5 min. So, the filter loss at time 7.5 min is

$$2 = \frac{10-x}{\sqrt{15}-\sqrt{7.5}} \times \sqrt{7.5}$$

Solving for x will yield 6.69 cc.

API filter loss is

$$V_{30} = 2V_{7.5} - V_{sp} = 2 \times 6.69 - 10 = 3.38 \text{ cc}$$

249. Problem 6.30

A kick was taken at 10,000 ft while drilling with a 14 ppg mud. Stabilized shut-in drillpipe pressure = 600 psi. Determine the amount of barite to be added to 800 bbl of original mud in order to contain formation pressure such that a differential pressure of 300 psi is achieved after killing the well.

Solution:

Equivalent mud weight needed to control the kick with the excess differential pressure is

$$\frac{600 + 0.052 \times 14 \times 10,000(7280) + 300}{0.052 \times 10,000} = 15.73$$

The amount of barite required is

$$m_B = 42 \times 28.02 \frac{(15.73-14)}{(28.02-15.73)} = 132,593 \text{ lbm}$$

Assuming 100 lbm per sack, the number of sacks of barite required is 1325.

250. Problem 6.31

Acidity-Alkalinity Calculations

pH is given as the negative logarithm of [H⁺] or [OH⁻] and is a measurement of the acidity of a solution. It is easy to compare by expressing it as below:

$$pH = -\log([H^+])$$

$$pH = -\log([OH^-]),$$

where

[H⁺] or [OH⁻] = hydrogen and hydroxide ion concentrations, respectively, in moles/liter.

Also, at room temperature, pH + pOH = 14

For other temperatures,

$$pH + pOH = pK_w,$$

where

K_w = ion product constant at that particular temperature.

At room temperature, the ion product constant for water is 1.0×10^{-14} moles/liter (mol/L or M).

A solution in which [H⁺] > [OH⁻] is acidic

A solution in which [H⁺] < [OH⁻] is basic.

Table 6.3 provides the ranges of acidity/alkalinity.

Table 6.3 Acidity/alkalinity ranges

pH	H⁺	Solution
< 7	> 1.0×10^{-7} M	Acid
> 7	< 1.0×10^{-7} M	Basic
7	= 1.0×10^{-7} M	Neutral

Moles per liter of hydroxide ion concentration required to change from one pH_1 to another pH_2 of a solution can be given as

$$\Delta[OH^-] = 10^{(pH_2 - 14)} - 10^{(pH_1 - 14)} \text{ mol/L}$$

The amount of material required = $\Delta[OH^-] \times$ MW gm/L,
where
MW = molecular weight of the material.

251. Problem 6.32

An aqueous potassium hydroxide completion fluid has a pH of 9. Determine the hydrogen ion concentration of this solution?

Solution:

From the pH equation $pH = -\log([H^+])$, hydrogen ion concentration can be given as

$$[H^+] = 10^{-pH} = 10^{-10} = 1 \times 10^{-9} \text{ mol/L (M)}$$

252. Problem 6.33

An aqueous hydrochloric acid solution has a pH of 5.24. Estimate the mass of the hydrochloric acid present in 1 liter of this solution. Use a molar mass of HCL = 36.5 g/mol.

Solution:

From the pH equation $pH = -\log([H^+])$, the hydrogen ion concentration can be given as

$$[H^+] = 10^{-pH} = 10^{-5.24} = 5.8 \times 10^{-6} \text{ mol/L}$$

Mass of hydrochloric acid = molar mass (g/mol) × hydrogen ion concentration (mol/L)

Mass of hydrochloric acid = 36.5 × 5.8 × 10⁻⁶ = 2.1 × 10⁻⁴ hydrogen ion concentration (mol/L)

253. Problem 6.34

An aqueous solution has a pH of 12.3. Determine the concentration of H^+ and OH^- in mole per liter.

Solution:

Hydrogen ion concentration is

$$[H^+] = 10^{-pH} = 10^{-12.3} = 2.0 \times 10^{-13} \text{ mol/L}$$

Using the relationship,

$$pH + pOH = 14$$

$$pOH = 14 - 12.3 = 1.3$$

Therefore, OH^- concentration is

$$[OH^-] = 10^{-pOH} = 10^{-1.3} = 0.05012 \text{ mol/L}$$

254. Problem 6.35

Calculate the amount of caustic required to raise the pH from 8 to 10. The molecular weight of caustic is 40.

Solution:

The moles of hydroxide concentration for increasing the pH can be given as

$$\Delta[OH^-] = 10^{(pH_2 - 14)} - 10^{(pH_1 - 14)}$$

Substituting the values,

$$\Delta[OH^-] = 10^{(10-14)} - 10^{(8-14)} = 9.9 \times 10^{-5} \text{ mol/L}$$

The weight of caustic required per liter of solution = $40 \times 9.9 \times 10^{-5}$
= 0.00396 g/L

255. Problem 6.36

An aqueous potassium hydroxide completion fluid has a pH of 9. Determine the hydrogen ion concentration of this solution.

Solution:

pH is given as the negative logarithm of [H⁺] or [OH⁻] and is a measurement of the acidity of a solution. It is easy to compare by expressing it as below

$$pH = -\log([H^+])$$
$$pH = -\log([OH^-]),$$

where
[H⁺] or [OH⁻] = hydrogen and hydroxide ion concentration respectively in moles/liter.

From the pH equation $pH = -\log([H^+])$ hydrogen ion concentration can be given as

$$[H^+] = 10^{-pH} = 10^{-10} = 1 \times 10^{-9} \text{ mol/L (M)}.$$

256. Problem 6.37
Marsh Funnel Calculations

The time for fresh water to drain = 26 sec ± 0.5 sec per quart for API water at 70°F + 0.5°F.

257. Problem 6.38
Common Terms

What are the common terms used in drilling fluids?

Solution:

1. Mud Weight
   The term "weight" is used in connection with mud more often than "density", even though density is a more correct term.
   Most tests of mud density are reported in pounds per gallon. Density can also be reported in pounds per cubic foot (lb/cu ft), grams per cc (gr/cc), specific gravity (sp/g), or pressure gradient (psi/ft).

2. Viscosity
   Mud viscosity is defined as the internal resistance of a fluid to flow. Only Newtonian fluids, such as water, have a true viscosity that can be defined by a single term. Non-Newtonian fluids, on the other

hand, require more than one component of viscosity to describe their flow behavior.

It is generally expressed as plastic viscosity (PV) with the unit cP.

3. Gel Strength

   It is a measurement of the electrochemical forces within the fluid under static conditions. Field unit is same as the yield strength and reported in lbf/100 sq.ft.

4. Yield Point (YP)

   Yield stress is the part of flow resistance of the fluid caused by electrochemical forces within the fluid. It is also called Yield Stress and expressed in lbf/100 sq.ft.

5. pH

   Ranges from 0 to 14

   7 – Neutral

   > Alkaline

   < Acidic

6. Spurt Loss – initial fluid loss

7. Resistivity

   Fresh Water – high resistivity

   Salt Water – low resistivity

8. Measured Depth (MD)

9. True Vertical Depth (TVD)

10. Hydrostatic Pressure
    - Normal pressure – the pressure gradient is approximately 0.433 psi/ft of depth.
    - Abnormal pressure – the pressure gradient is greater than 0.433 psi/ft of depth.
    - Subnormal pressure – the pressure gradient is less than 0.433 psi/ft of depth.

11. EMW

    Equivalent Mud Density

12. ESD

    Equivalent Static Density

13. ECD

    Equivalent Circulating Density

258. Problem 6.39
Common Weighting Materials

What are the common weighting materials?

Solution:

Average weight of the commonly used weighting materials are given Table 6.4.

Table 6.4 Density of commonly used weighting materials

Weighting materials	Specific gravity (gm/cm³)	Density (lbm/gal – ppg)	Density (lbm/bbl – ppb)
Barite Pure Grade	4.5	37.5	
Barite API – drilling grade	4.2	35	1470
Bentonite	2.6	21.7	910
Calcium carbonate	2.7	22.5	945
Calcium chloride	1.96	16.3	686
Sodium chloride	2.16	18	756
Water	1	8.33	1001
Diesel	0.86	7.2	300
Galena	6.6	55	6007

CHAPTER 7

HYDRAULICS

This chapter focuses on the different basic calculations involved in rig hydraulics and associated operations.

259. Problem 7.1
U-Tube Model

Find the pressure P_6 shown in Figure 7.1 with fluid densities.

Solution:

Applying pressure balance in both the legs of the manometer at the points 2 and 3

$$P_2 = P_1 + 0.052 \times \rho_1 \times h_1$$

$$P_3 = P_6 + 0.052 \times \rho_1 \times (h_1 - h_2 - h_3) + 0.052 \times \rho_3 \times h_3 + 0.052 \times \rho_2 \times h_2$$

Since $P_3 = P_2$

$$P_6 = P_1 + 0.052 \times h_2 (\rho_1 - \rho_2) + 0.052 \times h_3 (\rho_1 - \rho_3)$$

260. Problem 7.2

Find the pressure P_{bh} in terms of pressure at point 6 shown in Figure 7.2.

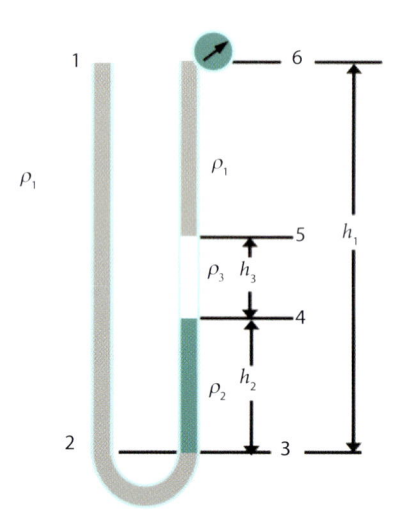

Figure 7.1 Problem 7.1

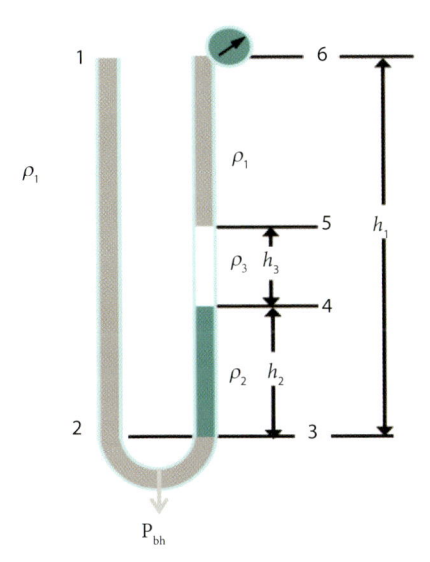

Figure 7.2 Problem 7.2

Applying pressure balance in both the legs of the manometer at points 2 and 3

$$P_3 = P_6 + 0.052 \times \rho_1 \times (h_1 - h_2 - h_3) + 0.052 \times \rho_3 \times h_3 + 0.052 \times \rho_2 \times h_2$$

Pressure at the bottom can be given as

$$P_{bh} = 0.052 \times \rho_1 \times x + P_6 + 0.052 \times \rho_1 \times (h_1 - h_2 - h_3) \\ + 0.052 \times \rho_3 \times h_3 + 0.052 \times \rho_2 \times h_2$$

261. Problem 7.3

Develop equations to calculate the bottomhole pressure at both annulus as well as pipe side. Assume

- no circulation
- different fluid densities inside pipe and annulus sides
- densities uniform, both inside and outside

Let the vertical length of the column from the surface to bottom be D_v.

Solution:

Since the pipe and annulus side are open, the U-tube has to balance.

Annulus Side:

The bottomhole pressure P_{bh} can be given as

$$P_{bh} = 0.052 \times \rho_a \times D_{av}$$

Pipe Side:

The bottomhole pressure P_{bh} can be given as

$$P_{bh} = 0.052 \times \rho_p \times D_{pv},$$

where
D_{av} = vertical height in the annulus in ft.
ρ_a = annulus fluid density in ppg.
D_{pv} = vertical height in the pipe in ft.
ρ_p = pipe side fluid density in ppg.

262. Problem 7.4

Develop equations to calculate the bottomhole pressure at both annulus as well as pipe side. Assume

- no circulation
- fluid densities remaining same both inside the pipe as well as annulus side of the wellbore

Let the vertical length of the column from the surface to bottom be D_v.

Solution:

Annulus Side:

The bottomhole pressure P_{bh} in psi can be given as

$$P_{bh} = 0.052 \times \rho \times D_v$$

Pipe Side:

The bottomhole pressure P_{bh} can be given as

$$P_{bh} = 0.052 \times \rho \times D_v,$$

where

D_v = vertical depth in ft.
ρ = fluid density in ppg.

263. Problem 7.5
Common Terms

What are the common terms used in drilling fluids?

Solution:

1. Mud Weight
   The term "weight" is used in connection with mud more often than "density", even though density is a more correct term.
   Most tests of mud density are reported in pounds per gallon. Density can also be reported in pounds per cubic foot (lb/cu ft), grams per cc (gr/cc), specific gravity (sp/g), or pressure gradient (psi/ft).
2. Viscosity
   Mud viscosity is defined as the internal resistance of a fluid to flow. Only Newtonian fluids, such as water, have a true viscosity that can be defined by a single term. Non-Newtonian fluids, on the other hand, require more than one component of viscosity to describe their flow behavior.
   It is generally expressed as Plastic Viscosity (PV) with the unit cP.
3. Gel Strength
   It is a measurement of the electrochemical forces within the fluid under static conditions. Field unit is same as the yield strength and reported in lbf/100 sq.ft.
4. Yield Point (YP)
   Yield stress is the part of flow resistance of the fluid caused by electrochemical forces within the fluid. It is also called Yield Stress expressed in lbf/100 sq.ft.
5. pH
   Ranges from 0 to 14
   7 – Neutral
   \> Alkaline
   < Acidic

6. Spurt Loss – initial fluid loss
7. Resistivity
    Fresh Water – high resistivity
    Salt Water – low resistivity
8. Measured Depth (MD)
9. True Vertical Depth (TVD)
10. Hydrostatic Pressure
    - Normal pressure - the pressure gradient is approximately 0.433 psi/ft of depth.
    - Abnormal pressure - the pressure gradient is greater than 0.433 psi/ft of depth.
    - Subnormal pressure - the pressure gradient is less than 0.433 psi/ft of depth.
11. EMW
    Equivalent Mud Density
12. ESD
    Equivalent Static Density
13. ECD
    Equivalent Circulating Density

264. Problem 7.6
Equivalent Mud Weight

The pressure in the wellbore or formation can be expressed in terms of equivalent mud weight (EMW). This is a convenient way to compare the pressures at any depth. EMW is calculated as:

$$\text{EMW} = \frac{P_h}{0.052 \times L_{tvd}} \text{ ppg,}$$

where
L_{tvd} = true vertical depth (TVD), ft.
P_h = pressure, psi.

If the well is deviated α deg from the vertical, the EMW is given by

$$\text{EMW} = \frac{P_h}{0.052 \times D_h \cos \alpha} \text{ ppg,}$$

where
D_h = measured depth, ft.

265. Problem 7.7
Equivalent Circulating Density

Equivalent circulating density results from the addition of the equivalent mud weight, due to the annulus pressure loss (Δp_a), to the original mud weight (ρ_m). This is calculated as:

$$\text{ECD} = \rho_m + \frac{\Delta p_a}{0.052 \times L_{tvd}} \text{ ppg}$$

In deviated wells vertical depth should be used, and the equation for multiple sections is given by:

$$\text{ECD} = \rho_m + \left(\frac{\sum_{i=1}^{n} \Delta p_a}{0.052 \times \sum_{i=1}^{n} \Delta L_{tvd}} \right) \text{ ppg,}$$

where
n = number of wellbore sections.

266. Problem 7.8

Calculate the equivalent mud weight at a depth of 10,000 ft (TVD) with an annulus back pressure of 500 psi. The mud density in the annulus is 10 ppg.

Solution:

$$\text{EMW} = \rho_m + \frac{500}{0.052 \times 10,000} = 10.96 \text{ ppg}$$

Alternatively, it can be calculated as follows:
Pressure exerted at the bottom is

$$P = 0.052 \times 10,000 \times 10 + 500 = 5700 \text{ psi}$$

Therefore,

$$\text{EMW} = \frac{5700}{0.052 \times 10,000} = 10.96 \text{ ppg}$$

267. Problem 7.9

Calculate the ECD while drilling a vertical well at a depth 10,000 ft. The mud density is 10 ppg. Stand pipe pressure is 3500 psi. Frictional pressure loss gradient in the drill string is 0.04137 psi/ft. Bit pressure loss is 1337 psi.

Solution:

$$\text{ECD} = 10 + \frac{3500 - 0.04137 \times 10{,}000 - 1337}{0.052 \times 10{,}000} = 13.36\,\text{ppg}$$

268. Problem 7.10

What are the conditions where the pressure losses calculated in a wellbore is affected?

Solution:

Temperature
Pressure
Eccentricity
Pipe roughness
Cuttings concentration
Cuttings density
Compressibility of the fluid
Besides other rheological properties

269. Problem 7.11

An engineer while analyzing a circulation operation expects ECD at the top of the well as shown in Figure 7.3 against "Expected" but the calculated value which is correct is shown against "Calculated". Explain the reason and justify your conclusion.

Solution:

The engineer expected that there is no back pressure based on an assumption.

Back pressure is applied on the annulus side. When circulating, the frictional pressure loss in the system should be considered. The pump pressure can be given as

$$P_p = P_{fs} + P_{fpdp} + P_{fpdc} + \Delta P_{mm} + \Delta P_{m/pwd} + \Delta P_b + P_{fadp} + P_{fadc} + P_{bp}$$

Hydraulics

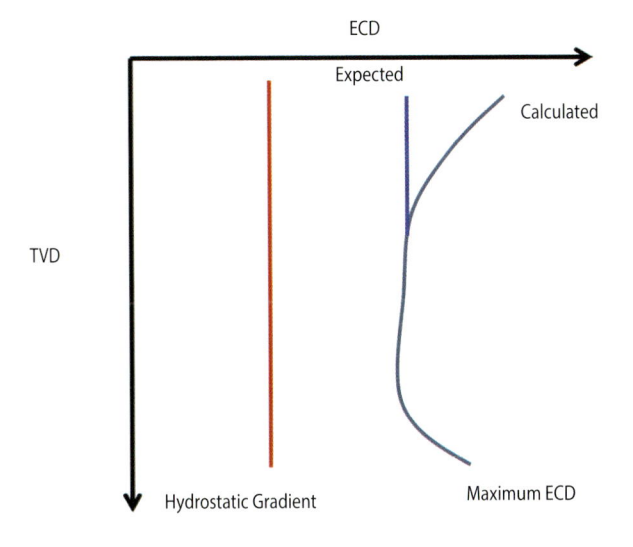

Figure 7.3 ECD vs TVD

Annulus Side:

The bottomhole pressure P_{bh} can be given as

$$P_{bh} = 0.052 \times \rho_m \times D_v + P_{fadp} + P_{fadc} + P_{bp}$$

ECD at any depth can be calculated as

$$\text{ECD} = \frac{0.052 \times \rho_m \times D_v + P_{fadp} + P_{fadc} + P_{bp}}{0.052 \times D_{TVD}}$$

This will result in the deviation of the ECD line to move to the right as shown in the calculated value.

270. Problem 7.12

For the annular flow the velocity profile cannot be mathematically defined and so the frictional loss formulations are _____ in nature.

Solution:
Empirical

271. Problem 7.13
Flow Velocity Calculations
Velocity is
$$V = \frac{\text{Flow Rate}}{\text{Cross-sectional Area}} = \frac{Q}{A}$$

When the flowrate is in gallons per minute and the cross sectional area is in sq.in.
$$V = \frac{19.25 \times Q}{A} \text{ ft/min}$$

When the flowrate is in barrels per minute and the cross sectional area is in sq.in.
$$V = \frac{808.5 \times Q}{A} \text{ ft/min}$$

272. Problem 7.14
If the desired annular velocity is 100 ft/min against 5" drillpipe while drilling 12 ¼" hole calculate the pump output flowrate in gal/min and bbl/min.

Solution:

When the flowrate is in gallons per minute and the cross-sectional area is in sq.in.
$$V = \frac{19.25 \times Q}{A} \text{ ft/min}$$

The annular cross-sectional area $A = \frac{\pi}{4}(12.25^2 - 5^2)$

So the flowrate in gpm $= \dfrac{V \times A}{19.25} = \dfrac{100 \times \frac{\pi}{4}(12.25^2 - 5^2)}{19.25} = 510$ gpm

When the flowrate is in barrels per minute and the cross sectional area is in sq.in.
$$V = \frac{808.5 \times Q}{A} \text{ ft/min}$$

So the flowrate in bpm $= \dfrac{V \times A}{808.5} = \dfrac{100 \times \frac{\pi}{4}(12.25^2 - 5^2)}{808.5} = 12.15$ bpm

273. Problem 7.15

Calculate the fluid velocity inside the pipe as well as in the annulus with the dimensions as follows for a flowrate of 350 gpm (4.762 bpm):

- Pipe inside diameter = 3 in.
- Pipe outside diameter = 4.5 in.
- Hole diameter = 8.5 in.

Solution:

Velocity inside pipe using flowrate in gpm:

$$V_p = \frac{\text{Flow Rate}}{\text{Cross-sectional Area}} = \frac{19.25 \times 200}{\frac{\pi}{4}(3^2)} = 544.7 \text{ fpm}$$

Velocity in the annulus using flowrate in gpm:

$$V_a = \frac{19.25 \times 200}{\frac{\pi}{4}(8.5^2 \quad 4.5^2)} = 94.3 \text{ fpm}$$

Velocity inside pipe using flowrate in bpm:

$$V_p = \frac{808.5 \times 4.762}{\frac{\pi}{4}(3^2)} = 544.7 \text{ fpm}$$

Velocity in the annulus using flowrate in bpm:

$$V_a = \frac{808.5 \times 4.762}{\frac{\pi}{4}(8.5^2 - 4.5^2)} = 94.3 \text{ fpm}$$

274. Problem 7.16

Calculate the ECD for the following data:
- Inclination of the well = 30°
- Measured depth = 5000 ft
- Calculated true vertical depth based on minimum curvature method = 4330 ft

- Annular pressure loss gradient = 0.03 psi/ft
- Mud weight = 9.2 ppg

Solution:

Total annular pressure loss = $0.03 \times 5000 = 150$ psi

$$\Delta p_a = 0.036 (\text{psi/ft}) \times 4500 (\text{ft}) = 162 \text{ psi}$$

Using equation from Problem 7.7

$$\text{ECD} = 9.2 + \frac{150}{0.052 \times 4330} = 9.866 \text{ ppg}$$

275. Problem 7.17

Hydraulics: Basic Calculations
Critical Velocity

Critical velocity is the velocity at which the flow regime changes from laminar to turbulent. It can be determined for the Bingham plastic model.

$$V_c = \frac{1.08 PV + 1.08\sqrt{PV^2 + 12.34 \rho_m D_i^2 YP}}{\rho D_i}$$

or

$$V_c = \frac{1.08 \mu_p + 1.08\sqrt{\mu_p^2 + 12.34 \rho_m D_i^2 \tau_y}}{\rho D_i} \text{ ft/sec,}$$

where
$PV = \mu_p$ = plastic viscosity, cP.
$YP = \tau_y$ = yield point, lbf/100 ft².
ρ_m = mud density, ppg.
Critical flowrate is

$$Q_c = 2.448 \times V_c \times D_i^2 \text{ gpm,}$$

where
D_i = inside diameter of the pipe, in.

276. Problem 7.18

Using the following data, calculate the critical flowrate at which laminar flow changes to turbulent flow.

- Drillpipe ID = 3 ¾"
- Mud density 10 ppg, $\theta_{600} = 37$; $\theta_{300} = 25$
- Target depth = 10,000 ft (TVD)

Solution:

Using the Fann reading, the yield and plastic viscosity are, respectively,

$$\tau_y = 2 \times 25 - 37 = 13 \text{ lbf/100 ft}^2$$

$$\mu_p = 37 - 25 = 12 \text{ cP}$$

Using equation from Problem 7.17, the critical velocity inside the pipe is calculated as

$$V_c = \frac{1.08\mu_p + 1.08\sqrt{\mu_p^2 + 12.34\rho_m D_i^2 \tau_y}}{\rho D_i}$$

$$V_c = \frac{1.08 \times 12 + 1.08\sqrt{12^2 + 12.34 \times 3.75^2 \times 12 \times 10}}{10 \times 3.75} = 4.69 \text{ ft/sec}$$

Critical flowrate = velocity × flow area
= 4.69 × 2.448 × 3.375² = 161.3 gpm

277. Problem 7.19

Pump Calculations

Pump pressure can be given as

$$P_p = \Delta p_b + P_d \text{ psi,}$$

where
Δp_b = bit pressure drop, psi.
P_d = frictional pressure losses.
Hydraulic horsepower (HHP) of the bit is

$$\frac{Q \times \Delta p_b}{1714} \text{ hp,}$$

where
Δp_b = bit pressure drop, psi.
Q = flowrate, gpm.

278. Problem 7.20

Given the following friction pressure loss-flow rate relationship:

$$P_f = cQ^m,$$

where
P = friction pressure loss.
C = constant.
m = flow exponent.
Q = flow rate.

Find an expression to determine m if the following measurements are known:

$P_f = P_{fi}$ at $Q = Q_i$

$P_f = P_{fij}$ at $Q = Q_j$

Solution:

For $P_f = P_{fi}$ at $Q = Q_i$

$P_{fi} = cQ_i^m$

For $P_f = P_{fij}$ at $Q = Q_j$

$P_{ji} = cQ_j^m$

Dividing the equations and taking the logarithm $m = \dfrac{\log\left(\dfrac{P_j}{P_i}\right)}{\log\left(\dfrac{Q_j}{Q_i}\right)}$

279. Problem 7.21

A 10 ppg-liquid is being circulated at 540 gpm through the system shown in Figure 7.1. A triplex-single-acting pump, having a volumetric efficiency of 85%, is being used. What must be the pump hydraulic horsepower requirement for the above operating conditions? Assume the friction pressure loss gradient in the circulating system to be 0.06 psi/ft. It is given that the hydrostatic pressure of a liquid column is given by $P = 0.052h$, where

$$P = 0.052h\rho,$$

where
h = column height, in feet.
ρ = liquid density, in ppg.

Solution:

Total head acting
(6800 − 6000) = 800 ft
Hydrostatic pressure is given as
P = 0.052 × 10 × 800 = 416 psi
Frictional pressure loss in the pipe
= (6000 + 1000 + 6000 + 800 + 8000) 0.06 = 1308 psi
Total pressure pump at the pump = 416 + 1308 = 1724 psi
Actual flowrate = 540 gpm
Theoretical flowrate = 540/0.85 = 635 gpm

$$\text{Hydraulic horsepower} = \frac{Q \Delta P}{1714} = \frac{635 \times 1724}{1714} = 640 \text{ hp}$$

280. Problem 7.22

A bit currently has 3 × 12 nozzles. The driller has recorded that when 10 ppg mud is pumped at a rate of 500 gpm, a pump pressure of 3000 psi is observed. When the pump is slowed to a rate of 250 gpm, a pump pressure of 800 psi is observed. The pump is rated at 2000 hp and has an overall efficiency of 90%. The minimum flowrate to lift the cuttings is 240 gpm. The maximum allowable surface pressure is 5000 psi.

A. Determine the pump operating conditions and bit nozzle sizes for maximum bit horsepower for the next bit run.
B. What bit horsepower will be obtained at the conditions selected?

Solution:

A. Bit pressure drop is 2097 psi for 500 gpm. Therefore, the frictional pressure losses can be calculated using the following equation:

$$P_{f1} = P_p - P_b = 3000 - 2097 = 903 \text{ psi}$$

The flow index m can be found as

$$m = \frac{\log\frac{903}{275.5}}{\log\frac{500}{250}} = 1.71$$

$$Q_{max} = 1714 \times 0.9 \times \frac{2000}{5000} = 617 \text{ gpm}$$

Find the optimum friction pressure loss using the maximum hydraulic horsepower criterion:

$$P_{fopt} = \frac{1}{m+1} P_{pmax} = 1845 \text{ psi}$$

$$P_{bopt} = 5000 - 1845 = 3155 \text{ psi}$$

$$Q_{opt} = 500a \log\left[\frac{1}{1.71}\log\frac{1845}{900}\right] = 761 \text{ gpm}$$

The calculated optimum flowrate has to be greater than the lowest limit of 240 gpm. Use the maximum flowrate as it cannot be exceeded, and calculate the area of the nozzle and the nozzle sizes, 12-12-12.

B. The nozzle pressure drop can be calculated as

$$\Delta P_b = \frac{8.3 \times 10^{-5} \rho \times Q^2}{C_d^2 \times A_n^2} = \frac{8.3 \times 10^{-5} \times 10 \times 617^2}{0.95^2 \times 0.3313^2} = 3189 \text{psi}$$

The bit hydraulic horsepower is

$$HHP_{bit} = \frac{Q\Delta P_b}{1714} = \frac{617 \times 3189}{1714} = 1148 \text{ hp}$$

281. Problem 7.23
Bingham Plastic Model
Reynolds Number

For the pipe side, the Reynolds number is calculated as

$$N_{Rep} = \frac{928 \rho_m v_p D_i}{\mu_{ep}},$$

where
ρ_m = mud weight in ppg.
v_p = velocity of the fluid in fps.

$$v_p = \frac{Q}{2.448 D_i^2} \text{ fps,}$$

where
Q = flowrate, gpm.
D_i = inside diameter of the pipe, in.
μ_{ep} = equivalent viscosity of the fluid and is calculated as:

$$\mu_{ep} = \mu_p + \frac{20 D_i \tau_y}{3 v_p},$$

where
μ_p = plastic viscosity, cP.
τ_y = yield point, lbf/100 ft².

The pressure gradient for unit length dL and for laminar flow is calculated as:

$$\left(\frac{dp_f}{dL}\right) = \frac{\mu_p v_p}{1500 D_i^2} + \frac{\tau_y}{225 D_i} \text{ psi/ft}$$

For the annulus side

$$N_{Rea} = \frac{757 \rho_m v_a (D_2 - D_p)}{\mu_{ep}}$$

$$\mu_{ep} = \mu_p + \frac{5(D_2 - D_p)\tau_y}{v_a},$$

where
D_2 = annulus diameter, in.
D_p = outside diameter of the pipe, in.
v_a = velocity of the fluid in fps.

$$v_a = \frac{Q}{2.448(D_2^2 - D_p^2)} \text{ ft/sec.}$$

The pressure gradient for laminar flow is calculated as:

$$\left(\frac{dp_f}{dL}\right) = \frac{\mu_p v_a}{1000(D_2 - D_p)^2} + \frac{\tau_y}{200(D_2 - D_p)} \text{ psi/ft}$$

282. Problem 7.24

Calculate the change in the bottomhole pressure when the yield value is 5 lb/100 ft² and the viscosity is 30 cP. Use the following data:

- Hole size = 9 ⅞"
- Depth = 10,000 ft
- Pipe OD = 4 ½"
- Flowrate = 400 gpm
- Mud weight = 12 ppg
- Yield point = 60 lb/100 ft²
- Plastic viscosity = 40 cP

Solution:

The annular velocity of the fluid is

$$v_a = \frac{400}{2.448(9.875^2 - 4.5^2)} = 2.11 \text{ ft/sec}$$

The equivalent viscosity is

$$\mu_{ep} = 40 + \frac{5 \times 60(9.875 - 4.5)}{2.11} = 802.5 \text{ cP}$$

The Reynolds number on the annulus side is

$$N_{Rea} = \frac{757 \times 12 \times 2.11 \times (9.875 - 4.5)}{802.5} = 128$$

Since the Reynolds's number is less than 2100, the flow is laminar. The equivalent viscosity with new YP and PV is

$$\mu_{ep} = 5 + \frac{5 \times 30(9.875 - 4.5)}{2.11} = 93.5 \text{ cP}$$

and the Reynolds's number is 1104. Therefore, the flow is still laminar.

Hydraulics

Using equation from Problem 7.23

$$\left(\frac{dp_f}{dL}\right) = \frac{(40-5)v_a}{1000(9.875-4.5)^2} + \frac{(60-30)}{200(9.875-4.5)} = 0.051895 \text{ psi/ft}$$

Total pressure change = 0.051895 × 10,000 = 519 psi

283. Problem 7.25

Two speed viscometer readings are $\theta_{600} = 52$ and $\theta_{300} = 35$. Mud density = 10 ppg. The hole size is 8 ¾", and the OD of the drillstring is 4 ½".

A. Using the Bingham plastic model, find out whether the flow is laminar or turbulent for a flowrate of 400 gpm.
B. Find the velocity at which the flow changes the regime.

Solution:

Viscosity is

$$\mu_p = \theta_{600} - \theta_{300} = 52 - 35 = 17 \text{ cP}$$

Yield point is

$$\tau_y = 35 - 17 = 18 \text{ lb}/100 \text{ ft}^2$$

Annular velocity of the fluid is

$$v_a = \frac{400}{2.448(8.75^2 - 4.5^2)} = 2.90 \text{ ft/sec}$$

The equivalent viscosity is

$$\mu_{ep} = 17 + \frac{5(8.75-4.5)18}{2.90} = 148.9 \text{ cP}$$

The Reynolds number is

$$N_{Rea} = \frac{757 \times 10 \times 2.9 \times (8.75-4.5)}{148.9} = 626$$

The flow is laminar.

284. Problem 7.26

At certain depth while drilling 12 ¼" hole, pump pressure is 3000 psi and parasitic pressure loss is 1500 psi at a circulation rate of 500 gpm with the mud weight of 12.5 ppg. Calculate the flowrate to achieve a bit HSI of 1.2, with the mud weight remaining same; $C_d = 0.95$.

Solution:

Pump pressure is given

$$P_p = P_b + P_d; \quad P_b = 3000 - 1500 = 1500 \text{ psi}$$

The bit pressure drop can be given as

$$\Delta P_b = \frac{8.311 \times 10^{-5} \rho Q^2}{C_d^2 A_n^2}$$

But

$$\frac{\Delta P_{b_1}}{\Delta P_{b_2}} = \frac{\rho_1 Q_1^2}{\rho_2 Q_2^2}$$

so,

$$\Delta P_{b_1} = \frac{1500 \times Q_1^2}{500^2}$$

$$\text{HSI} = \frac{\text{HHP}}{\frac{\pi}{4} \times 12.25^2} = 1.2 = \frac{\Delta P_{b_1} \times Q_1}{1714 \times \frac{\pi}{4} \times 12.25^2} = \frac{\frac{1500 \times Q_1^2}{500^2} \times Q_1}{1714 \times \frac{\pi}{4} \times 12.25^2}$$

$$= \frac{1500 \times Q_1^3}{500^2 \times 1714 \times \frac{\pi}{4} \times 12.25^2}$$

285. Problem 7.27

Two speed viscometer readings are $\theta_{600} = 52$ and $\theta_{300} = 34$. Mud density = 10 ppg. The hole size is 8 ½", and the OD of the drillstring is 4 ½".

Hydraulics

A. Using the Bingham plastic model, find out whether the flow is laminar or turbulent for a flowrate of 400 gpm.
B. Find out the percentage increase in the pressure loss gradient when the mud density is 11 ppg.

Solution:

$$\mu_p = \theta_{600} - \theta_{300} = 52 - 34 = 18 \text{ cP}$$

$$\tau_y = 35 - 18 = 16 \text{ lb}/100 \text{ ft}^2$$

$$V_a = \frac{400}{2.448(8.5^2 - 4.5^2)} = 3.14 \text{ ft/sec}$$

$$\mu_{ep} = 18 + \frac{5(8.5 - 4.5)16}{3.14} = 119.9 \text{ cP}$$

$$N_{Rea} = \frac{757 \times 10 \times 3.14 \times (8.5 - 4.5)}{119.9} = 792$$

Flow is laminar.

286. Problem 7.28

Determine the pump hydraulics horsepower requirement in order to drill to a target depth of 10,000 ft (TVD). Assume a pump volumetric efficiency of 85%. Calculate the bottomhole pressure and equivalent circulating density (ECD). Use the Bingham plastic model and the following data:

- Drillpipe: 4.5" OD, 3 ¾" ID, 18.10 ppf
- Drillcollar: 1200 ft, 7" OD, 3 ¾" ID, 110 ppf
- Surface connections: 250' (length) × 4" (inside diameter)
- Last intermediate casing details: Setting depth 5000 (TVD) and 9 ⅞" casing outside diameter; 0.3125 casing wall thickness
- Next hole size: 8 ½" tri-cone roller, 3 × 14 jets
- Mud density: 10 ppg, $\theta_{600} = 37$; $\theta_{300} = 25$
- For annular hole cleaning, 120 fpm of fluid velocity is required

Solution:

Use the thickness of the pipe and calculate the flowrate required to clean the hole against the drillpipe:

$$Q = v_{min} A_{adp} = 120 \times \frac{\pi}{4}\left(\left(\frac{9.25}{12}\right)^2 - \left(\frac{4.5}{12}\right)^2\right) \times 7.48 = 320 \text{ gpm}$$

$$\tau_y = 2 \times 25 - 37 = 13 \frac{\text{lbf}}{100 \text{ ft}^2}$$

$$\mu_p = 37 - 25 = 12 \text{ cP}$$

Pipe side calculation: drillpipe of 14,300 ft

$$v_{dp} = \frac{320}{2.448 \times 3.75^2} = 9.3 \text{ ft/sec}$$

$$N_{He} = \frac{37,100 \times 10 \times 13 \times 3.75^2}{12^2} = 47,100$$

$$N_{Re\,cr} \cong 13,000$$

$$N_{Re\,p} = \frac{928 \times 10 \times 9.3 \times 3.75}{12} = 26,970$$

Since $N_{Re\,cr} < N_{Re}$, the flow regime is turbulent and the frictional pressure drop can be found to be $\Delta p_{fdp} = 790$ psi.

Similarly, the following pressure loss can be calculated:

For drill collar of length 1200 ft, the flow regime is turbulent and the pressure drop is $\Delta p_{fdc} = 66$ psi.

Similarly, for surface connections the flow is turbulent and the pressure drop is $\Delta p_{fdc} = 10$ psi.

Bit pressure drop can further estimated to be

$$\frac{8.31 \times 10^{-5} \times 10 \times 320^2}{0.95^2 \times \left(3 \times \frac{\pi}{4}\left(\frac{14}{32}\right)^2\right)^2} = 463 \text{ psi}$$

Annulus side calculation:
Annulus 1 (hole/DC) length is 1200 ft
Flow is turbulent and the pressure drop is $\Delta p_{afdc} = 110$ psi
Annulus 2 (hole/DP) the length is 8800 ft
Flow is laminar and the pressure is $\Delta p_{afdp} = 160$ psi
Annulus 3 (casing/DP) the length is 5500 ft
Flow is laminar and the pressure drop is $\Delta p_{afdp} = 80$ psi

$$P_p = 790 + 66 + 10 + 463 + 110 + 160 + 80 = 1680 \text{ psi}$$

$$\text{HHP}_p = \frac{1680 \times 320}{1714} = 314 \text{ hp}$$

The bottomhole pressure is

$$P_h + \Delta p_a = 0.052 \times 10 \times \text{TVD} + \Delta p_a = 0.052 \times 10 \times 10,000 + 350 = 5550 \text{ psi}$$

$$\text{Equivalent circulating density} = \frac{\text{bottomhole pressure}}{0.052 \times \text{TVD}}$$

$$= \frac{5550}{0.052 \times 10,000} = 10.67 \text{ ppg}$$

287. Problem 7.29
Power Law Model

The power law constant for a pipe is calculated as

$$n_p = 3.32 \log\left(\frac{R_{600}}{R_{300}}\right)$$

The fluid consistency index (K_p) for a pipe is given by

$$K_p = \frac{5.11 R_{300}}{1022^{n_p}} \quad \frac{\text{dyne} \times \sec^n}{\text{cm}^2}$$

$$K_p \cong \frac{510 R_{300}}{511^{n_p}} \text{ eq.cP}$$

The equivalent viscosity is calculated as

$$\mu_{ep} = 100 K_p \left(\frac{96 V_p}{D}\right)^{n_p-1} \left(\frac{3 n_p + 1}{4 n_p}\right)^{n_p}$$

The Reynolds number for a pipe is

$$N_{\text{Re}_p} = \frac{928 D V_p \rho}{\mu_{ep}}$$

The friction factor in a pipe can be calculated as follows:

For $N_{\text{Re}_p} < 2100$

$$f_p = \frac{16}{N_{\text{Re}_p}}$$

For $N_{\text{Re}_p} > 2100$

$$f_p = \frac{a}{N_{\text{Re}_p}^b},$$

where

$$a = \frac{\log n_p + 3.93}{50}.$$

$$b = \frac{1.75 - \log n_p}{7}.$$

The friction pressure loss gradient in the pipe is calculated as

$$\left(\frac{dP}{dL}\right)_{dp} = \frac{f_p V_p^2 \rho}{25.81 d} \quad \frac{\text{psi}}{\text{ft}}$$

The power-law constant in the annulus is calculated as follows:

$$n_a = 0.657 \log\left(\frac{R_{100}}{R_3}\right)$$

$$K_a = \frac{5.11 R_{100}}{170.2^{n_a}} \quad \frac{\text{dyne} \times \text{sec}^n}{\text{cm}^2}$$

$$K_a = \frac{510 R_{100}}{511^{n_a}} \text{ eq.cP}$$

$$\mu_{ea} = 100 K_a \left(\frac{144 V_a}{d_2 - d_1}\right)^{n_a - 1} \left(\frac{2n_a + 1}{3n_a}\right)^{n_a}$$

The friction factor in the annulus can be calculated as follows:
For $N_{Re_p} < 2100$

$$f_a = \frac{24}{N_{Re_a}}$$

For $N_{Re_p} > 2100$

$$f_a = \frac{a}{N_{Re_a}^b},$$

where

$$a = \frac{\log n_a + 3.93}{50}.$$

$$b = \frac{1.75 - \log n_a}{7}.$$

The friction pressure loss gradient in the pipe is calculated as

$$\left(\frac{dP}{dL}\right)_a = \frac{f_a V_a^2 \rho}{25.81(d_2 - d_1)} \frac{\text{psi}}{\text{ft}}$$

288. Problem 7.30

Three speed viscometer reading are as follows: $\theta_{600} = 52$, $\theta_{300} = 34$, and $\theta_3 = 3$. Mud density = 10 ppg. The hole size is 8 ¾" and OD of the drillstring is 4 ½".

A. Using the Power Law model find out whether the flow is laminar or turbulent for a flowrate of 400 gpm.
B. Find out the percentage increase in the pressure loss gradient when the Reynolds number is 2400 and the average velocity is 4.69 ft/sec.

Solution:

Since OD of the pipe and hole diameter are given, we are concerned only with annulus:

$$\theta_{600} = \theta_{300} - \frac{2(52-34)}{3} = 22; n_a = 0.5685; K = 6.065$$

$$\mu_a = 100 \times 6.065 \left(\frac{144 \times 2.9}{4.25}\right)^{0.5685-1} \left(\frac{2 \times 0.5685+1}{3 \times 0.5685}\right)^{0.5685} = 95 \text{ cP}$$

Reynolds number is

$$R_e = \frac{928 \times 4.25 \times 2.9 \times 10}{95} = 1200$$

Since it is less than 2100 the flow regime is laminar

Assuming 2400 and 4.69 ft/sec

Calculation will result in $a = 0.737$; $b = 0.2850$; $fa2 = 0.0080$; $fa1 = 0.20$. By taking the ratio it can be seen that there will be a decrease of 5% pressure loss gradient.

289. Problem 7.31

Using the power law, calculate the change in the bottomhole pressure when the 600 rpm reading is 60. Use the following data:

- Hole size = 9 ⅞"
- Depth = 10,000 ft
- Pipe OD = 4 ½"
- Flowrate = 400 gpm
- Mud weight = 12 ppg
- Viscometer readings are $\theta_{600} = 55$, $\theta_{300} = 37$, and $\theta_3 = 3$

Solution:

$$\theta_{100} = \theta_{300} - \frac{2(55-37)}{3} = 25$$

$$n_a = 0.61$$

$$K = 5.7$$

Hydraulics

$$\mu_a = 100 \times 5.7 \left(\frac{144 \times 2.1}{5.375}\right)^{0.61-1} \left(\frac{2 \times 0.61+1}{3 \times 0.61}\right)^{0.61} = 130 \text{ cP}$$

$$R_e = \frac{928 \times 5.375 \times 2.1 \times 12}{130.6} = 968$$

Since the Reynolds number is less than 2100, the flow is laminar:

$$f_a = \frac{24}{968} = 0.025$$

$$\Delta p_a = 95 \text{ psi}$$

Use the same calculations for the second case

$$R_e = \frac{928 \times 5.375 \times 2.1 \times 12}{121} = 1040$$

The flow is laminar:

$$f_a = \frac{24}{1040} = 0.023$$

$$\Delta P_a = 88 \text{ psi}$$

290. Problem 7.32

A drilling engineer plans to calculate the hydrualic requirements while drilling at TD (Fig. 7.4).

Fann data

$$\theta_{600} = 65; \theta_{300} = 40$$

Bit nozzle sizes: 12-12-12.
Drillpipe: 4.5" × 3.958"
Heavy weight drillpipe: 4.5" × 2.75"
Drillcollar length 900 ft, 6" OD, 72.16 lb/ft in air
Surface connections: 500 ft × 4" ID
Use an uniform hole size of 8.5"

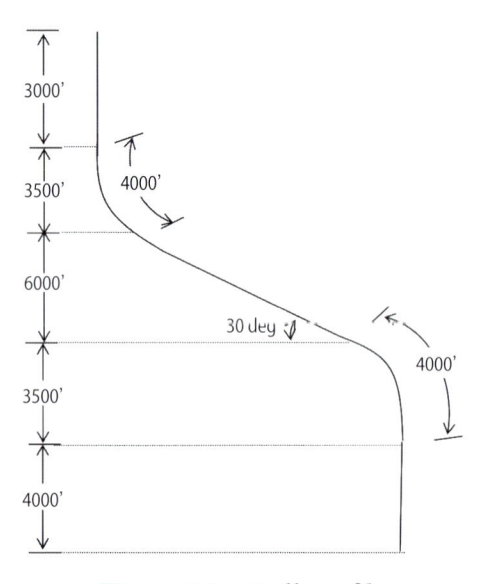

Figure 7.4 Well profile

For Normal Forward Circulation

a. Determine the pump hydraulics horsepower requirement to drill to target depth. Assume pump volumetric efficiency of 85% and a maximum flow rate of 400 gpm with a mud density of 10.2 ppg.
b. Calculate the percentage of bit pressure loss.

Solution:
1. **Dimensions of the drillstring component**
Total measured depth of the well
= 3000 + 4000 + 6000/sin 30° + 4000 + 4000 = 27,000 ft
Drillcollar length = 900 ft, 6" OD and the ID is calculated as below:
Using the density of the steel as 489.4 lbm/ft³ and OD = 6", the drill collar ID can be calculated as $d_i = \sqrt{d_o^2 - \frac{72.16}{\rho} \times \frac{4}{\pi}} = 3$ in.

Hwdp = 630 ft, 4.5" OD 2.75" ID
Length of the drillpipe = 27,000 − 900 − 630 = 25,470 ft, 4.5" OD 3.958" ID

Hydraulics

Forward circulation without tooljoint pressure losses

Calculation can be started by calculating the pressure losses from the pipe side to the annulus side.

1. **Calculating the pressure losses inside the drillpipe**
   Viscosity

$$\mu_p = \theta_{600} - \theta_{300}$$

$$\mu_p = 65 - 40 = 25$$

Yield stress

$$\tau_y = 2 \times \theta_{300} - \theta_{600}$$

$$\tau_y = 2 \times 40 - 65 = 15$$

Alternatively regressing the three dial readings the yield value can be found as $\tau_y = 15.75$

$$v_p = \frac{q}{2.448 \times d^2}$$

$$v_p = \frac{400}{2.448 \times 3.958^2} = 10.43 \text{ ft/sec}$$

$$\mu_{ep} = \mu_p + \frac{20 \times d \times \tau_y}{3 \times v_p}$$

$$\mu_{ep} = 25 + \frac{20 \times 3.958 \times 15}{3 \times 10.43} = 62.95$$

$$N_{Re} = \frac{928 \times \rho \times v_p \times d}{\mu_{ep}}$$

$$N_{Re} = \frac{928 \times 10.2 \times 10.2 \times 3.958}{62.95}$$

$$N_{Re} = 6208$$

Therefore the flow regime is turbulent.

Pipe pressure loss is

$$\left(\frac{dp_f}{dL}\right)_p = \frac{\rho^{0.75} v_p^{1.75} \mu_p^{0.25}}{1800 \times d^{1.25}}$$

$$\left(\frac{dp_f}{dL}\right)_p = \frac{10.2^{0.75} 10.43^{1.75} 25^{0.25}}{1800 \times 3.958^{1.25}} = 0.07688 \text{ psi/ft}$$

$\Delta P_{dp} - 0.07688 \text{ psi/ft}$
$\Delta P_{dp} = \Delta P_{dp} \times L_{dp} = 0.07688 \times 25{,}470$
$\Delta P_{dp} = 1958.0 \text{ psi}$

2. **Calculating the pressure losses inside the heavy weight drillpipe**
   Calculating as before
   $\Delta P_{hwdp} = 0.4335 \text{ psi/ft}$
   $\Delta P_{hwdp} = \Delta P_{hwdp} \times L_{hwdp} = 0.4335 \times 630$
   $\Delta P_{hwdp} = 273 \text{ psi}$

3. **Calculating the pressure losses inside the drillcollar**
   Calculating as before
   $\Delta P_{dc} = 0.2909 \text{ psi/ft}$
   $\Delta P_{dc} = \Delta P_{dc} \times L_{dc} = 0.2909 \times 900$
   $\Delta P_{dc} = 262 \text{ psi}$

4. **Calculating the pressure losses in the annulus between drillpipe and borehole**

$$v_a = \frac{q}{2.448 \times \left(d_2^2 - d_1^2\right)}$$

$$v_a = \frac{400}{2.448 \times \left(8.5^2 - 4.5^2\right)} = 3.14 \text{ ft/sec}$$

$$\mu_{ep} = \mu_p + \frac{5 \times \left(d_2 - d_1\right) \times \tau_y}{v_a}$$

$$\mu_{ep} = 25 + \frac{5 \times \left(8.5 - 4.5\right) \times 5}{3.14} = 120.54$$

Hydraulics

$$N_{Re} = \frac{757 \times \rho \times v_a \times (d_2 - d_1)}{\mu_{ep}}$$

$$N_{Re} = \frac{757 \times 10.2 \times 3.14 \times (8.5 - 4.5)}{120.54}$$

$$N_{Re} = 804.55$$

Therefore the flow regime is laminar.

$$\left(\frac{dp_{f\,adp}}{dL_{adp}}\right)_a = \frac{\mu_p v_a}{1000 \times (d_2 - d_1)^2} + \frac{\tau_y}{200 \times (d_2 - d_1)}$$

$$= \frac{25 \times 3.14}{1000 \times (8.5 - 4.5)^2} + \frac{15}{200 \times (8.5 - 4.5)} = 0.0049 + 0.0188$$

$\Delta P_{adp} = 0.024$ psi/ft
$\Delta P_{udp} = \Delta P_{udp} \times L_{udp} = 0.024 \times 25{,}470$
$\Delta P_{adp} = 603.0$ psi

5. **Calculating the pressure losses in the annulus between HWDP and borehole**

$$v_a = \frac{q}{2.448 \times (d_2^2 - d_1^2)}$$

$$v_a = \frac{400}{2.448 \times (8.5^2 - 4.5^2)} = 3.14 \text{ ft/sec}$$

$$\mu_{ep} = \mu_p + \frac{5 \times (d_2 - d_1) \times \tau_y}{v_a}$$

$$\mu_{ep} = 25 + \frac{5 \times (8.5 - 4.5) \times 15}{10.21} = 120.5$$

$$N_{Re} = \frac{757 \times \rho \times v_a \times (d_2 - d_1)}{\mu_{ep}}$$

$$N_{Re} = \frac{757 \times 3.14 \times 10.21 \times (8.5 - 4.5)}{120.5}$$

$$N_{Re} = 806$$

Therefore the flow regime is laminar.

$$\left(\frac{dp_{f\,adc}}{dL_{adc}}\right)_a = \frac{\mu_p v_a}{1000 \times (d_2 - d_1)^2} + \frac{\tau_y}{200 \times (d_2 - d_1)}$$

$$= \frac{25 \times 3.14}{1000 \times (8.5 - 4.5)^2} + \frac{15}{200 \times (8.5 - 4.5)} = 0.0237$$

$\Delta P_{adc} = 0.0237$ psi/ft
$\Delta P_{adc} = \Delta P_{adc} \times L_{adc} = 0.0237 \times 630$
$\Delta P_{adc} = 14.93$ psi

6. **Calculating the pressure losses in the annulus between drillcollars and borehole**

$$v_a = \frac{q}{2.448 \times (d_2^2 - d_1^2)}$$

$$v_a = \frac{400}{2.448 \times (8.5^2 - 6^2)} = 4.5 \text{ ft/sec}$$

$$\mu_{ep} = \mu_p + \frac{5 \times (d_2 - d_1) \times \tau_y}{v_a}$$

$$\mu_{ep} = 25 + \frac{5 \times (8.5 - 6) \times 15}{10.21} = 66.6$$

$$N_{Re} = \frac{757 \times \rho \times v_a \times (d_2 - d_1)}{\mu_{ep}}$$

$$N_{Re} = \frac{757 \times 10.2 \times 10.21 \times (8.5 - 6)}{66.6}$$

$$N_{Re} = 1307$$

Hydraulics

Therefore the flow regime is laminar.

$$\left(\frac{dp_{fadc}}{dL_{adc}}\right)_a = \frac{\mu_p V_a}{1000\times(d_2-d_1)^2} + \frac{\tau_y}{200\times(d_2-d_1)}$$

$$= \frac{25\times 4.5}{1000\times(8.5-6)^2} + \frac{15}{200\times(8.5-6)} = 0.0480$$

$\Delta P_{adc} = 0.0480$ psi/ft
$\Delta P_{adc} = \Delta P_{adc} \times L_{adc} = 0.0480 \times 900$
$\Delta P_{adc} = 43.2$ psi

7. **Calculating the pressure losses across the bit**

$$P_b = \frac{8.311\times 10^{-5}\times \rho \times Q^2}{C_d^2 \times A_n^2}.$$

Assuming a coefficient of discharge: $C_d = 0.95$

$$A_n = \frac{\pi}{4}\left(d_1^2 + d_2^2 + d_3^2\right)$$

$$A_n = \frac{3.1416}{4}\left(0.375^2 + 0.375^2 + 0.375^2\right)$$

$$A_n = 0.331 \text{ in}^2$$

$$P_b = \frac{8.311\times 10^{-5}\times 10.2\times 400^2}{0.95^2 \times 0.331^2} = \frac{135.64}{0.099}$$

$\Delta P_b = 1370.1$ psi

8. **Calculating the pressure losses in the surface**

$$P_{surf} = E\times \rho^{0.8}\times Q^{1.8}\times \mu_p^{0.2}$$

As the E factor is unknown, the surface pressure losses can be calculated using the drilling pipe equation:

$$v_p = \frac{q}{2.448\times d^2}$$

$$v_p = \frac{400}{2.448 \times 4^2} = 10.2 \text{ ft/sec}$$

$$\mu_{ep} = \mu_p + \frac{20 \times d \times \tau_y}{3 \times v_p}$$

$$\mu_{ep} = 25 + \frac{20 \times 4 \times 15}{3 \times 10.2} = 64.22$$

$$N_{Re} = \frac{928 \times \rho \times v_p \times d}{\mu_{ep}}$$

$$N_{Re} = \frac{928 \times 10.2 \times 10.2 \times 4}{64.22}$$

$$N_{Re} = 6013.6$$

Therefore the flow regime is turbulent.

$$\left(\frac{dp_f}{dL}\right)_{surf} = \frac{\rho^{0.75} v_p^{1.75} \mu_p^{0.25}}{1800 \times d^{1.25}}$$

$$\left(\frac{dp_f}{dL}\right)_{surf} = \frac{10.2^{0.75} 10.2^{1.75} 25^{0.25}}{1800 \times 4^{1.25}} = \frac{742.99}{10,182.34}$$

$\Delta P_{surf} = 0.073$ psi/ft
$\Delta P_{surf} = \Delta P_{surf} \times L_{surf} = 0.073 \times 500$
$\Delta P_{surf} = 36.5$ psi

Estimating E factor with the pressure loss calculated:

$$E = \frac{\Delta P_{surf}}{\rho^{0.8} \times Q^{1.8} \times \mu_p^{0.2}} \quad E = \frac{36.5}{10.2^{0.8} \times 400^{1.8} \times 25^{0.2}}$$

$$E = 0.0000062$$

9. **Pump hydraulics horsepower requirement**

$$\Delta P_p = \Delta P_f + \Delta P_b$$

Hydraulics

$$P_f = \Delta P_{fdp}L_{fdp} + \Delta P_{fhwdp}L_{fhwdp} + \Delta P_{fdc}L_{fdc} + \Delta P_{fadp}L_{fadp}$$
$$+ \Delta P_{fahwdp}L_{fadwp} + \Delta P_{fadc}L_{fadc} + \Delta P_{fsurf}L_{fsurf}$$

$$P_f = 3190 \text{ psi}$$

$$\Delta P_p = 3190 + 1370$$

$$\Delta P_p = 4560 \text{ psi}$$

$$P_{Hp} = \frac{\Delta P_p \times q}{1714}$$

$$P_{Hp} = \frac{4560 \times 400}{1714}$$

$$P_{Hp} = 1065 \text{ hp}$$

As the efficiency of the pump is 85 %,
$P_{Hp} = 1065/0.85$
$P_{Hp} = 1253 \text{ hp}$

10. **Bottomhole pressure calculation**
$P_{bhp} = P_h + \Delta P_a$
$P_h = 0.052 \times \rho \times D$
$P_h = 0.052 \times 10.2 \times 20,000 \quad P_h = 10,608 \text{ psi}$
$\Delta P_a = 661 \, psi$
$P_{bhp} = 10,608 + 661$
$P_{bhp} = 11,269 \text{ psi}$

11. **Percentage of bit pressure loss**

$$\%\text{Bit Pressure Loss} = \frac{\text{Bit Pressure Loss}}{\text{Total Pressure Loss}} = \frac{1370}{4492} = 30\%$$

291. Problem 7.33

A drilling engineer plans to calculate the ECD (Fig. 7.5) with a mud density of 10.2 ppg.

Fann data
$\theta_{600} = 65; \theta_{300} = 40; \theta_3 = 15$

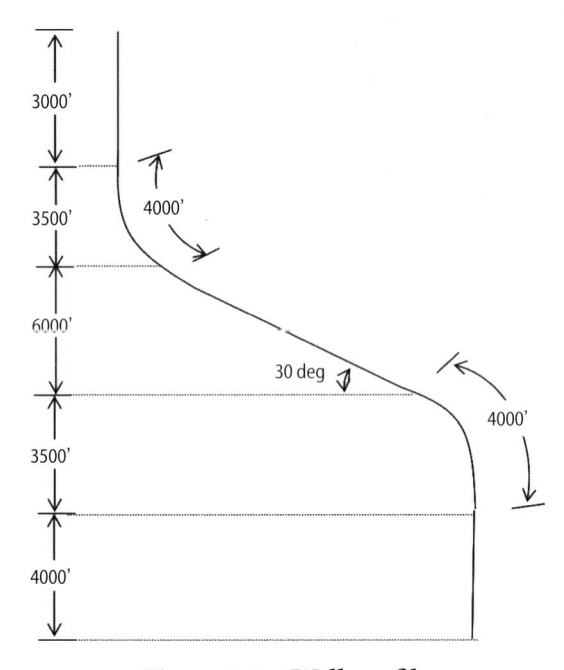

Figure 7.5 Well profile

Total nozzle area of the bit - .
Drillpipe: 4.5" × 3.958" and tool joint 6.00" × 3.25" - Length of tooljoint 0.71 ft on one side
Heavy Weight Drillpipe: 4.5" × 2.75" and tool joint 6.25" × 2.87" - Length of tooljoint 1.33 ft on one side
Drillcollar length 900 ft, 6" OD, 72.16 lb/ft in air.
Surface connections: 500 ft × 4" id
Hole: Last intermediate casing set: 8.5" ID set at 16,000' (TVD) openhole size 8.5".

For Normal Forward Circulation
Determine the pump hydraulics horsepower requirement to drill to target depth. Assume pump volumetric efficiency of 85% and a maximum flow rate of 400 gpm.

Using the data from Problem 7.32, calculate the percentage of bit pressure loss. Also calculate the bottomhole pressure and bottomhole circulating mud density (ECD) without tooljoint effects.

Solution:
1. **Equivalent Circulating Density at Target Depth (TD)**

$$\text{ECD} = \rho_m + \frac{P_{fa}}{0.052 \times D}$$

$$\text{ECD} = 10.2 + \frac{661}{0.052 \times 20{,}000}$$

$$\text{ECD} = 10.83 \text{ ppg}$$

2. **ECD at Shoe**

TVD at shoe is 16,000 ft and the measured depth is 23,000 ft. The frictional pressure loss up to shoe is against the drillpipe.
$\Delta P_{adp} = 0.024$ psi/ft
The total annular frictional pressure loss against the drillpipe is
$\Delta P_{adp} = \Delta P_{adp} \times L_{adp} = 0.024 \times 23{,}000$
$\Delta P_{adp} = 552$ psi

$$\text{ECD} = \rho_m + \frac{P_{fa}}{0.052 \times D}$$

$$\text{ECD} = 10.2 + \frac{552}{0.052 \times 16{,}000}$$

$$\text{ECD} = 10.86 \text{ ppg}$$

292. Problem 7.34

Using the data from Problem 7.32, plot the measured depth vs pressure and measured depth vs annular velocity for this condition.

Draw a pie-chart depicting the percentage of pressure losses through various components.

Solution:
1. Pie Chart of Inside Drillstring Pressure Losses (Fig. 7.6)

2. Pie Chart of Annular Pressure Losses (Fig. 7.7)

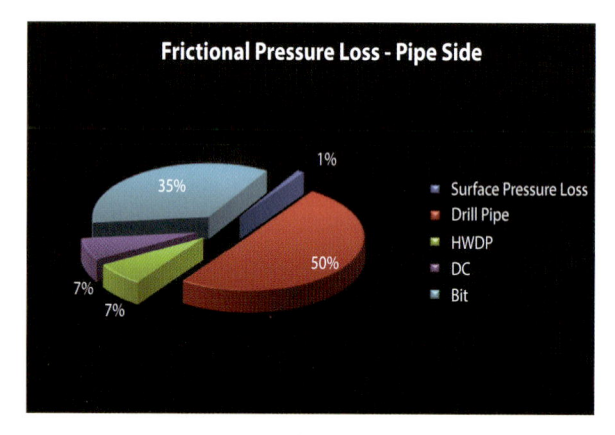

Figure 7.6 Pi-chart – pressure loss

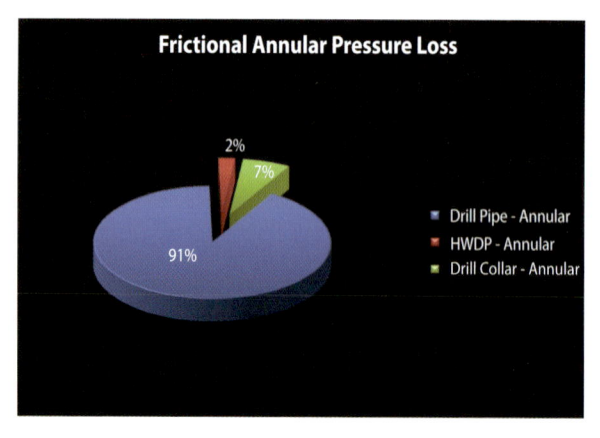

Figure 7.7 Pi-chart – annular pressure loss

Hydraulics

3. **Pie Chart of System Pressure Losses (Fig. 7.8)**

4. **Wellbore Pressure Profile (Fig. 7.9)**

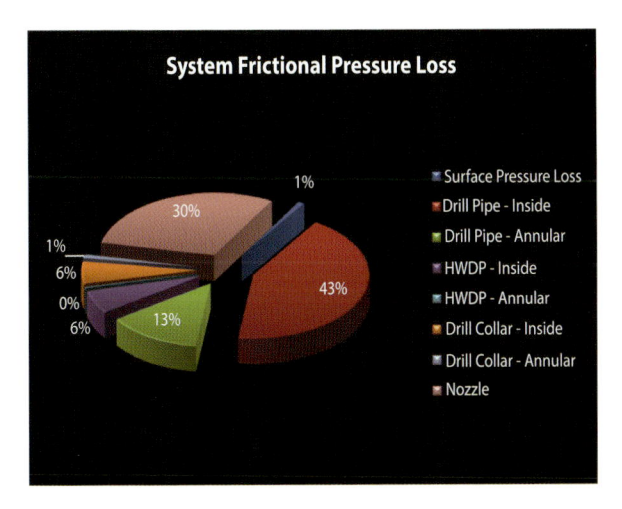

Figure 7.8   Pi-chart – system frictional pressure loss

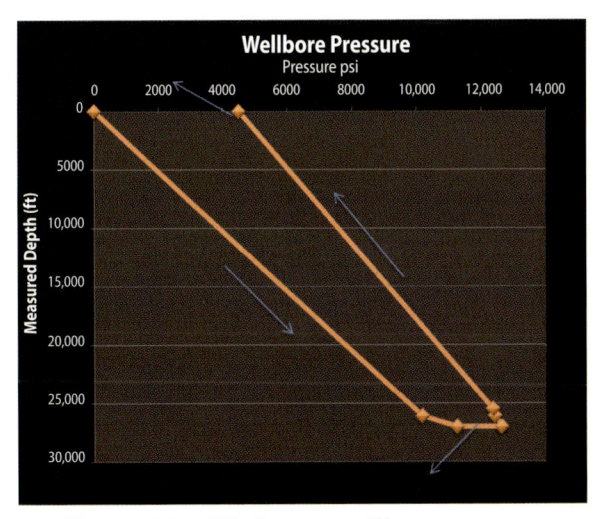

Figure 7.9   Pi-chart – wellbore pressure

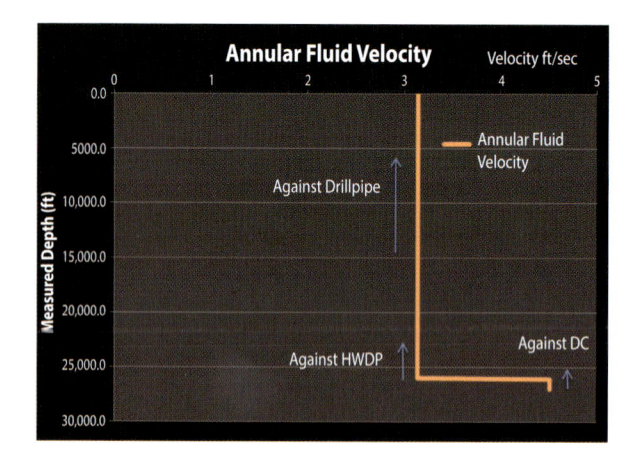

Figure 7.10 Annular velocity

5. *Annular Fluid Velocity Profile (Fig. 7.10)*

293. Problem 7.35

What are the causes of increased swabbing?

Solution:
- Tripping speed
- Acceleration and deceleration of the pipe
- Mud properties such as yield stress, gel strength
- Mud cake
- Clearance between the pipe and borehole
- Bit, stabilizer balling

294. Problem 7.36

Calculate the pressure required to break gel. Assume gel strength to be 15 lbf/100 ft². Determine whether the formation fracturing will occur during break circulation. Assume that the formation fracture gradient at the casing shoe is 0.545 psi/ft.

Solution:

Pressure to break the gel is given as

$$P_p + P_a = \frac{\tau_g L}{300 D_i} + \frac{\tau_g L}{300(D_h - D_p)} = 727 \text{ psi}$$

$$P_{pipe} = \frac{\tau_g L}{300 D_i} \text{ psi and annulus side } P_{ann} = \frac{\tau_g L}{300(D_h - D_p)} \text{ psi}$$

Since 10 sec or 10 min reading is not given, slow rpm reading of 3 is taken as reference gel strength reading.

To break the gel inside the drillpipe is

$$P_{pipe} = \frac{15 \times 25,470}{300 \times 3.958} = 322 \text{ psi}$$

To break the gel inside the heavy weight drillpipe is

$$P_{pipe} = \frac{15 \times 630}{300 \times 2.75} = 11.5 \text{ psi}$$

To break the gel inside the drillcollar is

$$P_{pipe} = \frac{15 \times 900}{300 \times 3} = 15 \text{ psi}$$

Total pressure in the pipe side = 349 psi
To break the gel in the annulus side of the drillpipe is

$$P_{pipe} = \frac{15 \times 25,470}{300 \times (8.5 - 4.5)} = 318 \text{ psi or } 0.0125 \text{ psi/ft}$$

To break the gel in the annulus side of the heavy weight drillpipe is

$$P_{pipe} = \frac{15 \times 630}{300 \times (8.5 - 4.5)} = 8 \text{ psi}$$

To break the gel in the annulus side of the drillcollar is

$$P_{pipe} = \frac{15 \times 900}{300 \times (8.5 - 6)} = 18 \text{ psi}$$

Total pressure in the pipe side = 344 psi
Total pressure both inside and outside = 349 + 344 = 693 psi

The pressure to break circulation at the shoe is the pressure required to break the gel up to the casing shoe plus the hydrostatic and is given as

$$P_{pipe} = \frac{15}{300 \times (8.5 - 4.5)} = 0.0125 \text{ psi/ft}$$

The pressure at the casing shoe at 23,000 ft (measured depth) when the gel is broken

$$P_{pipe} = \frac{15 \times 23,000}{300 \times (8.5 - 4.5)} + 0.052 \times 10.2 \times 16,000 = 8774 \text{ psi or } 10.54 \text{ ppg}$$

The fracture pressure at shoe is

$$P = 0.545 \times 16,000 = 8720 \text{ psi}$$

or

$$P = \frac{8720}{0.052 \times 16,000} = 10.48 \text{ ppg}$$

So while initiating circulation to breaking the gel will exceed the fracture pressure at shoe. To reduce the initial pressure requirements the drillstring can be rotated to aid in the breaking the gel or alternatively the drillstring is rotated simultaneously slowly pumping the fluid.

ECD at Shoe

TVD at shoe is 16,000 ft and the measured depth is 23,000 ft.
The frictional pressure loss up to shoe is against the drillpipe.
$\Delta P_{adp} = 0.024$ psi/ft
The total annular frictional pressure loss against the drillpipe is
$\Delta P_{adp} = \Delta P_{adp} \times L_{adp} = 0.024 \times 23,000$
$\Delta P_{adp} = 552$ psi

$$ECD = \rho_m + \frac{P_{fa}}{0.052 \times D}$$

$$ECD = 10.2 + \frac{552}{0.052 \times 16,000}$$

$$ECD = 10.86 \text{ ppg}$$

Hydraulics

Using the calculation from the previous section it can be seen that the ECD at shoe is 10.86 ppg which is also greater than the fracture gradient at shoe. This will result in fracture and thus loss.

295. Problem 7.37

Using the data from Problem 7.32, calculate the pressure losses when reverse circulating.

Solution:

Reverse Circulation without Tooljoint Pressure Losses (Fig. 7.11)
Calculating the pressure losses from the annulus side to the pipe side

1. **Calculating the pressure losses in the annulus between drillpipe and borehole**

$$v_a = \frac{q}{2.448 \times \left(d_2^2 - d_1^2\right)}$$

$$v_a = \frac{400}{2.448 \times \left(8.5^2 - 4.5^2\right)} = 3.14 \text{ ft/sec}$$

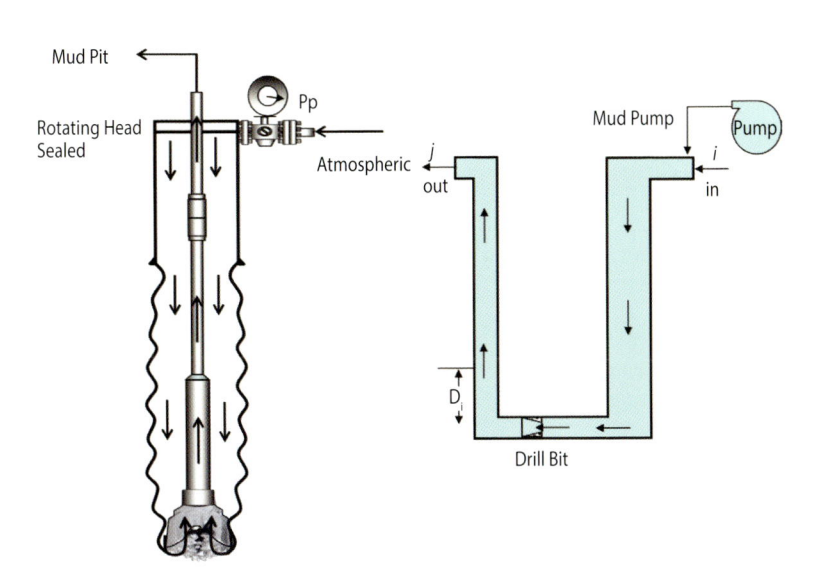

Figure 7.11 Reverse circulation flow path

$$\mu_{ep} = \mu_p + \frac{5 \times (d_2 - d_1) \times \tau_y}{v_a}$$

$$\mu_{ep} = 25 + \frac{5 \times (8.5 - 4.5) \times 15}{3.14} = 120.54$$

$$N_{Re} = \frac{757 \times \rho \times v_a \times (d_2 - d_1)}{\mu_{ep}}$$

$$N_{Re} = \frac{757 \times 10.2 \times 3.14 \times (8.5 - 4.5)}{120.54}$$

$$N_{Re} = 804.55$$

Therefore the flow regime is laminar.

$$\left(\frac{dp_{f\,adp}}{dL_{adp}}\right)_a = \frac{\mu_p v_a}{1000 \times (d_2 - d_1)^2} + \frac{\tau_y}{200 \times (d_2 - d_1)}$$

$$= \frac{25 \times 3.14}{1000 \times (8.5 - 4.5)^2} + \frac{15}{200 \times (8.5 - 4.5)} = 0.0049 + 0.0188$$

$\Delta P_{adp} = 0.024$ psi/ft
$\Delta P_{adp} = \Delta P_{adp} \times L_{adp} = 0.024 \times 25{,}470$
$\Delta P_{adp} = 603.0$ psi

2. **Calculating the pressure losses in the annulus between HWDP and borehole**

$$v_a = \frac{q}{2.448 \times (d_2^2 - d_1^2)}$$

$$v_a = \frac{400}{2.448 \times (8.5^2 - 4.5^2)} = 3.14 \text{ ft/sec}$$

$$\mu_{ep} = \mu_p + \frac{5 \times (d_2 - d_1) \times \tau_y}{v_a}$$

Hydraulics

$$\mu_{ep} = 25 + \frac{5 \times (8.5 - 4.5) \times 15}{10.21} = 120.5$$

$$N_{Re} = \frac{757 \times \rho \times v_a \times (d_2 - d_1)}{\mu_{ep}}$$

$$N_{Re} = \frac{757 \times 3.14 \times 10.21 \times (8.5 - 4.5)}{120.5}$$

$$N_{Re} = 806$$

Therefore the flow regime is laminar.

$$\left(\frac{dp_{fadc}}{dL_{adc}}\right)_a = \frac{\mu_p v_a}{1000 \times (d_2 - d_1)^2} + \frac{\tau_y}{200 \times (d_2 - d_1)}$$

$$= \frac{25 \times 3.14}{1000 \times (8.5 - 4.5)^2} + \frac{15}{200 \times (8.5 - 4.5)} = 0.0237$$

$\Delta P_{adc} = 0.0237$ psi/ft
$\Delta P_{adc} = \Delta P_{adc} \times L_{adc} = 0.0237 \times 630$
$\Delta P_{adc} = 14.93$ psi

3. **Calculating the pressure losses in the annulus between drillcollars and borehole**

$$v_a = \frac{q}{2.448 \times (d_2^2 - d_1^2)}$$

$$v_a = \frac{400}{2.448 \times (8.5^2 - 6^2)} = 4.5 \text{ ft/sec}$$

$$\mu_{ep} = \mu_p + \frac{5 \times (d_2 - d_1) \times \tau_y}{v_a}$$

$$\mu_{ep} = 25 + \frac{5 \times (8.5 - 6) \times 15}{10.21} = 66.6$$

$$N_{Re} = \frac{757 \times \rho \times v_a \times (d_2 - d_1)}{\mu_{ep}}$$

$$N_{Re} = \frac{757 \times 10.2 \times 10.21 \times (8.5 - 6)}{66.6}$$

$$N_{Re} = 1307$$

Therefore the flow regime is laminar.

$$\left(\frac{dp_{f\,adc}}{dL_{adc}}\right)_a = \frac{\mu_p v_a}{1000 \times (d_2 - d_1)^2} + \frac{\tau_y}{200 \times (d_2 - d_1)}$$

$$= \frac{25 \times 4.5}{1000 \times (8.5 - 6)^2} + \frac{15}{200 \times (8.5 - 6)} = 0.0480$$

$\Delta P_{adc} = 0.0480$ psi/ft
$\Delta P_{adc} = \Delta P_{adc} \times L_{adc} = 0.0480 \times 900$
$\Delta P_{adc} = 43.2$ psi

4. **Calculating the pressure losses across the bit**

$$P_b = \frac{8.311 \times 10^{-5} \times \rho \times Q^2}{C_d^2 \times A_n^2}$$

Assuming a coefficient of discharge: $C_d = 1.1$

$$A_n = \frac{\pi}{4}\left(d_1^2 + d_2^2 + d_3^2\right)$$

$$A_n = \frac{3.1416}{4}\left(0.375^2 + 0.375^2 + 0.375^2\right)$$

$$A_n = 0.331 \text{ in}^2$$

$$P_b = \frac{8.311 \times 10^{-5} \times 10.2 \times 400^2}{1.1^2 \times 0.331^2} = \frac{135.64}{0.099}$$

$$\Delta P_b = 1021 \text{ psi}$$

5. Calculating the pressure losses inside the drillcollar

Calculating as before

$\Delta P_{dc} = 0.2909$ psi/ft

$\Delta P_{dc} = \Delta P_{dc} \times L_{dc} = 0.2909 \times 900$

$\Delta P_{dc} = 262$ psi

6. Calculating the pressure losses inside the heavy weight drillpipe

$$\mu_p = \theta_{600} - \theta_{300}$$

$$\mu_p = 65 - 40 = 25$$

$$\tau_y = 2 \times \theta_{300} - \theta_{600}$$

$$\tau_y = 2 \times 40 - 65 = 15$$

Alternatively regressing the three dial readings the yield value can be found as $\tau_y = 15.75$

$$v_p = \frac{q}{2.448 \times d^2}$$

$$v_p = \frac{400}{2.448 \times 3.958^2} = 10.43 \text{ ft/sec}$$

$$\mu_{ep} = \mu_p + \frac{20 \times d \times \tau_y}{3 \times v_p}$$

$$\mu_{ep} = 25 + \frac{20 \times 3.958 \times 15}{3 \times 10.43} = 62.95$$

$$N_{Re} = \frac{928 \times \rho \times v_p \times d}{\mu_{ep}}$$

$$N_{Re} = \frac{928 \times 10.2 \times 10.2 \times 3.958}{62.95}$$

$$N_{Re} = 6208$$

Therefore the flow regime is turbulent.

$$\left(\frac{dp_f}{dL}\right)_p = \frac{\rho^{0.75} v_p^{1.75} \mu_p^{0.25}}{1800 \times d^{1.25}}$$

$$\left(\frac{dp_f}{dL}\right)_p = \frac{10.2^{0.75} 10.43^{1.75} 25^{0.25}}{1800 \times 3.958^{1.25}} = 0.07688 \text{ psi/ft}$$

$\Delta P_{hwdp} = 0.4335$ psi/ft
$\Delta P_{hwdp} = \Delta P_{hwdp} \times L_{hwdp} = 0.4335 \times 630$
$\Delta P_{hwdp} = 273$ psi

7. **Calculating the pressure losses inside the drillpipe**
$\Delta P_{dp} = 0.07688$ psi/ft
$\Delta P_{dp} = \Delta P_{dp} \times L_{dp} = 0.07688 \times 25{,}470$
$\Delta P_{dp} = 1958.0$ psi

8. **Calculating the pressure losses in the surface**

$$P_{surf} = E \times \rho^{0.8} \times Q^{1.8} \times \mu_p^{0.2}$$

As the E factor is unknown, the surface pressure losses can be calculated using the drilling pipe equation:

$$v_p = \frac{q}{2.448 \times d^2}$$

$$v_p = \frac{400}{2.448 \times 4^2} = 10.2 \text{ ft/sec}$$

$$\mu_{ep} = \mu_p + \frac{20 \times d \times \tau_y}{3 \times v_p}$$

$$\mu_{ep} = 25 + \frac{20 \times 4 \times 15}{3 \times 10.2} = 64.22$$

$$N_{Re} = \frac{928 \times \rho \times v_p \times d}{\mu_{ep}}$$

Hydraulics

$$N_{Re} = \frac{928 \times 10.2 \times 10.2 \times 4}{64.22}$$

$$N_{Re} = 6013.6$$

Therefore the flow regime is turbulent.

$$\left(\frac{dp_f}{dL}\right)_{surf} = \frac{\rho^{0.75} v_p^{1.75} \mu_p^{0.25}}{1800 \times d^{1.25}}$$

$$\left(\frac{dp_f}{dL}\right)_{surf} = \frac{10.2^{0.75} 10.2^{1.75} 25^{0.25}}{1800 \times 4^{1.25}} = \frac{742.99}{10{,}182.34}$$

$\Delta P_{surf} = 0.073$ psi/ft
$\Delta P_{surf} = \Delta P_{surf} \times L_{surf} = 0.073 \times 500$
$\Delta P_{surf} = 36.5$ psi

Estimating E factor with the pressure loss calculated:

$$E = \frac{\Delta P_{surf}}{\rho^{0.8} \times Q^{1.8} \times \mu_p^{0.2}}$$

$$E = \frac{36.5}{10.2^{0.8} \times 400^{1.8} \times 25^{0.2}}$$

$$E = 0.0000062$$

9. **Pump hydraulics horsepower requirement**

$$\Delta P_p = \Delta P_f + \Delta P_b$$

$$P_f = \Delta P_{fdp} L_{fdp} + \Delta P_{fhwdp} L_{fhwdp} + \Delta P_{fdc} L_{fdc} + \Delta P_{fadp} L_{fadp}$$
$$+ \Delta P_{fahwdp} L_{fadwp} + \Delta P_{fadc} L_{fadc} + \Delta P_{fsurf} L_{fsurf}$$

$$P_f = 603 + 273 + 43.2 + 262 + 15 + 1958 + 36.5 = 3191 \text{ psi}$$

$$\Delta P_p = 3191 + 1021$$

$$\Delta P_p = 4212 \text{ psi}$$

$$P_{Hp} = \frac{\Delta P_p \times q}{1714}$$

$$P_{Hp} = \frac{4212 \times 400}{1714}$$

$$P_{Hp} = 983 \text{ hp}$$

As the efficiency of the pump is 85%, I require additional 186.7 hp to be into the range, so:

$P_{Hp} = 983/0.85$

$P_{Hp} = 1157 \text{ hp}$

10. **Bottomhole pressure calculation**

$$P_{bhp} = P_h + \Delta P_p + \Delta P_b$$

$$P_h = 0.052 \times \rho \times D$$

$P_h = 0.052 \times 10.2 \times 20{,}000 \qquad P_h = 10{,}608 \text{ psi}$

$$\Delta P_p = 273 + 262 + 1958 + 1021 = 3514 \text{ psi}$$

$$P_{bhp} = 10{,}608 + 3514$$

$$P_{bhp} = 14{,}122 \text{ psi}$$

If the annular side pressure losses are used

$$P_{bhp} = P_h - \Delta P_a + P_p$$

$P_h = 0.052 \times 10.2 \times 20{,}000 \qquad P_h = 10{,}608 \text{ psi}$

$$\Delta P_a = 603 + 43.2 + 15 + 36.5 = 697 \text{ psi}$$

$$P_{bhp} = 10{,}608 - 697 + 4212 = 14{,}122 \text{ psi}$$

11. **Defining the equivalent circulating density**

$$\text{ECD} = \rho_m + \frac{P_{fa}}{0.052 \times D}$$

$$\text{ECD} = 10.2 + \frac{3514}{0.052 \times 20{,}000}$$

$$\text{ECD} = 13.57 \text{ ppg}$$

Hydraulics

Using the bottomhole pressure

$$ECD = \frac{14{,}122}{0.052 \times 20{,}000} = 13.57 \text{ ppg}$$

12. **Percentage of bit pressure loss**

$$\%\text{Bit Pressure Loss} = \frac{\text{Bit Pressure Loss}}{\text{Total Pressure Loss}} = \frac{1021}{4212} = 24\%$$

296. Problem 7.38

The hole section, drillstring, wellpath, and fluid data for a directional are given below. Calculate the bottomhole pressure and ECD while drilling at TMD (total measured depth) of 17,279 ft for a given flowrate of 600 gpm. Estimate the pump hydraulic horsepower requirement to drill to target depth. Assume a pump volumetric efficiency of 90%.

Other given data:

PDM performance is given as

$$\Delta p_{mm} = -0.0027Q^2 + 1.9457Q - 18.286 \text{ psi}$$

for a reference density of 8.3 ppg. Assume $C_d = 0.95$ for the nozzles. Use Bingham plastic model.

Plot the following:

Plot the measured depth vs pressure and measured depth vs annular velocity for this condition.

Also plot the TVD (true vertical depth) vs pressure.

Calculate the new ECD when a surface back pressure of 300 psi is applied at the top of the annulus and draw your conclusion.

Bit: Tri-cone Roller Cone Jet Bit with 12-12-12 nozzles.
Surface connections: 500 ft × 4" id (inside diameter).
Fluid Details:

Base Density (ppg)	Plastic Viscosity (cP)	Yield Point (Tau 0) (lbf/100 ft²)
11.90	22.97	19.926

Solution:

a) Well Schematic (Fig. 7.12)

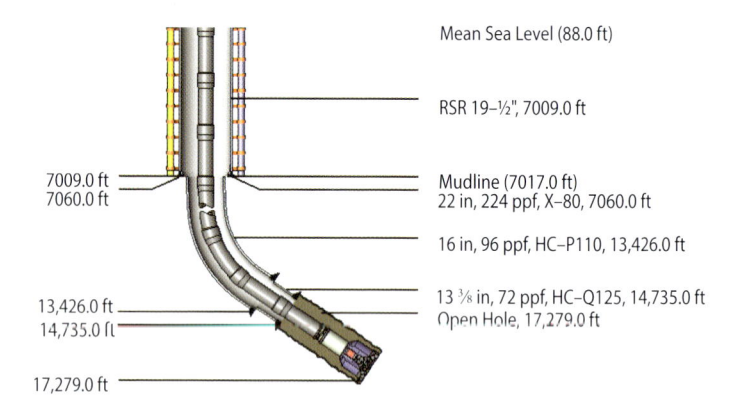

Figure 7.12 Well profile

MD/INC/DIR/TVD Calculation

MD (ft)	Delta MD (ft)	INC (°)	Delta TVD (ft)	TVD (ft)
0	0.0	0.00	0.0	0
88	88.0	0.00	88.0	88
7050	6962.0	0.00	6962.0	7050
7412	362.0	1.01	361.9	7412
7506	94.0	1.29	94.0	7506
7601	95.0	1.70	95.0	7601
7696	95.0	2.05	94.9	7696
7799	103.0	2.42	102.9	7799
7886	87.0	2.60	86.9	7886
7983	97.0	2.76	96.9	7983
8075	92.0	3.29	91.8	8074
8172	97.0	4.16	96.7	8171
8267	95.0	3.85	94.8	8266
8363	96.0	2.48	95.9	8362

Hydraulics

8455	92.0	1.75	92.0	8454
8549	94.0	2.60	93.9	8548
8646	97.0	2.94	96.9	8645
8740	94.0	4.25	93.7	8738
8835	95.0	5.62	94.5	8833
8930	95.0	6.59	94.4	8927
9024	94.0	8.08	93.1	9020
9118	94.0	9.45	92.7	9113
9214	96.0	10.63	94.4	9207
9310	96.0	12.17	93.8	9301
9407	97.0	13.80	94.2	9395
9503	96.0	15.69	92.4	9488
9599	96.0	17.54	91.5	9579
9695	96.0	19.43	90.5	9670
9784	89.0	20.77	83.2	9753
9880	96.0	22.29	88.8	9842
9974	94.0	23.61	86.1	9928
10,000	26.0	23.32	23.9	9952
10,097	97.2	22.23	90.0	10,042
10,100	2.8	22.26	2.6	10,044
10,200	100.0	23.57	91.7	10,136
10,300	100.0	24.89	90.7	10,227
10,400	100.0	26.20	89.7	10,317
10,500	100.0	27.53	88.7	10,405
10,600	100.0	28.85	87.6	10,493
10,700	100.0	30.18	86.4	10,579
10,800	100.0	31.51	85.3	10,665
10,900	100.0	32.84	84.0	10,749

11,000	100.0	34.18	82.7	10,831
11,100	100.0	35.51	81.4	10,913
11,200	100.0	36.85	80.0	10,993
11,300	100.0	38.19	78.6	11,071
11,400	100.0	39.53	77.1	11,148
11,500	100.0	40.87	75.6	11,224
11,600	100.0	42.21	74.1	11,298
11,700	100.0	43.56	72.5	11,371
11,800	100.0	44.90	70.8	11,441
11,900	100.0	46.25	69.2	11,511
12,000	100.0	47.59	67.4	11,578
12,100	100.0	48.94	65.7	11,644
12,200	100.0	50.28	63.9	11,708
12,300	100.0	51.63	62.1	11,770
12,400	100.0	52.98	60.2	11,830
12,457	57.3	53.75	33.9	11,864
12,469	11.9	53.90	7.0	11,871
12,500	30.8	54.28	18.0	11,889
12,519	19.2	54.52	11.1	11,900
12,569	50.0	55.14	28.6	11,929
12,600	30.8	55.52	17.4	11,946
12,619	19.2	55.76	10.8	11,957
12,669	50.0	56.38	27.7	11,984
12,672	3.2	56.42	1.8	11,986
12,700	27.6	56.42	15.3	12,001
12,800	100.0	56.42	55.3	12,057
12,900	100.0	56.42	55.3	12,112
13,000	100.0	56.42	55.3	12,167

Hydraulics

13,100	100.0	56.42	55.3	12,223
13,200	100.0	56.42	55.3	12,278
13,300	100.0	56.42	55.3	12,333
13,400	100.0	56.42	55.3	12,389
13,500	100.0	56.42	55.3	12,444
13,600	100.0	56.42	55.3	12,499
13,700	100.0	56.42	55.3	12,555
13,800	100.0	56.42	55.3	12,610
13,900	100.0	56.42	55.3	12,665
14,000	100.0	56.42	55.3	12,720
14,107	106.5	56.42	58.9	12,779
14,200	93.5	56.42	51.7	12,831
14,300	100.0	56.42	55.3	12,886
14,400	100.0	56.42	55.3	12,942
14,500	100.0	56.42	55.3	12,997
14,600	100.0	56.42	55.3	13,052
14,700	100.0	56.42	55.3	13,108
14,800	100.0	56.42	55.3	13,163
14,900	100.0	56.42	55.3	13,218
15,000	100.0	56.42	55.3	13,274
15,100	100.0	56.42	55.3	13,329
15,200	100.0	56.42	55.3	13,384
15,300	100.0	56.42	55.3	13,440
15,400	100.0	56.42	55.3	13,495
15,500	100.0	56.42	55.3	13,550
15,553	52.8	56.42	29.2	13,579
15,600	47.2	56.42	26.1	13,605
15,700	100.0	56.42	55.3	13,661

15,800	100.0	56.42	55.3	13,716
15,900	100.0	56.42	55.3	13,771
16,000	100.0	56.42	55.3	13,827
16,100	100.0	56.42	55.3	13,882
16,200	100.0	56.42	55.3	13,937
16,300	100.0	56.42	55.3	13,993
16,400	100.0	56.42	55.3	14,048
16,430	29.7	56.42	16.4	14,064
16,500	70.3	56.42	38.9	14,103
16,600	100.0	56.42	55.3	14,159
16,627	26.8	56.42	14.8	14,173
16,700	73.2	56.42	40.5	14,214
16,800	100.0	56.42	55.3	14,269
16,900	100.0	56.42	55.3	14,324
17,000	100.0	56.42	55.3	14,380
17,100	100.0	56.42	55.3	14,435
17,200	100.0	56.42	55.3	14,490
17,279	**79**	**56.42**	**55.3**	**14,533**
17,400	100.0	56.42	55.3	14,601

b) MD vs Inclination of the Well
c) Total Annular Frictional Pressure Losses = 210 psi
d) Total Pipe Frictional Pressure Losses = 640 psi
e) Bit Pressure Drop = 3594 psi
f) Motor Pressure Drop
The reference and the required flowrate are same.
$$\Delta p_{mm} = \left(-0.0027 \times 600^2 + 1.9457 \times 600 - 18.286\right)\frac{11.9}{8.3} = 254 \text{ psi}$$

g) Surface Connection Pressure
Use the regular pipe side equations = 85 psi

h) Standpipe Pressure
   = 210 + 640 + 3594 + 254 = 4698 psi
   This is the pressure measured at the rig and will be less than the surface connection pressure losses.
i) Total Pump Pressure
   = 4685 + 85 = 4785 psi
   This includes all the frictional pressure losses including the surface connections.
j) Bottomhole Pressure – Using Annulus Side Frictional Pressure Loss
   $P_{bh} = 0.052 \times 11.9 \times 14{,}533 + 210 = 9203$ psi
k) ECD Calculation
   $$\text{ECD} = 11.9 + \frac{210}{0.052 \times 14{,}533} = 12.2 \text{ ppg} \quad \text{or}$$

   $$\text{ECD} = \frac{9203}{0.052 \times 14{,}533} = 12.2 \text{ ppg}$$
l) Hydraulic Horsepower Requirement
   Using the efficiency to adjust the actual flowrate to the theoretical flowrate
   $$\text{HHP} = \frac{QP_p}{1714 \times \eta} = \frac{600 \times 4785}{1714 \times 0.9} = 1860 \text{ hp}$$
m) Pressure vs Depth Plot (Fig. 7.13)

297. Problem 7.39

Open hole size: 8 ⅞" (washed out to 9 ⅞")
to TMD (total measured depth) of 19,250 ft (Fig. 7.14)
Drillpipe: 5" OD; 3" ID
Drillcollar: 7 ½" OD; 3" ID; 800 ft length
Mud: Bingham Plastic $\gamma_m = 13$ ppg
Fann VG dial reading $\theta_{600} = 40$, $\theta_{300} = 25$
Last casing setting depth: 13,000 ft (TMD)
Pump: $P_{max} = 4300$ psi (allowed)

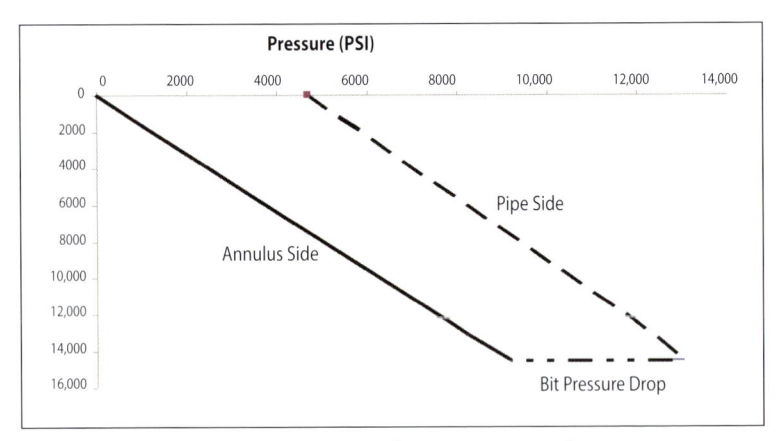

Figure 7.13 Depth vs pressure plot

Minimum annular fluid velocity required: 90 fpm

Neglect surface connections.

Calculate the equivalent circulating density at the bottom for minimum flowrate.

Determine the maximum flow rate while drilling at TMD of 19,250 ft that will not cause any formation fracturing during circulation.

Assume a formation fracture gradient of 0.77 psi/ft at 19,250 ft.

Solution:

Calculating viscosity and yield stress

$$\mu_p = 15 \text{ cP}; \tau_y = 10 \frac{\text{lbf}}{100 \text{ ft}^2}$$

Minimum flowrate is against the biggest annulus area

$$Q = 90 \times \frac{\pi}{4 \times 144}(9.875^2 - 5^2) \times 7.48 = 266 \text{ gpm}$$

Pipe Side:

Since the drillpipe and drillcollar has the same id 3", pressure loss is calculated the same way as the pressure loss inside the string.

$$\text{Hedstrom number} = \frac{37,100 \times 13 \times 10 \times 9}{225} = 1,92,920$$

Hydraulics

Critical Re. number is aprox. 9000

$$\text{Re. number} = \frac{928 \times 13 \times 12 \times 3}{15} = 29{,}147$$

Re, number is lower than Hedstrom number and so the flow is turbulent.

$$\Delta p_p = \frac{13^{0.75} \times 12.08^{1.75} \times 15^{0.25}}{1800 \times 3^{1.25}} 19{,}250 = 2857 \text{ psi}$$

Annulus side has to account for the change in the annulus profile.

Annulus Side:

There are three annulus sections.

Surface to casing shoe depth i.e., 13,000 ft. Since casing size is not explicitly given it can be assumed that the bit size is 8.875" (bit size). So annulus OD is 8.875" and the pipe OD is 5", length = 13,000 ft.

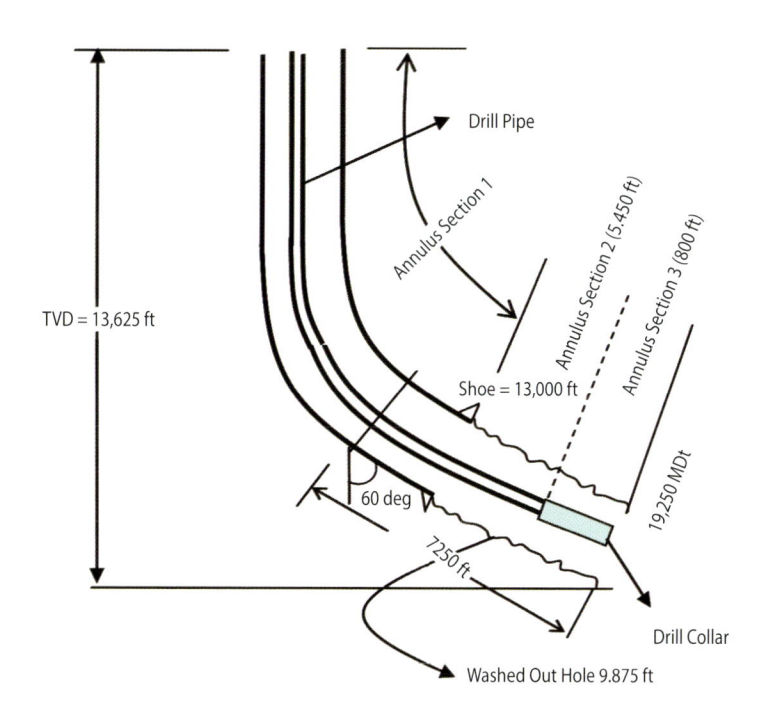

Figure 7.14   Well Schematic

Calculate Hedstrom and Reynolds number, flow will be laminar, pressure loss = 194 psi.

Second annulus section is shoe depth to the top of the drillcollar (i.e., drillpipe section). Since wash out is assumed, annulus od is 9.875", drillpipe od is 5", and length is 5450 ft.

Calculate Hedstrom and Reynolds number, flow will be laminar, pressure loss = 61 psi.

Third section is the drillcollar section in the openhole. Since wash out is assumed, annulus OD is 9.875", drillpipe OD is 7.5", length = 800 ft.

Calculate Hedstrom and Reynolds number, flow will be laminar, pressure loss = 22 psi.

Total pressure loss
= 194 + 61 + 22 + 2857 = 3134 psi; maximum pump pressure of 4300 psi

ECD at Minimum Flowrate

$$ECD = 13 + \frac{194 + 61 + 22}{0.052 \times TVD}$$

Using trig. relation between TVD and MD at 19,250 ft is 13,625 ft.
So ECD = 13.391 ppg

Maximum Flowrate:

With the given data the fracture pressure, i.e., $0.77 \times 13,625 = 10,491$ psi.

13 ppg mud is acting at the bottom and the hydrostatic pressure at the bottom is

$13 \times 0.052 \times 13,625 = 9.210$ psi.

So the extra annulus pressure that can withstand is
1281 psi (10,491 − 9210)

So we have to find out the flowrate that can cause 1281 psi pressure loss in the annulus.

When calculating the drillpipe, pressure also increases.

There is another condition given in the problem that total pressure loss should not be more than 4.300 psi.

Either using goal seek in excel or using iteration, the flowrate that can satisfy these conditions can be calculated and will be approximately 325 gpm.

CHAPTER 8

DRILLBIT HYDRAULICS

This chapter focuses on the calculations related to drillbit pressure losses, nozzle selection, and optimization of nozzle pressure losses and size.

298. Problem 8.1

Bit Pressure Drop Calculation

It is desired to place an orifice above the mud motor shaft so that flow can be diverted through the shaft. Derive an equation starting from Bernoulli's principle for the pressure drop across the orifice plate. Assume steady-state, incompressible laminar flow. Also neglect the frictional pressure losses due to the thickness of the orifice plate.

Solution:

Based on Bernoulli's principle and by ignoring the friction pressure loss

$$P_1 + \rho g h_1 + \frac{\rho v_1^2}{2} = P_2 + \rho g h_2 + \frac{\rho v_2^2}{2}$$

Since the thickness of the plate is small and the hydrostatic pressure difference can be neglected $\rho g h_1 \approx \rho g h_2$, the pressure drop across the orifice plate is

$$\Delta P = P_1 - P_2 \approx \frac{\rho}{2}(v_2^2 - v_1^2)$$

Now, $v_2 \gg v_1$ hence we can write

$$\Delta P = P_1 - P_2 \approx \frac{\rho}{2} v_2^2$$

i.e., $v_2 = \sqrt{\dfrac{2\Delta p}{\rho}}$

299. Problem 8.2

Pump and Bit Pressure Drop Calculations

Pump pressure can be given as

$$P_p = \Delta p_b + P_d \text{ psi}$$

From the bit pressure drop can be calculated

$$\Delta p_b = P_p - P_d \text{ psi,}$$

where
Δp_b = bit pressure drop, psi.
P_d = frictional pressure losses, psi.
Hydraulic horsepower (HHP) of the bit is

$$\frac{Q \times \Delta p_b}{1714} \text{ hp,}$$

where
Δp_b = bit pressure drop, psi.
Q = flowrate, gpm.

300. Problem 8.3
Basic Calculations

Total nozzle flow area is calculated as

$$A_n = \frac{\pi \times S_n^2}{64} \text{ in}^2$$

Nozzle velocity is calculated as

$$V_n = 0.3208 \frac{Q}{A_n} \text{ ft/sec}$$

The pressure drop across bit is calculated as

$$\Delta p_b = \frac{8.311 \times 10^{-5} \rho_m Q^2}{C_d^2 A_n^2} \text{ psi,}$$

where
C_d = discharge coefficient (usually 0.95 used).

The pressure drop across bit can also be calculated with 0.95 discharge coefficient as

$$\Delta p_b = \frac{\rho_m \times Q^2}{10,858 \times A_n^2} \text{ psi}$$

The pressure drop across bit is calculated with nozzle velocity as

$$\Delta p_b = \frac{\rho_m \times V_n^2}{1120} \text{ psi}$$

The percentage of pressure drop across bit is calculated as

$$\frac{\Delta p_b}{\Delta p} \times 100$$

The bit hydraulic power (BHHP) is calculated as

$$HP_b = \frac{Q \times \Delta p_b}{1714} \text{ hp}$$

Hydraulic bit (jet) impact force (IF) is calculated as

$$F_{imp} = 0.01823 \times C_d \times Q\sqrt{\rho_m \times \Delta p_b} \text{ lbf}$$

Hydraulic bit (jet) impact force (IF) can also be written as

$$F_{imp} = \frac{Q \times \sqrt{\rho_m \times \Delta p_b}}{57.66} \text{ lbf}$$

Hydraulic bit (jet) impact force (IF) with nozzle velocity is given as

$$F_{imp} = \frac{\rho_m \times Q \times V_n}{1930} \text{ lbf}$$

Impact force per square inch of the bit area is calculated as

$$\frac{F_{imp}}{\left(\frac{\pi \times D_b^2}{4}\right)} \text{ lbf/in}^2$$

Impact force per square inch of the hole area is calculated as

$$\frac{F_{imp}}{\left(\frac{\pi \times D_h^2}{4}\right)} \text{ lbf/in}^2$$

301. Problem 8.4

A trip is planned to run in with a tricone bit with 3 × 13 nozzles. The flowrate used is based on the minimum annular fluid velocity of 120 ft/min

Drillbit Hydraulics

against the 4.5 inch OD drillpipe inside 9.25 inch hole. Calculate the pressure drop across the bit for a mud density of 10 ppg.

Solution:

Flowrate needed based on the minimum annular fluid velocity

$$Q = v_{min} A_{adp} = 120 \frac{\pi}{4}(9.25^2 - 4.5^2) = 320 \text{ gpm}$$

Total area of the 3 nozzles

$$A_n = \left(3 \times \frac{\pi}{4}\left(\frac{13}{32}\right)^2\right) = 0.151215$$

Bit pressure drop: $\dfrac{8.31 \times 10^{-5} \times 10 \times 320^2}{0.95^2 \times (0.151215)^2} = 623 \text{ psi}$

302. Problem 8.5

A pump is capable of a pressure of 2000 psi at a circulation rate of 400 gpm. Frictional pressure loss is estimated to be 900 psi with 10 ppg mud. It is found that an increase in mud weight results in an increase of 55 bit HHP for the same flowrate. Estimate the new mud weight.

Solution:

The pump pressure can be given as

$$P_p = \Delta p_b + P_d,$$

from which the bit pressure drop can be written as

$$\Delta p_b = P_p - P_d$$

Substituting the values, the bit pressure drop is

$$\Delta p_b = 2000 - 900 = 1100 \text{ psi}$$

The hydraulic horsepower of the bit is

$$\frac{400 \times 1100}{1714} = 256.7 \text{ hp}$$

New HP = 256.7 + 55 = 311.7 hp

Bit pressure drop for the new hydraulic horsepower is

$$\Delta p_{b_1} = 311.7 \times \frac{1714}{400} = 1335.7 \text{ psi}$$

Using the relationship

$$\frac{\Delta p_{b_1}}{\Delta p_{b_2}} = \frac{P_1 Q_1^2}{P_2 Q_1^2} = 12.14 \text{ ppg}$$

$$\frac{1100}{1335.7} = \frac{10 \times 400^2}{P_2 \times 400^2}$$

Therefore,

$$P_2 = \frac{10 \times 400^2}{\left(\frac{1100}{1335.7}\right) \times 400^2} = 12.14 \text{ ppg}$$

303. Problem 8.6

HSI Calculation

Bit hydraulic horsepower per square inch of bit (HSI) is

$$HSI = \frac{HHP}{\frac{\pi}{4} \times D_b^2} \text{ HP/in.}^2$$

where
D_b = diameter of the bit, in.

Bit hydraulic horsepower per square inch of hole drilled (HSI) is

$$HSI = \frac{HHP}{\frac{\pi}{4} \times D_h^2} \text{ HP/in.}^2$$

where
D_h = diameter of the hole, in.

304. Problem 8.7

While drilling, the drilling engineer wants to increase the bit hydraulic horsepower at the bit to 1.1 times the existing HHP at the bit for the same mud weight. Calculate the flowrate needed to achieve his objective if the original flowrate is 400 gpm. Bit pressure drop is given as $\Delta P_b = \dfrac{K\rho Q^2}{A_n^2}$ psi where K is unit conversion constant and other variables in conventional oilfield units.

Solution:

$$HHP_1 = \frac{\Delta P_{b1} Q_1}{1714}$$

and new HP is

$$HHP_2 = \frac{\Delta P_{b2} Q_2}{1714}$$

But $0.9 \dfrac{\Delta P_{b1} Q_1}{1714} = \dfrac{\Delta P_{b2} Q_2}{1714}$

Substituting the bit pressure drop

$$0.9 \frac{\dfrac{K\rho Q_1^2}{A_n^2} Q_1}{1714} = \frac{\dfrac{K\rho Q_2^2}{A_n^2} Q_2}{1714}$$

which reduces to $0.9 Q_1^3 = Q_2^3$
So, $Q_2 = 0.965 Q_1$
Since Q_1 is 400 gpm
$Q_2 = 0.965 \times 400 = 386$ gpm

305. Problem 8.8

At a certain depth while drilling a 12 ¼" hole, the pump pressure is 3000 psi and parasitic pressure loss is 1500 psi at a circulation rate of 500 gpm with the mud weight of 12.5 ppg. Calculate the flowrate to achieve a bit HSI of 1.2 with the mud weight remaining the same. $C_d = 0.95$.

Solution:
Using the pump pressure drop equation
$P_p = \Delta p_b + P_d$, bit pressure drop can be calculated as
$\Delta p_b = 3000 - 1500 = 1500$ psi

$$\text{HSI} = \frac{\text{HHP}}{\frac{\pi}{4} \times 12.25^2} = 1.2$$

Substituting the equation for HHP and other respective values, it can be written as

$$1.2 = \frac{\Delta p_{b_1} \times Q_1}{\frac{1714}{\frac{\pi}{4} \times 12.25^2}} = \frac{\frac{1500 \times Q_1^2}{500^2} \times Q_1}{\frac{1714}{\frac{\pi}{4} \times 12.25^2}} = \frac{1500 \times Q_1^3}{500^2 \times 1714 \times \frac{\pi}{4} \times 12.25^2}$$

Solving for Q, flowrate will be 343 gpm.

306. Problem 8.9

At certain depth the pressure drop with 3 × 13 nozzles is estimated to be 1490 psi. Mud density 10 ppg. $C_d = 0.95$. Calculate the velocity of the mud in the annulus of 5 in. OD drillpipe inside 10 in. hole.
Solution:

Total nozzle area $A_n = \left(3 \times \frac{\pi}{4}\left(\frac{13}{32}\right)^2\right) = 0.151215$

Bit pressure drop: $\dfrac{8.31 \times 10^{-5} \times 10 \times Q^2}{0.95^2 \times (0.151215)^2} = 1490$ psi

Solving the above equation the flowrate is calculated to be
$Q = 350$ gpm

$$Q = vA_{adp} = \left(\frac{v}{144 \times 7.48}\right)\frac{\pi}{4}(10^2 - 5^2) = 350 \text{ gpm}$$

So, the velocity of the fluid in the annulus = 114.5 ft/min

Drillbit Hydraulics

307. Problem 8.10

At a certain depth during drilling, the pump pressure is 3000 psi, and parasitic pressure loss is 1500 psi at a circulation rate of 500 gpm. The mud weight is 12.5 ppg. Calculate the available bit hydraulic horsepower if the mud weight is increased to 14.5 ppg and the circulation rate remains same at 500 gpm.

Solution:

$$P_p = \Delta p_b + P_d$$

$$\Delta p_b = 3000 - 1500 = 1500 \text{ psi}$$

$$\Delta p_b = \frac{8.311 \times 10^{-5} \rho Q^2}{C_d^2 A_n^2}$$

$$\frac{\Delta p_{b_1}}{\Delta p_{b_2}} = \frac{\rho_1 Q_1^2}{\rho_2 Q_1^2}$$

$$\Delta p_{b_1} = 1500 \times \frac{14.5}{12.5} = 1740 \text{ psi}$$

$$\text{Bit HHP} = \frac{1740 \times 500}{1714} = 507 \text{ hp}$$

308. Problem 8.11

Optimization Calculations

Flow index is

$$m = \left(\frac{\log\left(\frac{(\Delta p_d)_j}{(\Delta p_d)_i} \right)}{\log\left(\frac{Q_j}{Q_i} \right)} \right)$$

For maximum bit hydraulic horsepower, the optimum bit pressure drop is

$$\Delta P_{bopt} = \frac{m}{m+1} \Delta p_{pmax}$$

and the optimum flowrate is

$$Q_{opt} = Q_a \times a\log\left[\frac{1}{m}\log\left(\frac{\Delta p_{d_{opt}}}{\Delta p_{d_{Q_a}}}\right)\right].$$

309. Problem 8.12
Limitation 1—Available Pump Horsepower
For maximum impact force, the optimum bit pressure drop is

$$\Delta p_{b_{opt}} = \frac{m+1}{m+2}\Delta p_{p_{opt}}$$

The optimum flowrate is

$$Q_{opt} = \left(\frac{2 \times \Delta p_{p\max}}{c(m+2)}\right)^{\frac{1}{m}}$$

310. Problem 8.13
Limitation 2—Surface Operating Pressure
For maximum impact force, the optimum bit pressure drop is

$$\Delta p_{b_{opt}} = \frac{m}{m+2}\Delta p_{p\max}$$

311. Problem 8.14

Estimate the optimum nozzle size and optimum flowrate for the following conditions:

- $m = 1.66$
- Maximum allowed operating pressure = 5440 psi
- Frictional pressure loss = 2334 psi for a flowrate of 300 gpm
- Volumetric efficiency of the pump = 80%
- Mud weight = 15.5 ppg
- Minimum flowrate required for holecleaning = 265 gpm

Use limitation condition 2 for the calculations.

Drillbit Hydraulics

Solution:

Using limitation condition 2,

$$\Delta p_{d\,opt} = \frac{2}{m+2}\Delta p_{p\,max}$$

$$\Delta p_{d\,opt} = \frac{2}{1.66+2} \times 5440 = 2972.7 \text{ psi}$$

$$\Delta p_{b\,opt} = \Delta p_{P\,max} - \Delta p_{d\,opt}$$

$$\Delta p_{b\,opt} = 5440 - 2975 = 2467.3 \text{ psi}$$

Using the corresponding equation for the optimized flowrate

$$Q_{opt} = Q_a \text{alog}\left[\frac{1}{m}\log\left(\frac{\Delta p_{d\,opt}}{\Delta p_{d\,Q_a}}\right)\right]$$

$$Q_{opt} = 300 a \log\left[\frac{1}{1.66}\log\left(\frac{2975}{2334}\right)\right] = 359 \text{ gpm}$$

$$\Delta p_{bopt} = \frac{8.3 \times 10^{-5} \times 15.5 \times 347^2}{0.95^2 \times A_{nopt}^2} = 2465 \text{ psi}$$

$$A_{opt} = 0.2728 \text{ in}^2$$

The average diameter of the nozzle can be calculated from the total optimum nozzle as

Three nozzles: 11-11-11.

312. Problem 8.15

Given the following friction pressure loss-flow rate relationship:

$$P_f = cQ^m,$$

where
P = friction pressure loss.
C = constant.
m = flow exponent.
Q = flow rate.

Find an expression to determine m if the following measurements are known:

$$P_f = P_{fi} \text{ at } Q = Q_i$$
$$P_f = P_{fij} \text{ at } Q = Q_j$$

Solution:

For $P_f = P_{fi}$ at $Q = Q_i$
$$P_{fi} = cQ_i^m$$
For $P_f = P_{fij}$ at $Q = Q_j$
$$P_{ji} = cQ_j^m$$
Dividing the equations and taking the logarithm

$$m = \frac{\log\left(\dfrac{P_j}{P_i}\right)}{\log\left(\dfrac{Q_j}{Q_i}\right)}$$

313. Problem 8.16

Calculate the optimal nozzle size using the maximum bit hydraulic horsepower criterion with the following data:

- Frictional pressure loss is $P_d = 2400$ psi for $Q = 350$ gpm
- $m = 1.86$
- Maximum allowed pump operating pressure = 5000 psi
- Pump hydraulic horsepower = 1600 hp
- Pump volumetric efficiency = 80%
- Minimum flowrate required for holecleaning = 200 gpm
- Mud density = 10 ppg

Solution:

$$p_{d_{opt}} = \left(\frac{1}{m+1}\right) p_{p_{max}} = \frac{1}{1.86+1} \times 5000 = 1748 \text{ psi}$$

$$\Delta p_{b_{opt}} = \frac{1.86}{2.86} \times 5000 = 3252 \text{ psi}$$

Drillbit Hydraulics

$$Q_{opt} = 350 \times a\log\left[\left(\frac{1}{1.86}\log\frac{1748}{2400}\right)\right] = 295 \text{ gpm}$$

$$Q_{max} = \frac{1714 \times 1500 \times 0.85}{5000} = 437 \text{ gpm}$$

Optimum flowrate check is

$$200 \text{ gpm} < Q_{opt} < 403 \text{ gpm}$$

From the pressure drop equation

$$\Delta p_b = \frac{8.311 \times 10^{-5} \rho_m Q^2}{C_d^2 A_n^2}$$

The area of the nozzle can be estimated as

$$A_{nopt}^2 = \frac{8.3 \times 10^{-5} \times 10 \times 295^2}{0.95^2 \times 3252} = 0.1568 \text{ in}^2$$

314. Problem 8.17

Calculate the optimal flowrate optimal nozzle area using the bit impact force criterion (limitation 2) with the following data:

- Maximum allowed pump operating pressure = 5000 psi
- Minimum flowrate required for holecleaning = 300 gpm
- Mud density = 10 ppg
- Pump pressures of 4000 psi and 2500 psi were recorded while circulating mud at 500 gpm and 387 gpm, respectively, with 11-11-11 nozzles.

Solution:

$$m = \frac{\log\left(\frac{p_{d2}}{p_{d1}}\right)}{\log\left(\frac{Q_2}{Q_1}\right)} = \frac{\log\left(\frac{2970}{1779}\right)}{\log\left(\frac{500}{387}\right)} = 1.80$$

Using the bit impact force criterion (limitation 2),

$$p_{dopt} = \left(\frac{2}{m+2}\right) p_{pmax} = \left(\frac{2}{1.8+2}\right) \times 5000 = 2630 \text{ psi}$$

Optimum bit pressure drop can be estimated as

$$p_{bopt} = p_{max} - p_{dopt} = 5000 - 2630 = 2369 \text{ psi}$$

Using the optimum values, the optimum flowrate can be calculated as

$$Q_{opt} = Q_a a \log\left[\frac{1}{m}\log\left(\frac{\Delta p_{dopt}}{\Delta p_{da}}\right)\right] = 387 \times a \log\left[\frac{1}{1.8}\log\left(\frac{2630}{1779}\right)\right] = 435 \text{ gpm}$$

With the pressure drop equation and using the optimum flowrate, the corresponding optimum nozzle area can be calculated as

$$\Delta p_{b_{opt}} = \frac{8.3 \times 10^{-5} \times 10 \times 435^2}{0.95^2 \times A_n^2} = 2369 \text{ psi}$$

$A_n^2 = 0.07337$ in^4, and the total nozzle area is

$$A_n = \sqrt{0.07337} = 0.2708 \text{ in.}^2$$

315. Problem 8.18

Calculate the maximum flowrate at which the driller can pump if the maximum hydraulic horsepower is 1500 hp. The maximum bit hydraulic horsepower criterion yields an optimum flowrate of 350 gpm. Frictional pressure loss is $p_d = 1975$ psi for $Q = 300$ gpm. Calculate the HSI at optimum condition for a 12¼-in. hole. Use $m = 1.6$ and $C_d = 0.95$

Solution:

Using the maximum bit hydraulic horsepower criterion

$$\Delta p_{d_{opt}} = 1975 \times a \log\left[1.6\log\left(\frac{350}{300}\right)\right] = 2527 \text{ psi}$$

$$p_{d_{opt}}(m+1) = p_{p_{max}} = (1.6+1) \times 2527 = 6570 \text{ psi}$$

$$\Delta p_{d_{opt}} = 1975 \times a \log\left[1.6\log\left(\frac{350}{300}\right)\right] = 2527 \text{ psi}$$

$$Q_{max} = \frac{1714 \times 1500}{6570} = 391 \text{ gpm}$$

The optimum bit pressure drop is

$$\Delta P_{b_{opt}} = 6570 - 2527 = 4044 \text{ psi}$$

The optimum HSI is

$$HSI_{bit} = \frac{Q_{opt} \times \Delta P_{b_{opt}}}{1714 \times \frac{\pi}{4} d_b^2} = \frac{350 \times 4044}{1714 \times \frac{\pi}{4} 12.25^2} = 7 \text{ hp/in.}^2$$

316. Problem 8.19

A bit currently has 3 × 12 nozzles. The driller has recorded that when 10 ppg mud is pumped at a rate of 500 gpm, a pump pressure of 3000 psi is observed. When the pump is slowed to a rate of 250 gpm, a pump pressure of 800 psi is observed. The pump is rated at 2000 hp and has an overall efficiency of 90%. The minimum flowrate to lift the cuttings is 240 gpm. The maximum allowable surface pressure is 5000 psi.

A. Determine the pump operating conditions and bit nozzle sizes for maximum bit horsepower for the next bit run.
B. What bit horsepower will be obtained at the conditions selected?

Solution:
A.
Bit pressure drop is 2097 psi for 500 gpm. Therefore, the frictional pressure losses can be calculated using the following equation:

$$P_{fl} = P_p - P_b = 3000 - 2097 = 903 \text{ psi}$$

The flow index m can be found as

$$m = \frac{\log \frac{903}{275.5}}{\log \frac{500}{250}} = 1.71$$

$$Q_{max} = 1714 \times 0.9 \times \frac{2000}{5000} = 617 \text{ gpm}$$

Find the optimum friction pressure loss using the maximum hydraulic horsepower criterion:

$$P_{fopt} = \frac{1}{m+1} P_{pmax} = 1845 \text{ psi}$$

$$P_{bopt} = 5000 - 1845 = 3155 \text{ psi}$$

$$Q_{opt} = 500 \, a \log\left[\frac{1}{1.71} \log \frac{1845}{900}\right] = 761 \text{ gpm}$$

The optimum flowrate calculated should be between the minimum flowrate to clean the hole and maximum pump capacity.

Lowest limit and is 240 gpm.

Since the maximum flowrate 617 gpm, optimum flowrate has to be adjusted to 617 gpm.

Using this 617 gpm the calculated nozzle sizes are, 12-12-12.

B.

The nozzle pressure drop can be calculated as

$$\Delta p_b = \frac{8.3 \times 10^{-5} \rho \times Q^2}{C_d^2 \times A_n^2} = \frac{8.3 \times 10^{-5} \times 10 \times 617^2}{0.95^2 \times 0.3313^2} = 3189 \text{ psi}$$

The bit hydraulic horsepower is

$$HHP_{bit} = \frac{Q \Delta P_b}{1714} = \frac{617 \times 3189}{1714} = 1148 \text{ hp}$$

317. Problem 8.20

Using the data from Problem 316, calculate the optimum nozzle sizes for the maximum impact force (limitation 2) criteria for the next bit run.
Solution:

If the bit pressure drop is 2097 psi for 500 gpm, then the frictional pressure losses can be calculated using the following equation:

$$P_{fl} = P_p - \Delta p_b = 3000 - 2097 = 903 \text{ psi}$$

Drillbit Hydraulics

The flow index m can be obtained as

$$m = \frac{\log \frac{903}{275.5}}{\log \frac{500}{250}} = 1.71$$

$$Q_{max} = 1714 \times 0.9 \times \frac{2000}{5000} = 617 \text{ gpm}$$

$$Q_{min} = 240 \text{ gpm}$$

Find the optimum friction pressure loss using the maximum hydraulic horsepower criterion:

$$P_{fopt} = \frac{2}{m+2} P_{pmax} = \frac{2}{3.71} \times 5000 = 2695.4 \text{ psi}$$

$$\Delta P_{bopt} = 5000 - 2695.4 = 2304.6 \text{ psi}$$

$$Q_{opt} = 500a \log \left[\frac{1}{1.71} \log \frac{2695.4}{900} \right] = 950 \text{ gpm}$$

Since the calculated 950 gpm is greater than the maximum flowrate, the optimum flowrate is 617 gpm. This flowrate is greater than the required minimum flowrate.

Using this flowrate, the optimum nozzle area can be found as

$$A_n = \sqrt{\frac{8.3 \times 10^{-5} \rho \times Q^2}{C_d^2 \times \Delta P_b}} = \sqrt{\frac{8.3 \times 10^{-5} \times 10 \times 617^2}{0.95^2 \times 2304}} = 0.39 \text{ in.}^2$$

Nozzle sizes can be 13-13-13.

318. Problem 8.21

At certain depth during drilling, pump pressure is 3000 psi and parasitic pressure loss is 1500 psi at a circulation rate of 500 gpm. The mud weight is 12.5 ppg. Calculate the available bit hydraulic horsepower if the mud weight is increased to 14.5 ppg and the circulation rate remains same at 500 gpm.

$$\text{BHHP} = \frac{Q \Delta p}{1714}$$

Solution:

$P_p = P_b + P_d$; $P_b = 3000 - 1500 = 1500$ psi

The bit pressure drop is given as

$$\Delta P_b = \frac{8.311 \times 10^{-5} \rho Q^2}{C_d^2 A_n^2}$$

So, the ratio of the bit pressure drops for two different flowrate and densities for the same nozzle size can be given as

$$\frac{\Delta P_{b_1}}{\Delta P_{b_2}} = \frac{\rho_1 Q_1^2}{\rho_2 Q_1^2}$$

Substituting the values for the densities and the bit pressure drop is calculated as

$$\Delta P_{b_1} = 1500 \times \frac{14.5}{12.5} = 1740 \text{ psi}$$

$$\text{BHHP} = \frac{1740 \times 500}{1714} = 507 \text{ hp}$$

319. Problem 8.22

Determine the optimum nozzle sizes to be used for the next depth interval using the bit hydraulic horsepower criterion, from the data given below. Calculate the HSI at the optimum flowrate.

Pump: National Duplex (double acting) – Maximum allowed operating pressure = 5440 psi Hydraulic horsepower = 1600 hp

Volumetric efficiency = 80%

Drilled cuttings: Average minimum required cuttings rise velocity for annular holecleaning = 65 ft/min

Mud density: 15.5 ppg

8 ⅞" Tri-cone Roller Cone Jet Bit with 14-14-14 nozzles to be used to drill the next phase starting from 12,000 ft

Field data: At 12,000 ft while using 8 ⅞" bit with 3 × 14 nozzles:

$Q_1 = 300$ gpm; $P_{p1} = 2966$ psi
$Q_1 = 400$ gpm; $P_{p1} = 4883$ psi

Solution:

a. Pressure drop through the bit at flow rates 300 and 400 gpm are 631 psi and 1123 psi, respectively. Then the frictional pressure losses can be calculated using the equation

$$P_{fl} = P_p - P_b = 2966 - 632 = 2334 \text{ psi}$$
$$P_{fl} = P_p - P_b = 4883 - 1123 = 3760 \text{ psi}$$

Find m

$$m = \frac{\log \frac{3760.2}{2334.4}}{\log \frac{400}{300}} = 1.657$$

Find optimum friction pressure loss using the maximum hydraulic horsepower criterion;

$$P_{fopt} = \frac{1}{m+1} P_{pmax} = 2047 \text{ psi}$$
$$P_{bopt} = 5440 - 2047 = 3393 \text{ psi}$$
$$Q_{opt} = 300 \, a \log \left[\frac{1}{1657} \log \frac{2047}{2334} \right] = 277 \text{ gpm}$$

Check whether optimum is within min and max flowrate. Assume 4.5" drillpipe.

$$Q_{max} = 1714 \times 0.8 \times \frac{1600}{5440} = 403 \text{ gpm}$$
$$Q_{min} = 2.448 \left(d_h^2 - d_p^2 \right) \frac{65}{60} = 268 \text{ gpm}$$

Optimum area of the nozzle

$$A_{nopt} = \sqrt{\frac{8.311 \times 10^{-5} \times 15.5 \times 277^2}{0.95^2 \times 3393}} = 0.1797 \text{ in.}^2$$

This area is close to 9-9-9 nozzles:
b. HSI at the optimum flow rate

$$\text{HSI} = \frac{\text{HHP}}{A_b} = \frac{549}{\frac{\pi}{4} \times 8.875^2} = 8.87$$

320. Problem 8.23

Calculate the optimal flowrate and optimal nozzle size using the maximum bit hydraulic horsepower criterion with the following data:

frictional pressure loss is P_d = 2334 psi for Q = 300 gpm
$m = 1.66$
maximum allowed pump operating pressure = 5440 psi
pump hydraulic horsepower = 1600
pump volumetric efficiency = 80%
Minimum flowrate required for holecleaning = 200 gpm

Solution:

$$P_{d_{opt}} = \left(\frac{1}{m+1}\right) P_{p_{max}} = \frac{1}{1.66+1} \times 5440 = 2045 \text{ psi}$$

$$Q_{opt} = 300 \times a\log\left[\left(\frac{1}{1.66}\log\frac{2045}{2334}\right)\right] = 277 \text{ gpm}$$

$$Q_{max} = \frac{1714 \times 1600 \times 0.8}{5440} = 403 \text{ gpm}$$

$$200 \text{ gpm} < Q_{opt} < 403 \text{ gpm}$$

321. Problem 8.24

Use the following data for the drillpipe. Tooljoint effect for heavy weight drillpipe (HWDP) may be neglected.
Tooljoint OD = 8"
Tooljoint ID = 4.25"
Inside and outside tooljoint length = 1.48 ft

Drillbit Hydraulics

Calculate the new ECD and bottom hole pressure with tooljoints. Estimate the following for the bit:

a. Bit Impact Force
b. Bit Hydraulic Power
c. Percent Power at Bit
d. HSI
e. Bit Nozzle Velocity

With the following constraints, calculate the optimum nozzle sizes and flowrate for the HHP criterion.

a. Maximum Surface Pressure = 5000 psi
b. Maximum Pump Power = 2200 hp
c. Minimum Annular Velocity = 70 ft/min
d. Constant Slope of Hydraulics = 1.85

Solution:

Inside Pipe:
Internal pressure loss due to tooljoint is calculated as 0.45 psi/ft.
One stand of 90 ft will have 3 tooljoints.
Total number of tooljoints = 16,350/31 = 528
Total length of tooljoint is = 528 × 1.48 = 781 ft
Calculating the velocity, Reynolds's number, and the coefficient K_L as below:
N_{Re} = 3404 and since it is >3000 and <13,000, K_L = 0.95
Total tooljoint pressure loss = 0.45 × 528 = 237 psi

Outside Pipe:
External pressure loss due to tooljoint is calculated using the outside geometry of the tooljoint and the total tooljoint lengths are lumped together.
Lumping the tooljoints length and calculating the length of drill pipe for outside pressure loss calculation

$$16{,}350 - 781 = 15{,}569 \text{ ft}$$

Total annulus pressure loss = 212 psi

$$\text{ECD} = \rho_m + \frac{\Delta P_{fa}}{0.052 \times \text{TVD}} = 11.9 + \frac{212}{0.052 \times 14{,}533} = 12.2 \text{ ppg}$$

Bit Nozzle Velocity is calculated as

$$V_n = \frac{600}{2.448 \times 3\left(\frac{12}{32}\right)^2} = 581 \text{ fps}$$

Bit Impact Force $F_{bit} = 0.000516\,(1-53.4^{-0.122}) \times 11.9 \times 600 \times 581 = 823$ lbf

Bit Hydraulic Power $\text{BHHP} = \dfrac{3593 \times 600}{1714} = 1258$ hp

Percentage of power lost at the bit $= \dfrac{3593}{4850} = 75\%$

$$\text{HSI} = \frac{1258}{\dfrac{\pi}{4} \times 12.25^2} = 10.7 \text{ hp/in.}^2$$

Optimization:

Calculate the minimum annular flowrate based on min. annular velocity of 70 fpm for holecleaning against the drillpipe.

$$Q_{min} = 2.448 \times (12.25^2 - 6.625^2) \times 70/60 = 303 \text{ gpm}$$

Calculate the maximum flowrate for the high limit:

$$\frac{1714 \times 2200 \times 0.9}{5000} = 680 \text{ gpm}$$

Using the maximum HHP criterion:

$$\Delta P_{bopt} = \frac{m}{m+1} \Delta P_{max} = \frac{1.85}{1.85+1} \times 5000 = 3246 \text{ psi}$$

$$\Delta P_{dopt} = \Delta P_{max} - \Delta P_{dopt} = 5000 - 3246 = 1754 \text{ psi}$$

$$Q_{opt} = 600 \times \text{anti}\log\left(\frac{1}{1.85}\log\left(\frac{1754}{1260}\right)\right) = 717 \text{ gpm}$$

Since the calculated optimum flowrate is greater than the maximum allowed pump flowrate, the maximum pump flowrate becomes the optimum flowrate.

Using this optimum flowrate the optimum pressure drop across the bit can be calculated and the optimum nozzle size can be estimated to be 13-13-13.

322. Problem 8.25

Calculate the optimal nozzle size using the maximum bit hydraulic horsepower criterion with the following data:

Maximum permitted surface pressure = 3500 psi
Mud density = 13 ppg
The parasitic pressure loss = $0.001Q^{1.86}$ psi, where Q is flowrate in gpm.

Solution:

Optimum bit pressure drop is given as

$$\Delta p_b = \frac{m}{m+1} P_{p\max} = \frac{1.86}{1+1.86} \times 3500 = 2276.22 \text{ psi}$$

$$\Delta p_d = P_{p\max} - \Delta P_b = 3500 - 2276 = 1224$$

Since parasitic pressure loss is given as

$$\Delta p_d = 0.001 Q^{1.86} = 1224$$

Flow rate is 1875 gpm

$$P_b = \frac{8.311 \times 10^{-5} \times \rho \times Q^2}{C_d^2 \times A_n^2}$$

Assuming a coefficient of discharge: $C_d = 0.95$

$$A_n = \sqrt{\frac{8.311 \times 10^{-5} \times 13 \times 1875^2}{0.95^2 \times 2276}} = 1.35 \text{ in.}^2$$

323. Problem 8.26

An ECD of 14 ppg is calculated while drilling a vertical well of 12,000 ft using a mud density of 12 ppg with a flowrate of 550 gpm. Standpipe pressure observed is 3500 psi. Frictional pressure loss gradient inside the drillstring is 0.04821 psi/ft. Pressure drop of the mud motor in the drillstring is $\Delta p_{mm} = -0.002645Q^2 + 1.9457Q - 18.26$ psi for a reference fluid density of 8.33 ppg, where Q is in gpm. Estimate the total bit nozzle area. You may neglect the length of the motor and assume bit discharge coefficient $C_d = 0.95$.

Solution:

The bottomhole pressure is given by

$$P_{bh} = 0.052 \times \rho \times 12{,}000 + 3500 - 0.04821 \times 12{,}000$$

$$- \left(0.002645 Q^2 + 1.9457 Q - 18.26\right)\left(\frac{\rho_m}{\rho_r}\right) - \Delta p_b,$$

which can be written in terms of ECD as

$$\text{ECD} \times 0.052 \times 12{,}000 = 0.052 \times \rho \times 12{,}000 + 3500 - 0.04821 \times 12{,}000$$

$$- \left(-0.002645 Q^2 + 1.9457 Q - 18.26\right)\left(\frac{\rho_m}{\rho_r}\right) - \Delta p_b$$

$$14 \times 0.052 \times 12{,}000 = 12 \times 0.052 \times 12{,}000 + 3500 - 0.04821 \times 12{,}000$$

$$- \left(-0.002645 \times 550^2 + 1.9457 \cdot 550 - 18.26\right)\left(\frac{12}{8.33}\right) - \Delta p_b$$

$$8736 = 7488 + 3500 - 578.52 - 362.68 - \Delta p_b$$

So pressure drop is
$\Delta p_b = 1310$ psi

$$A_n = \sqrt{\frac{8.311 \times 10^{-5} \times 12 \times 550^2}{0.95^2 \times 1310}} = 0.2785$$

324. Problem 8.27

Nozzle Selection with Different Size, Calculations

Table 8.1 Nozzle selection

Nozzle size	Area of one nozzle (sq.in)	Area of two nozzles (sq.in)	Area of three nozzles (sq.in)
7	0.0376	0.0752	0.1127
8	0.0491	0.0982	0.1473
9	0.0621	0.1243	0.1864
10	0.0767	0.1534	0.2301
11	0.0928	0.1856	0.2784
12	0.1104	0.2209	0.3313

13	0.1296	0.2592	0.3889
14	0.1503	0.3007	0.451
15	0.1726	0.3451	0.5177
16	0.1963	0.3927	0.589
18	0.2485	0.497	0.7455
20	0.3068	0.6136	0.9204
22	0.3712	0.7424	1.1137
24	0.4418	0.8836	1.3254
26	0.5185	1.037	1.5555
28	0.6013	1.2026	1.804
30	0.6903	1.3806	2.0709
32	0.7854	1.5708	2.3562

325. Problem 8.28

Nozzle Area with Different Nozzle Sizes, Calculations

Table 8.2 Nozzle area calculation

Nozzle size	Nozzle size	Nozzle size	Total Flow Area (TFA) (sq.in)
7	7	7	0.1127
7	7	8	0.1243
7	8	8	0.1358
8	8	8	0.1473
8	8	9	0.1603
8	9	9	0.1733
9	9	9	0.1864
9	9	10	0.201
9	10	10	0.2155
10	10	10	0.2301
10	10	11	0.2462
10	11	11	0.2623
11	11	11	0.2784
11	11	12	0.2961
11	12	12	0.3137
12	12	12	0.3313
12	12	13	0.3505

12	13	13	0.3697
13	13	13	0.3889
13	13	14	0.4096
13	14	14	0.4303
14	14	14	0.451
14	14	15	0.4732
14	15	15	0.4955
15	15	15	0.5177
15	15	16	0.5415
15	16	16	0.5653
16	16	16	0.589
16	16	18	0.6412
16	18	18	0.6934
18	18	18	0.7455
18	18	20	0.8038
18	20	20	0.8621
20	20	20	0.9204
20	20	22	0.9848
20	22	22	1.0492
22	22	22	1.1137
22	22	24	1.1842
22	24	24	1.2548
24	24	24	1.3254
24	24	26	1.4021
24	26	26	1.4788
26	26	26	1.5555
26	26	28	1.6383
26	28	28	1.7211
28	28	28	1.804
28	28	30	1.8929
28	30	30	1.9819
30	30	30	2.0709
30	30	32	2.166
30	32	32	2.2611
32	32	32	2.3562

CHAPTER 9

DRILLING TOOLS

This chapter focuses on the different basic calculations using various downhole drilling tools.

326. Problem 9.1

Diameter of the hole: 12.25 in.
Bit nozzles: 3 × 12
Weight on bit: 20 kips
Bit rotation: 120 RPM
Coefficient of friction: 0.2
Rate of penetration: 20 fph
Torque: 1333 ft lbf
Flowrate: 400 gpm
Drilling fluid density: 10 ppg
Neglect the impact force.

Calculate the mechanical specific energy as well as hydromechanical specific energy.

Solution:

Hydromechanical specific energy is given as

$$E_s = \frac{W - \eta F_j}{A} + \frac{120\pi NT + 1154\eta \Delta p_b Q}{AR},$$

where
  W = weight on bit (lbs)
  N = revolutions in rpm
  T = torque (ft-lbf)
  Q = flowrate in gpm
  A = cross sectional area of the hole drilled, sq.in
  R = rate of penetration in ft/hr
  ΔP_b = pressure drop through the bit in psi
  F_j = jet impact force, lbf
  η = hydraulic jet efficiency

$$A = \frac{\pi}{4} \times 12.25^2 = 117.85 \text{ sq.in}$$

Since there is no flowrate the mechanical specific energy is given as

$$E_s = \frac{W}{A} + \frac{120\pi NT}{AR}$$

Drilling Tools

Substituting the values the Mechanical specific energy

$$E_s = \frac{20{,}000}{117.85} + \frac{2\pi \times 120 \times 1333 \times 60}{117.85 \times 20} = 25{,}752 \text{ psi}$$

Assume efficiency is 100%
Nozzle pressure drop

$$\Delta p_b = \frac{8.311 \times 10^{-5} \times 10 \times 400^2}{0.95^2 \times 0.33134^2} = 1342 \text{ psi}$$

$$E_s = \frac{W - 1 \times 8025}{117.85} + \frac{120\pi NT + 1154 \times 1 \times 1342 \times 400}{117.85 \times 20} = 28{,}8568 \text{ psi}$$

327. Problem 9.2

Drilling efficiency is given by $\eta_d = \dfrac{E_{S\min}}{E_S}$ where $E_{S\min}$ is roughly equal to the compressive strength of the formation being drilled and E_S is the mechanical specific energy.

A drilling engineer wants to achieve a drilling efficiency of 50% while drilling a 12 ¼" hole with at a certain depth with the formation compressive strength of 22.5 kpsi.

Weight on bit = 10 kips
Total footage drilled = 22.5 ft in 45 min
Bit Torque = 2100 ft-lbf

Solution:

$$E_S = \frac{22.5}{0.5} = 45{,}000 \text{ psi}$$

$$\text{ROP} = \frac{22.5}{45} \times 60 = 30 \text{ fph}$$

$$A_b = \frac{\pi}{4} \times 12.25^2 = 117.8 \text{ in}^2$$

$$45{,}000 = \frac{10{,}000}{117.8} + \frac{2\pi \times 60 \times 2100}{117.8 \times 30} N = 84.9 + 223.9N$$

Solving N = 200 RPM

328. Problem 9.3

Diameter of the hole: 12.25 in.
Bit nozzles: 3 × 12
Weight on bit: 20 kips
Bit rotation: 120 RPM
Coefficient of friction: 0.2
Rate of penetration: 20 fph
Torque: 1333 ft-lbf
Flowrate: 400 gpm
Fluid weight: 10 ppg

With the given data above calculate the hydromechanical specific energy.

Solution:

$$E_s = \frac{20,000}{\frac{\pi}{4} \times 12.25^2} + \frac{2\pi \times 120 \times 1333}{\frac{\pi}{4} \times 12.25^2 \times 20} = 25,752 \text{ psi}$$

Nozzle pressure drop

$$\Delta P_b = \frac{8.311 \times 10^{-5} 10 \times 400^2}{0.95^2 \left[\frac{3\pi}{4}\left(\frac{12}{32}\right)^2\right]^2} = 1342 \text{ psi}$$

Effective weight on bit

$$W_{eff} = 20,000 - \frac{400}{58}\sqrt{10 \times 1342} = 19,201 - \text{lbf}$$

The hydro-mechanical specific energy can be calculated

$$E_s = \frac{19,021}{\frac{\pi}{4} \times 12.25^2} + \frac{120\pi \times 120 \times 1333}{\frac{\pi}{4} \times 12.25^2 \times 20} + \frac{400 \times 1342}{\frac{\pi}{4} \times 12.25^2 \times 20} = 25,819 \text{ psi}$$

329. Problem 9.4

An ECD of 14 ppg is calculated while drilling a vertical well of 12,000 ft using a mud density of 12 ppg with a flowrate of 550 gpm. Standpipe pressure observed is 3500 psi. Frictional pressure loss gradient inside the drillstring is 0.04821 psi/ft. Pressure drop of the mud motor in the drillstring is $\Delta P_{mm} = -0.002645Q^2 + 1.9457Q - 18.26$ psi for a reference fluid density of 8.33 ppg where Q is in gpm. Estimate the total bit nozzle area. You may neglect the length of the motor and assume bit discharge coefficient $C_d = 0.95$.

Solution:

The bottomhole pressure is given by

$$P_{bh} = 0.052 \times \rho \times 12,000 + 3500 - 0.04821 \times 12,000$$

$$-\left(0.002645Q^2 + 1.9457Q - 18.26\right)\left(\frac{\rho_m}{\rho_r}\right) - \Delta p_b$$

Which can be written in terms of ECD as

$$ECD \times 0.052 \times 12,000 = 0.052 \times \rho \times 12,000 + 3500 - 0.04821 \times 12,000$$

$$-\left(-0.002645Q^2 + 1.9457Q - 18.26\right)\left(\frac{\rho_m}{\rho_r}\right) - \Delta p_b$$

$$14 \times 0.052 \times 12,000 = 12 \times 0.052 \times 12,000 + 3500 - 0.04821 \times 12,000$$

$$-\left(-0.002645 \times 550^2 + 1.9457 \cdot 550 - 18.26\right)\left(\frac{12}{8.33}\right) - \Delta p_b$$

$$8736 = 7488 + 3500 - 578.52 - 362.68 - \Delta P_b$$

So pressure drop is

$$\Delta P_b = 1310 \text{ psi}$$

$$A_n = \sqrt{\frac{8.311 \times 10^{-5} \times 12 \cdot 550^2}{0.95^2 \times 1310}} = 0.2785$$

330. Problem 9.5

A drilling engineer is planning to place a narrow passage in a MWD tool sub to have an orifice effect as shown in Figure 9.1 with respective

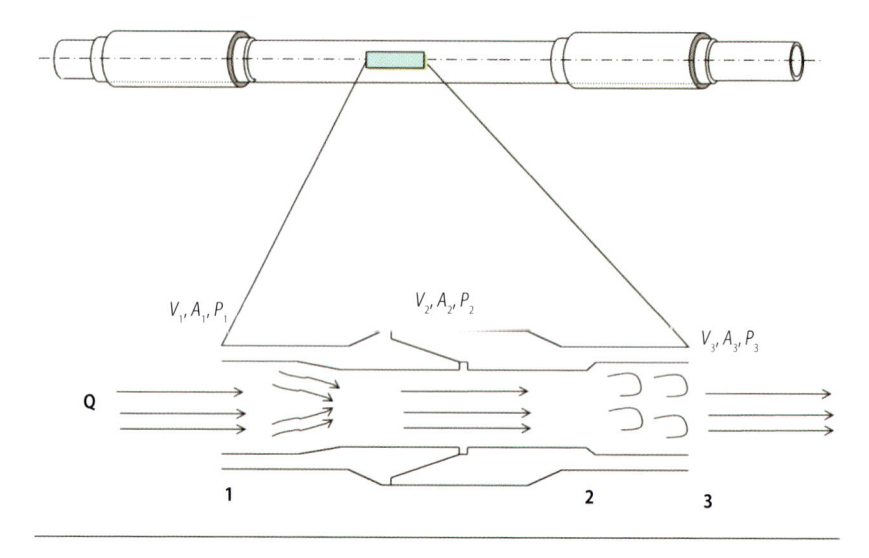

Figure 9.1 Pressures across the tooljoint

velocity, area, and pressure. Point 3 is where there is uniform flow in the downstream. You may neglect entry and exit losses.

Prove that

$$\Delta P = P_1 - P_3 = \frac{1}{2}\rho V_1^2 \left(\frac{A_1}{A_2} - 1\right)^2$$

Neglect the frictional pressure losses in the narrow region.

$$\Delta P = P_1 - P_3 = (P_1 - P_2) + (P_2 - P_3)$$

Using Bernoulli's principle

$$P_1 + \frac{1}{2}\rho V_1^2 = P_2 + \frac{1}{2}\rho V_2^2$$

$$P_1 - P_2 = \frac{1}{2}\rho\left(V_2^2 - V_1^2\right)$$

Using continuity

$$P_1 - P_2 = \frac{1}{2}\rho V_1^2 \left[\left(\frac{A_1}{A_2}\right)^2 - 1\right]$$

Change of system momentum is equated to the sum of forces acting on the mass inside the control volume between 2 and 3.

Momentum flux out is

$$P_2 A_3 - P_3 A_3 = A_3 (P_2 - P_3) = \rho A_3 V_3^2 - \rho A_2 V_2^2$$

Rewriting the above equation

$$A_3 (P_2 - P_3) = \rho A_3 V_2 \frac{A_2^2}{A_3^2} - \rho A_2 V_2^2 = \rho A_2 V_2^2 \left(\frac{A_2}{A_3} - 1 \right)$$

$$(P_2 - P_3) = \rho V_2^2 \frac{A_2}{A_3} \left(\frac{A_2}{A_3} - 1 \right)$$

Combining will result in

$$\Delta P = P_1 - P_3 = \frac{1}{2} \rho V_1^2 \left(\frac{A_1}{A_2} - 1 \right)^2$$

331. Problem 9.6

A drilling engineer desires to measure the temperature generated at the bit during drilling by embedding a small square measurement device at the surface of the bit with a dimension a mm and a heat transfer coefficient h in W/m²K. He desires to minimize the heat loss from the device to the surrounding. The heat transfer coefficient is given as

$$h = \frac{(2 + 12.5\sqrt{a}) \Delta T^{0.25}}{a},$$

where

ΔT = temperature difference in K deg.

From the laboratory result it is found that the bond strength between the device and the surface of the bit to be 6.1, which is a product of the dimension of the device and the temperature difference.

With the constraint being the strength of the bond, calculate the dimension a that would minimize the heat loss. Use both constrained and unconstrained methods.

Objective function is the heat loss which is
$$Q = hA\Delta T.$$

$$\text{So } Q = \frac{(2+12.5\sqrt{a})\Delta T^{0.25}}{a} \cdot a \times a\Delta T$$

Constraint is
$$a\Delta T = 6.1$$

Lagrangian
$$L(a, \Delta T, \lambda) = \left(2a + 12.5a^{\frac{3}{2}}\right)\Delta T^{1.25} + \lambda(6.1 - a\Delta T)$$

Taking the derivatives
$$\frac{\partial L}{\partial a} = 0, \quad \frac{\partial L}{\partial \Delta T} = 0, \text{ and } \frac{\partial L}{\partial \lambda} = 0$$

and solving will result in
$$a = 0.0256 \text{ mm}$$

332. Problem 9.7

Calculate the ECD while drilling a vertical well with the mud motor at a depth 10,000 ft. The mud density is 10 ppg. Stand pipe pressure is 3500 psi. Actual mud motor pressure drop corresponding to the flowrate and mud density is 400 psi. Frictional pressure loss gradient in the drill string is 0.04137 psi/ft. Bit pressure loss is 1337 psi. You may neglect the length of the motor.

Solution:

When fluid is circulating, the frictional pressure loss in the system (Fig. 9.2) should be considered. The pump pressure can be given as

$$P_p = P_{fs} + P_{fpdp} + P_{fpdc} + \Delta P_{mm} + \Delta P_{m/pwd} + \Delta P_b + P_{fadp} + P_{fadc},$$

where

P_{fdp}, P_{fdc} = pressure friction loss inside drillpipe and drill collar, respectively.

Drilling Tools

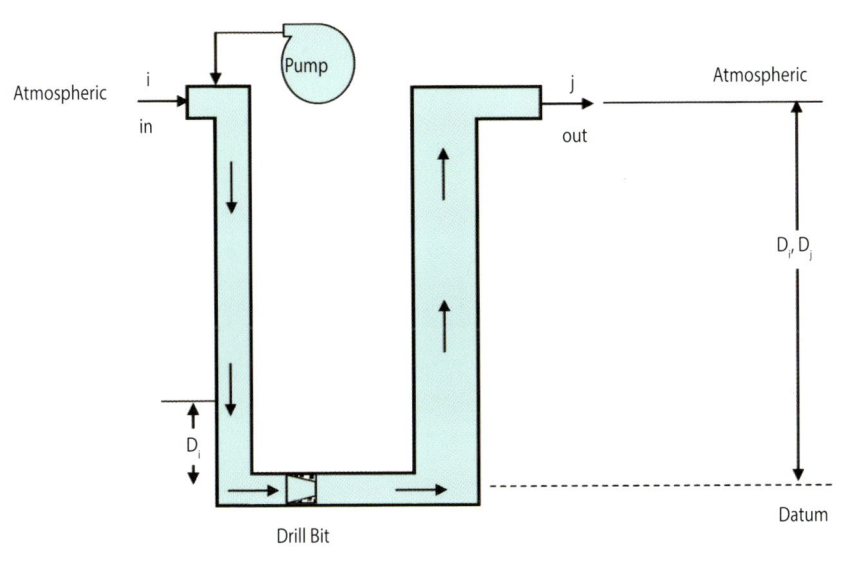

Figure 9.2 Pressure in the string and annulus

P_{fadp}, P_{fadc} = pressure friction loss in annulus around drillpipe and drill collar, respectively.

P_{fs} = pressure friction loss inside surface connections.
ΔP_b = dynamic pressure change across the bit.

Annulus Side:
The bottomhole pressure P_{bh} can be given as

$$P_{bh} = 0.052 \times \rho_m \times D_v + P_{fadp} + P_{fadc}$$

Pipe Side:
The bottomhole pressure P_{bh} can be given as

$$P_{bh} = P_p - P_{fpdp} - P_{fpdc} - \Delta P_{mm} - \Delta P_{m/pwd} - \Delta P_b + 0.052 \times \rho_m \times D_v$$

Since the details pertain to the pipe side, writing the pressure balance at the pipe side as

Pipe Side:
The bottomhole pressure P_{bh} can be given as

$$P_{bh} = P_p - P_{fpdp} - P_{fpdc} - \Delta P_{mm} - \Delta P_{m/pwd} - \Delta P_b + 0.052 \times \rho_m \times D_v$$

Substituting the values

$P_{bh} = 3500 - 0.04137 \times 10,000 - 400 - 1337 + 0.052 \times 10 \times 10,000$
$= 6549.3$

$$\text{Equivalent circulating density} = \frac{6549.3}{0.052 \times 10,000} = 12.59 \text{ ppg}$$

333. Problem 9.8
Bending Stress

For pure bending as shown in Figure 9.3, and if the material is elastic, the bending stress can be given as

$$\sigma_b = \frac{E \times y}{R},$$

where
E = Young's Modulus.
R = radius of curvature.
y = calculation point.
The bending moment can be given as

$$M = \frac{EI}{R},$$

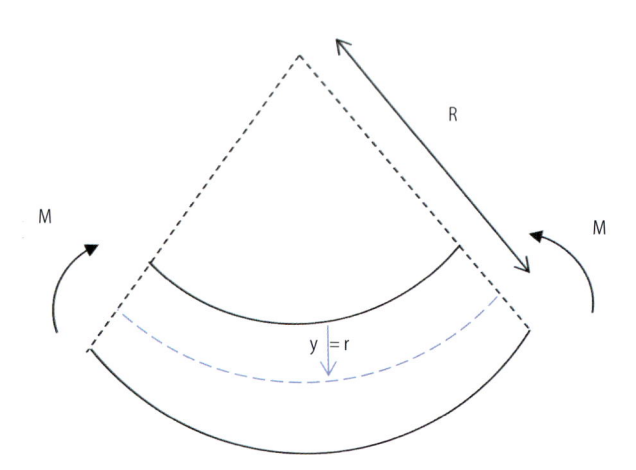

Figure 9.3 **Bending stress**

which can be written as
M = the bending moment = $EI\kappa$,
where
curvature $\kappa = 1/R$ or
radius of curvature $R = 1/\kappa$.
So the previous equation can be written as

$$\sigma_b = \frac{M \times y}{I},$$

where
M = bending moment = $EI\kappa$.
I = axial moment of inertia.
y = calculation point.

The stress obtains the maximum value at the outside diameter of the pipe from the neutral axis (center if it is an uniform pipe) as it bends. Using the outside diameter of the pipe the above equation can be written as

$$\sigma_b = \frac{M \times OD}{2I}$$

It can be written as

$$\sigma_b = \frac{M \times OD}{2I}$$

It can be written as

$$\sigma_b = \frac{E\kappa \times OD}{2I}$$

The bending stress can conveniently be expressed using the wellbore curvature as the pipe curvature as

$$\sigma_b = \frac{rE\kappa}{\frac{12 \times 100 \times 180}{\pi}} \text{ psi}$$

or

$$\sigma_b = \frac{rE\kappa}{68{,}754.9},$$

where
E = modulus of elasticity in psi.
r = radius of the pipe in inches.
κ = wellbore curvature as dogleg severity in °/100 ft.

Since the pipe bending takes different form of curvature along the wellbore such as no contact, point contact, and wrap contact the bending stress needs to be adjusted. So a factor, called contact factor, has to be applied and so the equation has to be modified as

$$\sigma_b = \frac{\xi r E \kappa}{68,754.9} \text{ psi,}$$

where
E = modulus of elasticity in psi.
r = radius of the pipe in inches.
κ = wellbore curvature as dogleg severity in °/100 ft.
ξ = bending stress contact factor.

The bending stresses at the outside and inside pipe are given as

$$\sigma_{bi} = \frac{\xi r_i E \kappa}{68,754.9}$$

$$\sigma_{bo} = \frac{\xi r_o E \kappa}{68,754.9}$$

Alternatively, the pipe dogleg severity can be estimated from the bending stress of the section of the pipe

$$\kappa = \frac{68,754.9 \times \sigma_{bo}}{\xi r_o E}$$

334. Problem 9.9
Bending Moment and Curvature Calculations

The bending moment can be related to curvature as below

$$M = \frac{\xi E I \kappa}{5729.6},$$

where
I = moment of inertia of the pipe section.

334. Problem 9.10
Curvature Calculation from Bending Moment
The dogleg severity can be estimated from the bending moment measured as below:

$$\kappa = \frac{5729.6 M}{\xi r_o EI} \text{ °}/100 \text{ ft},$$

where
 I = moment of inertia of the pipe section.
 M = bending moment in lbf-ft.

335. Problem 9.11
Bending Moment Sign Convention
The sign of R depends on the sign of M.
  i. $M > 0$, concave up, $R > 0$, i.e., center of curvature is above the beam; and
  ii. $M < 0$, concave down, $R < 0$, i.e., center of curvature is below the beam.

336. Problem 9.12
Bending Energy Calculation
Considering an elastic beam inside the wellbore as shown in Figure 9.4:

$$E_{bend} = \frac{1}{2EI} \int_0^L M^2 \, dx$$

Where bending moment is

$$M = EI\kappa(x).$$

That is

$$E_{bend} = \frac{EI}{2} \int_0^L \kappa(x)^2 \, dx$$

As the bending moment is constant, the curvature of the deflection curve is also constant. This becomes a constant curvature circular arc.

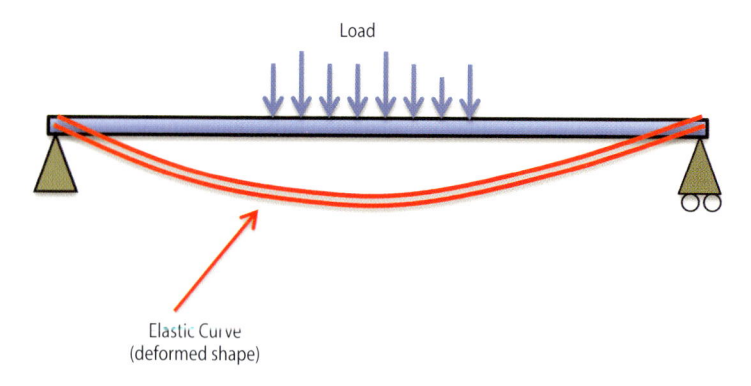

Figure 9.4   Elastic beam

337. Problem 9.13
Bending Stiffness Calculations

The product (EI) of Young's modulus (E) and moment of inertia (I) is called the *bending stiffness*. It relates to the amount of bending deformation (the curvature) to the internal force causing the bending moment.

Bending stiffness can be calculated as

$$EI = MR \text{ or } EI = \frac{M}{\kappa}$$

338. Problem 9.14

A drilling engineer is planning to calculate the bending stress in the dump value of the downhole motor. The expected dogleg close at the dump value is 9°/100 ft. The details of the dump sub is—body diameter = 8"; inside diameter = 4"

Assume $E = 30 \times 10^6$ psi

Solution:

The bending stress inside the dump sub is

$$\sigma_{bi} = \frac{r_i \kappa k}{68{,}754.9} = \frac{\left(\frac{8}{2}\right) 30{,}000{,}000 \times 9}{68{,}754.9} = 15{,}708 \text{ psi}$$

The bending stress outside the dump sub is

$$\sigma_{bo} = \frac{r_0 \kappa k}{68,754.9} = \frac{\left(\frac{4}{2}\right)30,000,000 \times 9}{68,754.9} = 7854 \text{ psi}$$

Therefore, the maximum bending stress occurs outside the body of the dump sub.

339. Problem 9.15
Wellbore Curvature Calculation from Bending Moment

Alternatively, the curvature of the pipe can be calculated from the bending moment estimation from the downhole measuring subs. Figure 9.5 shows different wellprofiles. It has to be noted that the bending stress calculated may be affected by buckling, twist, and other conditions.

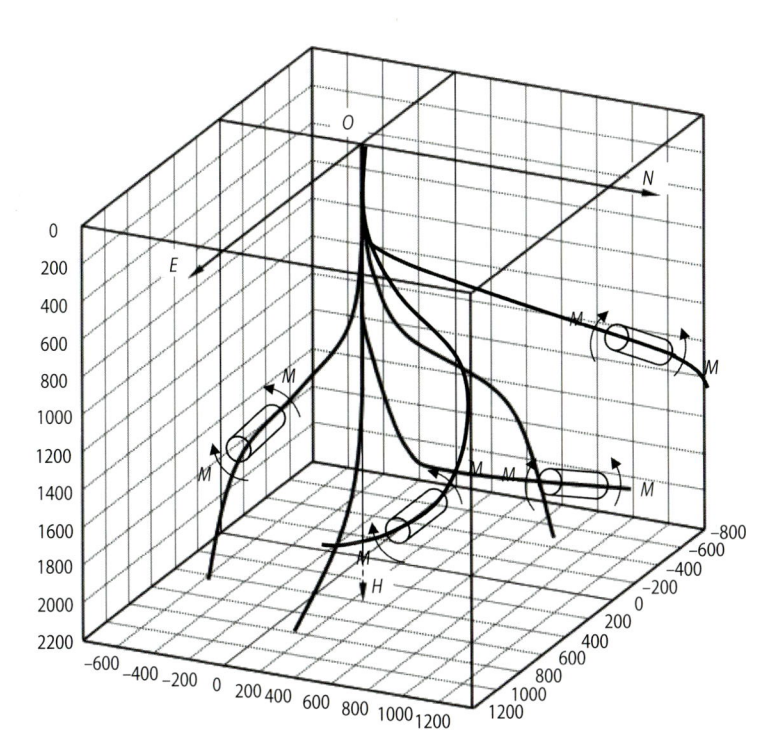

Figure 9.5 Well profiles

$$\kappa = \frac{5729.6M}{\xi EI} \text{°}/100 \text{ ft},$$

where

I = moment of inertia of the pipe section.
M = bending moment in lbf-ft.

340. Problem 9.16

Calculate the curvature of the section where the bending moment is 277 lbf-ft
- section of the pipe = 10 ft
- no buckling occurs
- modulus of elasticity = 30×10^6 psi
Drillpipe = 5" OD × 4.276" ID

Solution:

The average curvature is

$$\kappa = \frac{5729.6M}{\xi EI} \text{°}/100 \text{ ft}$$

Axial Moment of Inertia is

$$I = \frac{\pi(D_o^4 - D_i^4)}{64} = \frac{\pi(5^4 - 4.276^4)}{64} = 14.269 \text{ in.}^4$$

Substituting the respective values

$$\kappa = \frac{5729.6 \times 277}{1 \times 30{,}000{,}000 \times 14.269} = 0.0037\text{°}/100 \text{ ft}$$

Dogleg in the pipe section is 0.00037°/10 ft

341. Problem 9.17

It is assumed that the string follows the profile of the wellbore where the survey details between two survey stations are below:
Survey point 1 - Incl = 44°, Azm = 200°
Survey point 2 - Incl = 48°, Azm = 180°

Calculate the bending stress with the following assumption:
- distance between the survey point = 100 ft
- no buckling occurs
- modulus of elasticity = 30×10^6 psi

Determine the bending stress in the outside diameter of the 5" drillpipe.

Solution:

The average borehole curvature between the survey point is

$$\kappa = \frac{\beta}{\Delta L},$$

where

κ = average borehole curvature.
$\beta = a\cos(\cos\alpha_1 \cos\alpha_2 + \sin\alpha_1 \sin\alpha_2 \cos\Delta\phi)$.

Substituting the respective values

$$\kappa = \frac{a\cos(\cos 44 \cos 48 + \sin 44 \sin 48 \cos(200-180))}{100}$$

$= 0.00256$ rad/ft

Converting to °/100 ft $= 0.00256 \times 100 \times \dfrac{180}{\pi} = 14.88°/100$ ft

The bending stress outside the pipe is

$$\sigma_{bo} = \frac{r_o E \kappa}{12} = \frac{\left(\dfrac{5}{2}\right) 30{,}000{,}000 \times 0.00256}{12} = 1240 \text{ psi}$$

342. Problem 9.18

In drilling series and parallel flows are encountered due to various downhole tools such as port collars, holeopeners, before and after coring, and leakage in downhole motor seals. The correct estimation of the split flow and pressure drops are important.

For series flow the following relationships apply

Total pressure drop $\Delta P_T = \Delta P_1 + \Delta P_2 \ldots\ldots + \Delta P_n = \sum \Delta P_n$,

where

ΔP_n = pressure drop through individual path.

n = number of orifices.

Q_T = total flow.

Q_n = flow through individual path.

The flowrate

$$Q_T = Q_1 = Q_2 \ldots\ldots = Q_n$$

So the total pressure drop for total area

$$\Delta P_T = \frac{\rho Q_T^2}{2 C_{dT}^2 A_T^2},$$

where

C_{dT} = total discharge coefficient.

A_T = total nozzle area

Q_T = total flowrate.

ρ = fluid density.

The individual pressure drop through the orifice is

$$\Delta P_n = \frac{\rho Q_T^2}{2 C_{dn}^2 A_n^2}$$

So total pressure drop is

$$\Delta P_T = \sum \frac{\rho Q_T^2}{2 C_{dn}^2 A_n^2}$$

Comparing the total pressure drop, it can be written as

$$\frac{1}{C_{dT}^2 A_T^2} = \sum \frac{1}{C_{dn}^2 A_n^2}$$

Using this approach for parallel flow prove that

$$Q_T = \sum Q_n = \sum \frac{C_{dn} A_n}{C_{dT} A_T} Q_T$$

and

$$Q_n = \frac{C_{dn} A_n}{\sum C_{dn} A_n} Q_T$$

Solution:

When the flow is parallel through different orifices the total flow is

$$Q_T = Q_1 + Q_2 \ldots\ldots\ldots + Q_n = \Sigma Q_n$$

For parallel flow, the total pressure drop is same as through the individual orifices or nozzles

$$\Delta P_T = \Delta P_1 = \Delta P_2 \ldots\ldots\ldots = \Delta P_n$$

The total pressure drop can be given as

$$\Delta P_T = \frac{\rho Q_T^2}{2 C_{dT}^2 A_T^2}$$

The individual pressure drop can be given as

$$\Delta P_n = \frac{\rho Q_n^2}{2 C_{dn}^2 A_n^2}$$

Since the pressure drop is same it can be written as

$$\Delta P_T = \frac{\rho Q_T^2}{2 C_{dT}^2 A_T^2} = \frac{\rho Q_n^2}{2 C_{dn}^2 A_n^2}$$

so

$$Q_n = \frac{C_{dn} A_n}{C_{dT} A_T},$$

where
C_{dT} = total discharge coefficient.
A_T = total nozzle area.
Q_T = total flowrate.
ρ = fluid density.
Using the flowrate relationship the total flowrate can be written as

$$Q_T = \Sigma Q_n = \Sigma \frac{C_{dn} A_n}{C_{dT} A_T} Q_T,$$

where n is the number of nozzles.
Since

$$\sum \frac{C_{dn} A_n}{C_{dT} A_T} \equiv 1$$

Hence, it can be written as

$$Q_n = \frac{C_{dn} A_n}{\sum C_{dn} A_n} Q_T$$

343. Problem 9.19

A holeopener with 3 × 13 nozzles are placed above the pilot bit which has 3 × 12 nozzles. If the pressure drops through the holeopener and bits are Dp_h and Dp_b, respectively. Calculate the ratio of the pressure drop through the holeopener and pilot bit if the flow is split in the ratio 3:2 between holeopener and pilot bit. Assume same discharge coefficient for both the components' nozzles.

Solution:

Given ratio can be written as

$$\frac{Q_h}{Q_b} = \frac{3}{2}$$

Area of the bit

$$A_b^2 = \left(3 \times \frac{\pi}{4} \left(\frac{12}{32}\right)^2\right)^2 = 0.109786$$

The pressure drop across the bit

$$\Delta P_b = \frac{8.311 \times 10^{-5} \rho_m Q_b^2}{C_d^2 A_b^2}$$

Pressure drop across the holeopner

$$\Delta P_h = \frac{8.311 \times 10^{-5} \rho_m Q_h^2}{C_d^2 A_h^2}$$

The ratio is

$$\frac{\Delta P_h}{\Delta P_b} = \frac{\dfrac{8.311 \times 10^{-5} \rho_m Q_h^2}{C_d^2 A_h^2}}{\dfrac{8.311 \times 10^{-5} \rho_m Q_b^2}{C_d^2 A_b^2}}$$

$$\frac{\Delta P_h}{\Delta P_b} = \frac{\dfrac{Q_h^2}{A_h^2}}{\dfrac{Q_b^2}{A_b^2}} = \frac{\dfrac{\frac{9}{4} Q_b^2}{Q_b^2}}{} = \frac{0.151215}{0.109786} = 1.63$$

344. Problem 9.20

Plot the hydraulic horsepower at bit and holeopener versus nozzle sizes for different $\beta = \dfrac{Q_r}{Q_T}$ where Q_r is the flow through the reamer and Q_T is the total flow. Assume data.

Assign β values ranging from 0.1 to 0.9
Assume Q_T = 350 gpm Cd = 0.95, ρ = 15 ppg
Calculate Q_r using the following formula:
$Q_r = \beta \times Q_T$
Calculate Q_b using the following formula:
$Q_b = Q_T - Q_r$

Solution:

Assume that the bit and holeopener has three nozzles each and calculate the area using the following formula:

$$A_n = 3 \times \frac{\pi}{4} \left(\frac{d}{32}\right)^2$$

Calculate the pressure drop at the bit and holeopener using the following formula:

$$P_{bj} = \frac{8.311 \times 10^{-5} \rho_m Q_j^2}{C_d^2 A_n^2}$$

Calculate HHP using the following formula:

$$\text{HHP}_b = \frac{Q_r P_b}{1714}$$

Results are shown in the graphs below (Fig. 9.6).

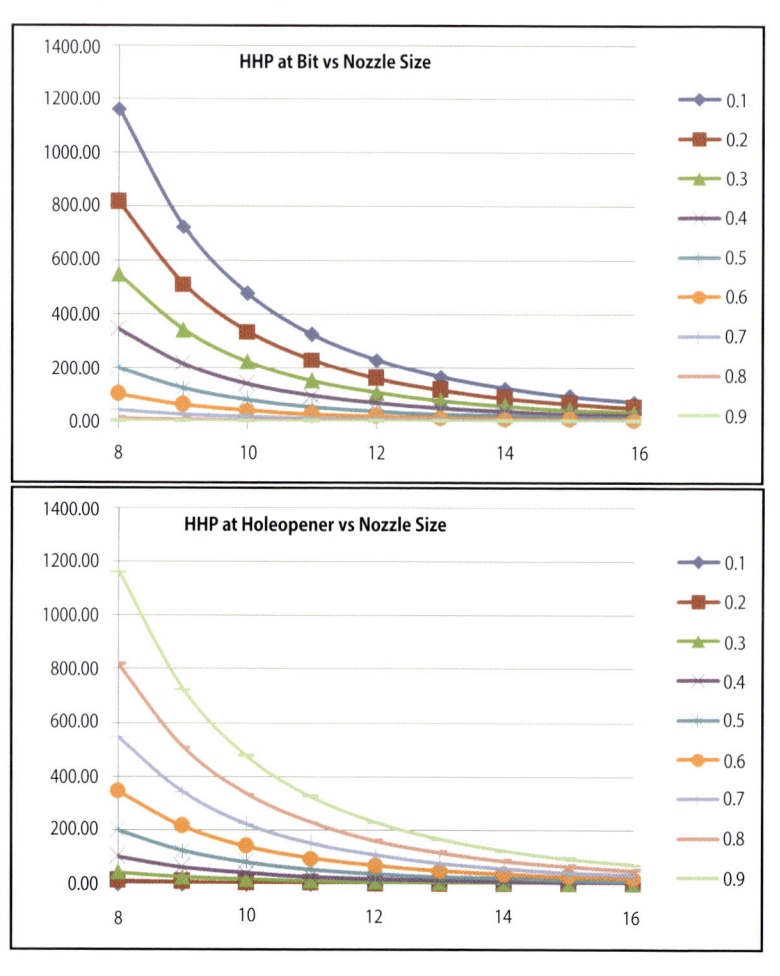

Figure 9.6 HHP vs nozzle size

CHAPTER 10

HOLE CLEANING

Hole cleaning or cuttings transport is an important part of well engineering and excessive cuttings concentration results in additional problems and decrease rate of penetration. This chapter concentrates on the calculations related to hole cleaning.

345. Problem 10.1

What are the basic functions of drilling fluids?
Solution:

- Drilled cuttings removal
- Containment of formation fluid pressures
- Hole stabilization
- Reduced heavy weight of casing string and drillstring
- Cooling and lubrication of drillstring and drillbit
- Suspension of desired solids
- Cooling and lubrication of drillstring and drill bit
- Drillbit and other tools such as stabilizers, reamers cleaning

346. Problem 10.2

What are the effects of inadequate holecleaning?
Solution:

- Pipe sticking
- Premature bit and tool wear
- Slow drilling rate
- Formation fracturing
- High torque and drag
- Overall increase in well cost

347. Problem 10.3

What are the factors that affect cuttings transport?
Solution:

- Cutting slip velocity
- Annular mud velocity
- Flow regime of fluid and cuttings slippage
- Annular velocity
- Cuttings bed formation
- Drillpipe rotational speed

- Rate of penetration (ROP)
- Drillpipe eccentricity
- Fluid Rheological properties
- Hole inclination
- Cuttings:
  - Size
  - Density
  - Shape
  - Concentration
  - Size distribution

348. Problem 10.4

What are the conditions where the pressure losses calculated in a wellbore is affected?
Solution:
    Temperature
    Pressure
    Eccentricity
    Pipe roughness
    Cuttings concentration
    Cuttings density
    Compressibility of the fluid
    Besides other rheological properties

349. Problem 10.5

What are terminal velocity of a particle and slip velocity of a particle and free settling velocity of a particle?
Solution:
    Terminal velocity of a particle in a still fluid medium is when particle attains a constant velocity and is called terminal settling velocity.
    Slip velocity of a particle is the velocity of the particle when the fluid is in motion.

Terminal velocity, also called free settling velocity, is measured using the following relationship:

$$\bar{V}_{se} = \frac{h_t - h_i}{t},$$

where
h_i = initial height.
h_t = height at time t.
t = time to reach the height.

350. Problem 10.6

If the annular mud velocity is V_a fpm and the transport ratio is R_t, what is the slip velocity?

Solution:

Transport velocity = Annular velocity − slip velocity

So the transport velocity has to be greater than zero else the cuttings cannot get transported.

Slip velocity goes to zero which is not possible

$$R_t = \frac{V_t}{V_a} = \frac{V_a - V_s}{V_a}$$

So V_s can be expressed as $V_s = V_a(1 - R_t)$

351. Problem 10.7

What is the condition between V_a and V_s for efficient hole cleaning?
where
V_a = annular velocity
V_s = slip velocity

Transport ratio is defined as the ratio between the transport velocity and the annular velocity.

Why 100% transport ratio is impossible?

Solution:

Transport ratio is given as

$$R_t = \frac{V_t}{V_a} = \frac{V_a - V_s}{V_a}$$

Hole Cleaning

If transport ratio is 100% then

$$R_t = \frac{V_a - V_s}{V_a} = 1$$

That is, $V_a - V_s = V_a$, which will result in $V_s = 0$
Slip velocity goes to zero which is not possible.

352. Problem 10.8

Transport ratio and transport efficiency are defined as below:

$$\text{Transport Ratio } R_t = \frac{\text{cuttings transport velocity}}{\text{annular fluid velocity}}$$

$$\text{Transport Efficiency } R_{t_{eff}} = \frac{\text{cuttings transport velocity}}{\text{annular fluid velocity}} \times 100\%,$$

where cuttings transport velocity is given as

$$\overline{V}_t = \overline{V}_a - \overline{V}_s,$$

where
$\overline{V}_a$ = annular fluid velocity.
$\overline{V}_s$ = slip velocity.

Calculate the transport ratio and transport efficiency for against the 5" drillpipe in a 8 ½" hole.

Use flowrate of 425 gpm. Slip velocity of the particle is 0.18 fps.

Solution:

The annular velocity is given as

$$v_a = \frac{q}{2.448(d_2^2 - d_1^2)} \text{ ft/sec}$$

$$v_a = \frac{425}{2.448(8.5^2 - 5^2)} = 3.67 \text{ ft/sec}$$

Transport velocity = $\overline{V}_t$ = 3.67 − 0.18 = 3.49 fps
Transport ratio = 3.49/3.67 = 95%

353. Problem 10.9

If the annular mud velocity is V_a fpm and the transport ratio is R_t, what is the slip velocity?

Transport ratio is given as

$$R_t = \frac{V_t}{V_a}$$

This can be written using the annular velocity as

$$R_t = \frac{V_a - V_s}{V_a}$$

Rearranging the equation the slip velocity is given as

$$V_s = V_a - R_t V_a \text{ or } V_s = V_a(1 - R_t)$$

354. Problem 10.10

Hydraulic Diameter Calculation

With the foregoing calculations of the wetted perimeter and flow area, the hydraulic diameter can be calculated using the equation:

$$\text{Hydraulic diameter} = \frac{4 \times \text{cross-sectional area}}{\text{wetted perimeter}}$$

The hydraulic diameter for annulus

$$D_{hyd} = D_h - D_p,$$

where
D_p = diameter of the pipe.
D_h = diameter of the hole.

355. Problem 10.11

Hydraulic Diameter Calculation

Equivalent diameter given by Lamb is:

$$D_e = \sqrt{D_h^2 + D_p^2 - \frac{D_h^2 - D_p^2}{\ln \frac{D_h^2}{D_p^2}}}$$

Hole Cleaning

Another simple and convenient form of expression to calculate the equivalent diameter is obtained by comparing the Lamb's equation and the slot flow approximation for annulus and is given as:

$$D_e = 0.816(D_h^2 - D_p^2)$$

Another expression based on empirical results

$$D_e = \frac{1}{2}\sqrt[4]{D_h^4 - D_p^4 - \frac{(D_h^2 - D_p^2)^2}{\ln\frac{D_h^2}{D_p^2}}} + \sqrt{D_h^2 - D_p^2}$$

Equivalent diameter by Jones and Leungs given as

$$D_e = \frac{D_l^2}{D_{hy}},$$

where

$$D_l = \sqrt{D_h^2 + D_p^2 - \frac{D_h^2 - D_p^2}{\ln\frac{D_h^2}{D_p^2}}}.$$

356. Problem 10.12

Calculate the particle Reynolds number for the following data:
Density of the fluid = 14 ppg
Viscosity = 109.2 cP
Diameter of the particle = 0.256 in.
Velocity of the particle = 13.9 ft/min
The Reynolds Number of the particle is calculated as always using the equivalent spherical diameter of the cutting in inches.
Solution:

$$Re_p = \frac{928\rho_f v_s d_s}{\mu_a}$$

$$Re_p = \frac{928(14)(\frac{13.9\,\text{ft/min}}{60})(0.256\,\text{in})}{109.2\,\text{cP}}$$

$$Re_p = 7.07$$

357. Problem 10.13

Calculate the volume rate of cuttings generated when drilling a hole of 8 ½" with a rate of penetration of (ROP) is 60 ft/hr.
Solution:
The rate at which cuttings are generated during drilling is

$$Q_c = \frac{\pi}{4} D_h^2 \times \text{ROP}$$

$$Q_c = \frac{\pi}{4}\left(\frac{8.5}{12}\right)^2 \times 60 = 23.64\,\text{ft}^3/\text{hr}$$

$$= 23.64 \times 7.48/60 = 2.95\,\text{gpm}$$

358. Problem 10.14

Calculate the volume rate of cuttings generated when drilling a hole of 12 ¼" with a rate of penetration (ROP) is 25 m/hr.
Solution:
Converting diameter of the hole = 12.25 × 0.0254 = 0.3115
The rate at which cuttings are generated during drilling is

$$Q_c = \frac{\pi}{4} D_h^2 \times \text{ROP}$$

$$Q_c = \frac{\pi}{4}(0.3115)^2 \times 25 = 1.900\,\text{m}^3/\text{hr}$$

$$= 1.900/60 = 0.031682\,\text{m}^3/\text{min}$$

359. Problem 10.15

Calculate the cuttings concentration in the annulus when drilling at 60 ft/hr a 12 ¼" hole. Assume 5" drillpipe.

Annular velocity = 100 ft/min
Slip velocity = 25 ft/min

Solution:

The cuttings concentration is given as

$$C_a = \frac{100}{60} \frac{D_h^2 \times ROP}{(V_a - V_s)(D_h^2 - D_p^2)}\%$$

Substituting the respective values

$$C_a = \frac{100}{60} \frac{12.25^2 \times 60}{(100-25)(12.25^2 - 5^2)}\% = 1.6\%$$

360. Problem 10.16

Calculate the maximum ROP that can be achieved if the cuttings concentration is to be limited to 5% in the annulus while drilling a 12 ¼" hole. Assume 5" drillpipe.

Annular velocity = 100 ft/min
Slip velocity 25 ft/min

Solution:

The ROP can be expressed using cuttings concentration as

$$ROP = \frac{60}{100} C_a \frac{(V_a - V_s)(D_h^2 - D_p^2)}{D_h^2}$$

Substituting the respective values

$$ROP = \frac{60}{100} \times 5 \times \frac{(100-25)(12.25^2 - 5^2)}{12.25^2}$$

$$= 188 \text{ ft/hr}$$

361. Problem 10.17

Calculate the cuttings concentration in the annulus when drilling at 60 ft/hr a 12 ¼" hole. Assume 5" drillpipe.

Annular velocity = 100 ft/min
Slip velocity 25 ft/min
Use bed porosity = 30%

Solution:

The cuttings concentration with bed porosity is given as

$$C_a = \frac{100}{60} \frac{D_h^2 \times ROP(1-\phi)}{(V_a - V_s)(D_h^2 - D_p^2)} \%$$

Substituting the respective values

$$C_a = \frac{100}{60} \frac{12.25^2 \times 60 \times 0.7}{(100-25)(12.25^2 - 5^2)} \% = 1.12\%$$

362. Problem 10.18

Calculate the maximum ROP that can be achieved if the cuttings concentration is to be limited to 5% in the annulus while drilling a 12 ¼" hole. Assume 5" drillpipe.

Annular velocity = 100 ft/min
Slip velocity 25 ft/min
Use bed porosity = 30%

Solution:

The ROP can be expressed using cuttings concentration as

$$ROP = \frac{60}{100} C_a \frac{(V_a - V_s)(D_h^2 - D_p^2)}{(1-\phi)D_h^2}$$

Substituting the respective values

$$ROP = \frac{60}{100} \times 5 \times \frac{(100-25)(12.25^2 - 5^2)}{0.7 \times 12.25^2}$$

$$= 268 \text{ ft/hr}$$

363. Problem 10.19

Calculate the volume of cuttings generated while drilling 12 ¼" hole with the rate of penetration of 50 ft/hr. Assume formation porosity = 30%

Hole Cleaning

Solution:

Using the equation
Volume of cuttings entering the mud system

$$V_c = \frac{(1-\phi)D_b^2 \times ROP}{1029} \text{ bbl/hr,}$$

where
ϕ = average formation porosity.
D_b = diameter of the bit, in.
ROP = rate of penetration, ft/hr.

$$V_c = \frac{(1-\phi)D_b^2 \times ROP}{24.49} \text{ gal/hr or } V_c = \frac{(1-\phi)D_b^2 \times ROP}{1469.4} \text{ gpm}$$

Substituting the values

$$V_c = \frac{(1-\phi)D_b^2 \times ROP}{1029} = \frac{\left(1-\frac{30}{100}\right) \times 12.5^2 \times 50}{1029} = 5.1 \text{ bbl/hr}$$

Volume of cuttings generated in barrels per hour

$$V_c = \frac{(1-\phi)D_b^2 \times ROP}{24.49} = \frac{\left(1-\frac{30}{100}\right) \times 12.5^2 \times 50}{24.49} = 214.5 \text{ gal/hr}$$

Volume of cuttings generated in barrels per hour

$$V_c = \frac{(1-\phi)D_b^2 \times ROP}{1469.4} = \frac{\left(1-\frac{30}{100}\right) \times 12.5^2 \times 50}{1469.4} = 3.57 \text{ gpm}$$

364. Problem 10.20

During drilling 12 ¼" at a depth 10,000 ft (True vertical depth -TVD), a wellbore pressure overbalance of 150 psi is observed with mud density 10 ppg. Maximum allowable overbalance pressure during drilling is 400 psi.

Determine:

1. The maximum rate of penetration for a flowrate of 500 gpm
2. Minimum flowrate required to maintain a penetration rate of 130 ft/hr

Solution:

The initial hydrostatic mud = 0.052 × 10 × 10,000 = 5200 psi
Allowed overbalance over the existing overbalance is = 400 − 250 = 150 psi
Total pressure allowed = 5200 + 150 = 5350 psi

Maximum allowed mud weight $= \dfrac{5350}{0.052 \times 1000} = 10.2884$ ppg

$$10.2884 = \dfrac{10 \times 500 + 141.4296 \times 10^{-4} \times \text{ROP} \times 12.25^2}{500 + 6.7995 \times 10^{-4} \times \text{ROP} \times 12.25^2}$$

Solving ROP = 129 ft/hr

With the given rate of penetration of 130 ft/hr the flowrate needed is 504 gpm.

365. Problem 10.21

Assume the pump is on for 1.5 min while a 30-foot joint is drilled down, (R) will be equal to 20.0 ft/min. Hole diameter is 17-½ in. diameter, the flow rate is 800 gpm, and the mud density is 9.0 ppg. Calculate the average density in the annulus.

Solution:

An estimate of the average annulus mud weight ($\gamma_{m,av.}$) can be formulated from the above equations and is given by

$$\gamma_m = \dfrac{\gamma_{ps} Q + 0.85 D_h^2 R}{Q + 0.0408 D_h^2 R},$$

where

$\gamma_{m,av.}$ = average annular mud weight (lb/gal).
Q = flow rate (gpm).
γ_{ps} = measured mud weight at pump suction (lb/gal).

Hole Cleaning

D_h = diameter of hole (in).
R = penetration rate, based on time the pump is on before, during and after joint is drilled down (fpm).

If the pump is on for 1.5 min while a 30-foot joint is drilled down, (R) will equal 20.0 ft/min.

$$\gamma_{m,av} = \frac{800 \times 9 + 0.85 \times 20 \times 17.5 \times 17.5}{800 + 0.0408 \times 17.5 \times 17.5 \times 20} = 11.8 \text{ ppg}$$

366. Problem 10.22

The fracture pressure at a critical depth is found to be 15 ppg. Determine whether the cuttings loading will result in fracturing of the formation when drilling at a rate of 100 ft/hr with the mud flowrate of 500 gpm.

Hole drilled 12 ¼". Assume density of cuttings to be 20 ppg.
Cuttings generated = 2 bbl/min
Density of cuttings = 20 ppg
Flowrate of mud = 500 gpm
Mud weight = 10 ppg

Solution

The rate at which cuttings are generated during drilling is

$$Q_c = \frac{\pi}{4} D_h^2 \times ROP$$

$$Q_c = \frac{\pi}{4} \left(\frac{12.25}{12}\right)^2 \times 100 = 10.20 \text{ gpm}$$

Weight of cuttings generated = 10.20 × 20 = 200.4 lbs/min
Weight of mud circulated = 500 × 12 = 6000 lbs/min
Volume rate of mud = 500 gpm
Mud weight in the annulus

$$\gamma_{ann} = \frac{(6000 + 200.4)}{(500 + 10.2)} = 15.68 \text{ ppg}$$

Cuttings loading will result in fracture of the formation.

367. Problem 10.23

Calculate the percent weight increase due to cuttings loading with the density of 20 ppg in a mud of 10 ppg.

Solution

Percent weight increase is given as

$$P = \frac{\frac{\rho_m - \rho_w}{\rho_m}}{\frac{\rho_c - \rho_w}{\rho_c}} \times 100$$

Substituting the values using density of water to be 8.33 ppg

$$P = \frac{\frac{10 - 8.33}{10}}{\frac{20 - 8.33}{20}} \times 100$$

$$= \frac{0.167}{0.5835} \times 100 = 29\%$$

368. Problem 10.24

Lag time from bit to surface is calculated as

$$\text{Lag time} = \frac{\text{Annuluar volume}}{\text{flow rate}}$$

In offshore when booster pumps is used the lag time is calculated as

$$\text{Lag time} = \frac{\text{Annuluar volume to seabed}}{\text{pump flowrate}}$$
$$+ \frac{\text{Riser volume}}{\text{pump flowrate} + \text{booster pump flowrate}}$$

369. Problem 10.25

Compute the transport ratio of a 0.375" cutting (both diameter and thickness) having a specific gravity of 2.5" in a 14 ppg mud being pumped at an annular velocity of 90 ft/min in a 6.5" × 3.5" annulus.

Hole Cleaning

The following data were obtained for the drilling fluid using a rotational viscometer.

Rotor speed (rpm)	Dial reading (degree)
3	4
6	6.6
100	26
200	44
300	60
600	100

a. Compute the transport ratio using Moore correlation.
b. Compute the transport ratio using Chien correlation.
c. Compute the transport ratio using the Walker and Mayes correlation.
d. Compute the transport ratio using Peden correlation.

370. Problem 10.26

Using Chein's correlation calculate the slip velocity using the following data:

Cuttings diameter = 0.125 in.
Cuttings density = 2.5 sp.gr.
Bed porosity = 36%
Mud Details:
Mud Density: 10.50 ppg
Plastic Viscosity: 26.24 cP
Annular Velocity 90 ft/min
Yield Point: 8.363 lbf/100 ft²

Solution:
Slip velocity using Chien's correlation is given as

$$v_s = 0.458\beta\left[\sqrt{\frac{36,800d_s}{\beta^2}\left(\frac{\rho_p - \rho_f}{\rho_f}\right) + 1} - 1\right]\left(\text{for } \beta < 10\right)$$

$$v_s = 86.4\sqrt{\left(\frac{\rho_p - \rho_f}{\rho_f}\right)d_p}\ (\text{for } \beta > 10)$$

$$\beta = \frac{\mu_a}{\rho_f d_p}$$

whereas the apparent viscosity is given as

$$\mu_a = \mu_p + \frac{300\tau_y d_p}{V_a},$$

where
μ_a = apparrent viscosity(cP).
ρ_f = mud density (ppg).
ρ_p = cuttings density (ppg).
τ_y = mud yield value (lb/100 ft²).
d_p = equivalent diameter of cuttings (in).
v_a = annular velocity (fpm).
Annular velocity is 90 ft/min

$$\mu_a = 26.24 + \frac{300 \times 8.363 \times 0.125}{90} = 29.72$$

Calculate beta

$$\beta = \frac{29.72}{10.5 \times 0.125} = 22.6$$

so slip velocity is calculated as

$$v_s = 86.4\sqrt{\frac{2.5 \times 8.33 - 10.5}{10.5} \times 0.125} = 0.18\ \text{fps}$$

371. Problem 10.27

While drilling at a certain depth it is estimated that the cuttings slip velocity is 14 ft/min. What would be the minimum flowrate needed to clean the wellbore against the 5" drillpipe in the 12 ¼" hole? Also calculate the transport velocity.

Solution:

Cuttings slip velocity = 14 ft/min

So the flow velocity has to be greater than the slip velocity.

With this annular velocity the flowrate needed in the annular section of 12 ¼" hole and 5" pipe is $A_a = \frac{\pi}{4}(12.25^2 - 5^2)$

$$\text{Flowrate} = V \times A_a = \frac{14}{60} \times 2.448 \times (12.25^2 - 5^2) = 71 \text{ gpm}$$

372. Problem 10.28

Compute the transport ratio of a 0.375" cutting (both diameter and thickness) having a specific gravity of 2.5 in a 14 ppg mud being pumped at an annular velocity of 90 ft/min in a 6.5" × 3.5" annulus. The following data were obtained for the drilling fluid using a rotational viscometer.

Rotor speed (rpm)	Dial Reading (degree)
3	4
6	6.6
100	26
200	44
300	60
600	100

a. Compute the transport ratio using Moore correlation

Solution:
Moore correlation

Calculate n and K using

$$n_a = 3.32 \cdot \log\left(\frac{\theta_{600}}{\theta_{300}}\right) \text{ and}$$

$$K = \frac{510 \times \theta_{300}}{511^n}$$

Or using regression fit of all the Fann data.

Hence,
n = 0.737
K = 308.8 eq. cP
The apparent viscosity in the annulus is:

$$\mu_a = \frac{K}{144}\left(\frac{d_h - d_p}{\frac{v_a}{60}}\right)^{1-n}\left(\frac{2+\frac{1}{n}}{0.0208}\right)^{n_p}$$

$$\mu_a = \frac{308.8}{144}\left(\frac{6.5-3.5}{90/60}\right)^{1-0.737}\left(\frac{2+\frac{1}{0.737}}{0.0208}\right)^{0.737} = 109 \text{ eq.cP}$$

Particle Re number is calculated as

$$N_{Re} = \frac{928 \times \rho_f \times v_s \times d_s}{\mu_a}$$

$$N_{Re} = \frac{928 \times 14.0 \times 0.59 \times 0.375}{109} = 26$$

So slip velocity is

$$v_s = \frac{174 \times 0.375(20.8-14.0)^{0.667}}{(20.8 \times 14.0)^{0.333}} = 0.60 \text{ fps}$$

373. Problem 10.29

Using the data from problem 372:
Compute the transport ratio using Chien correlation.
Solution:
Chien correlation
The apparent viscosity is given as

$$\mu_a = \mu_p + \frac{300\tau_y d_p}{V_a},$$

Hole Cleaning

where
μ_a = apparrent viscosity(cP).
ρ_f = mud density (ppg).
ρ_p = cuttings density (ppg).
τ_y = mud yield value (lb/100ft²)
d_p = equivalent diameter of cuttings (in.).
v_a = annular velocity (fpm).

$$\mu_a = 40 + \frac{300 \times 20 \times 0.375}{90} = 65$$

Calculate beta

$$\beta = \frac{\mu_a}{\rho_f d_p}$$

$$\beta = \frac{65}{14 \times 0.375} = 12.4$$

So, slip velocity using Chien's correlation is given as

$$v_s = 0.458\beta \left[\sqrt{\frac{36,800 \, d_s}{\beta^2}\left(\frac{\rho_p - \rho_f}{\rho_f}\right) + 1} - 1 \right] (\text{for } \beta < 10)$$

$$v_s = 86.4 \sqrt{\left(\frac{\rho_p - \rho_f}{\rho_f}\right) d_p} \quad (\text{for } \beta > 10)$$

$$v_s = 86.4 \sqrt{\frac{20.8 - 14}{14}} \times 0.375 = 0.62 \text{ fps}$$

374. Problem 10.30

Using the data from problem 372:
Compute the transport ratio using the Walker and Mayes correlation.

Solution:

Walker and Mayes

$\gamma = 0.99$

$\tau_p = 7.9\sqrt{h_s(\rho_s - \rho_f)} \quad \tau_p = 7.9\sqrt{0.375(20.8 - 14)} = 12.6 \text{ lbm}/100 \text{ ft}^2$

Note: thickness is same as diameter (given)

$$v_s = 1.22 \times 12.6 \sqrt{\left(\frac{20.8 \times 0.99}{\sqrt{14}}\right)} = 0.60 \text{ fps}$$

375. Problem 10.31

Using the data from Problem 372:
Compute the transport ratio using Peden correlation.

Peden correlation

In the laminar region:
a = 39.8 − 9 × 0.737 = 33.2; e = 1.2 − 0.47 × 0.737 = 0.85;
K = 0.654 lbf-s^n/100 ft^2
Assume disk shapes: F_{hb} = 1; ψ = 0.87; F_s = 1.065
You should note that the original equation is in SI units. You should use consistent unit for all variables: converting to oilfield units.

$$v_s = \left[\frac{4}{3} \times 9.81 \frac{\left[\left(10^{-3} \times 0.375 \times 25.40\right)^{1+0.85 \times 0.737} (20.8-14) \times 119.8\right]}{33.2 \times 1.065 \times (0.31532)^{0.85} \times \left(14 \times 0.1198\right)^{(1-0.85)}}\right]^{\frac{1}{[2-0.85(2-0.737)]}}$$

v_s = 0.41 fps

376. Problem 10.32

You are given the following well data:
Mud data
Yield point and plastic viscosity = 12 lbf/100 ft^2/cP
Density = 9 ppg
Hole = 9" Diameter, 70° inclination
Cuttings average size = 0.174"
Drillpipe = 3.5" OD
Bed Porosity = 40%
ROP = 60 ft/hr
Estimate the minimum flowrate needed to clean the hole.

Hole Cleaning

Solution:

The cutting rise velocity is given as

$$\bar{V}_{cr} = \frac{1}{\left[1-\left(\frac{D_{op}}{D_h}\right)^2\right]\left[0.64+\frac{18.16}{ROP}\right]}$$

$$\bar{V}_{cr} = \frac{1}{\left\{1-\left(\frac{3.5}{9}\right)^2\left[0.64+\left(\frac{18.16}{60}\right)\right]\right\}} = 1.166 \text{ ft/sec}$$

Apparent viscosity is calculated using the following conditions

$\mu_a = PV + 1.12YP(D_h - D_{op})$ $PV < 20cP$ $YP < 20\frac{\text{lb}}{100 \text{ ft}^2}$

$\mu_a = PV + 0.9YP(D_h - D_{op})$ $PV > 20cP$ $YP > 20\frac{\text{lb}}{100 \text{ ft}^2}$

$\mu_a = 12 + 1.12 \times 12(9 - 3.5) = 85.9$ cP
The slip velocity is $v_{cs} = 0.02554 \times 85.92 + 3.28 = 5.47$ ft/sec
The hole inclination correction is $c_{ang} = 0.0342 \times 70 - 0.000233 \times 70^2 - 0.213 = 1.039$
Cutting size correction is $c_{size} = -1.04 \times 0.174 + 1.286 = 1.105$
Thus cuttings average slip velocity is $\bar{v}_{cs} = 5.47 \times 1.039 \times 1.10504 \times 0.99 = 6.22$ ft/sec
Critical transport fluid velocity is $\bar{v}_{ca} = 1.1663 + 6.224 = 7.4$ ft/sec and flowrate = 1244 gpm

377. Problem 10.33

With the following data calculate the slip velocity, transport velocity, and transport ratio.

PV = 26 cP
YP = 13 lbf/100 ft²
Flow rate = 500 gpm

Hole diameter = 12 ¼"
Pipe outside diameter = 5"

Solution:

Using Chien correlation, the slip velocity can be calculated. Initially the apparent viscosity can be estimated as

$$\mu_a = 26 + \frac{300 \times 13 \times 0.28}{97.9} = 37.14$$

Calculate beta

$$\beta = \frac{37.14}{10 \times 0.28} = 13.26 \text{ As } \beta > 10$$

so, the slip velocity is:

$$v_s = 86.4\sqrt{\frac{20.8 - 10}{10}} \times 0.28 = 25.14 \text{ fpm}$$

The transport ratio is given by: $R_t = 1 - \dfrac{25.14}{97.99} = 0.7434$

378. Problem 10.34

After milling a certain portion of the well, the drilling engineer observes the cuttings in the form of flakes like swarfs (Fig.10.1) during the under-reaming operation. Approximate dimensions of the cuttings are shown and given below:

- Thickness ⅟₃₂" - ⅟₁₆"
- Width ~ ¼"
- Length ½" - 1"

Calculate the slip velocity using other relevant data from Problem 377.

Solution:

Because the cuttings are in the form of flakes like swarfs, Walker Mayes Correlation should be used. Because ranges of data were given for the dimensions, the mean values were used. The same process as used above is also used here.

Figure 10.1 Cuttings Shape

	Min	Max	Mean	
Thickness	0.03125	0.0625	0.04688	in
Width	0.25	0.25	0.25	in
Length	0.5	1	0.75	in

The shear stress of the particle falling in a static fluid is first estimated using the thickness of the particle h_s.

$$\tau_p = 7.9\sqrt{h_s(\rho_s - \rho_f)}$$

$$\tau_p = 7.9\sqrt{(0.046875)(20.8 - 14)}$$

$$\tau_p = 4.46 \text{ lb}_m/100 \text{ ft}^2$$

The corresponding Fann dial reading for this shear stress is then calculated

$$\text{reading} = \frac{\tau_p}{1.0599}$$

$$\text{reading} = \frac{4.46}{1.0599}$$

$$\text{reading} = 4.2°$$

The rotation speed for this reading then needs to be correlated. Because the provided data shows that a 4° corresponds with 3 RPM, this rotation speed is used and the data is not plotted. This RPM is then converted to a shear rate.

$$\gamma_p = 1.7034 \times RPM$$
$$\gamma_p = 1.7034(3)$$
$$\gamma_p = 5.11 \text{ sec}^{-1}$$

The apparent viscosity can then be calculated.

$$\mu_a = 511\frac{\tau_p}{\gamma_p}$$

$$\mu_a = 511\frac{4.46}{5.11}$$

$$\mu_a = 446 \text{ cP}$$

The slip velocity has two different equations depending on the Reynolds Number. For now assume that the Reynolds Number of the particle is less than 100. If this results in inaccurate numbers, change the equation used. The equation below uses equivalent spherical diameter, so it must be calculated. Start by calculating the volume of the particle.

$$V_p = TWL$$
$$V_p = (0.046875)(0.25)(0.75)$$
$$V_p = 0.008789 \text{ in}^3$$

Then calculate the diameter of a sphere with the same volume.

$$V_{sphere} = \frac{\pi d^3}{6}$$

$$0.008789 = \frac{\pi d^3}{6}$$

$$d_s = 0.256 \text{ in}$$

$$v_s = 1.22\tau_p\sqrt{\frac{d_s\gamma_p}{\sqrt{\rho_f}}} \text{ (for Re}_p < 100)$$

$$v_s = 1.22(12.62)\sqrt{\frac{(0.256)(5.11)}{\sqrt{14}}}$$

$$v_s = 2.02 \text{ ft/min}$$

The Reynolds Number of the particle is then calculated to double check the previous assumption.

When calculating the Reynolds Number, always use the equivalent spherical diameter of the cutting in inches. Also, the velocity should be in ft/sec.

$$Re_p = \frac{928 \rho_f v_s d_s}{\mu_a}$$

$$Re_p = \frac{928(14)(\frac{2.02 \text{ ft/min}}{60})(0.256 \text{ in})}{126.1 \text{ cP}}$$

$$Re_p = 0.251$$

The transport ratio can finally be calculated.

$$R_t = 1 - \frac{v_s}{v_a}$$

$$R_t = 1 - \frac{2.02}{90}$$

$$R_t = 0.977$$

This same process was repeated for the maximum and minimum possible size of the cuttings to make sure that the slip velocity of the largest cuttings exceeded the annular fluid velocity. All the equations remained unchanged, and the results are shown below.

Cuttings	Min	Max	Mean	Unit
Thickness	0.03125	0.0625	0.04688	in
Width	0.25	0.25	0.25	in
Length	0.5	1	0.75	in
Surface Area	0.29688	0.65625	0.46875	in²
Volume	0.00391	0.01563	0.00879	in³
Sphere Diameter	0.1954	0.31018	0.25604	in
Sphere Area	0.11995	0.30225	0.20596	in²
Sphericity (ψ)	0.40403	0.46057	0.43938	--

Cuttings	Min	Max	Mean	Unit
Thickness	0.03125	0.0625	0.04688	in
Width	0.25	0.25	0.25	in
Length	0.5	1	0.75	in
τ (Shear Stress)	3.64172	5.15017	4.46018	
Fann Dial Reading	3.43591	4.85911	4.20811	$lb_m/100\ ft^2$
RPM (from plot)	2.5	3.5	3	°
γ (Shear Rate)	4.2585	5.9619	5.1102	
μ_a	436.989	441.426	446	sec-1
v_s	1.31581	2.77407	2.02082	cP
Re_p	0.1274	0.42208	0.25121	ft/min
R_t	0.98538	0.96918	0.97755	--

379. Problem 10.35

Calculate the new slip velocity using sphericity and draw conclusions.

Solution:

Because the Moore correlation appeared to produce more accurate results, it will be used for this calculation as well. Sphericity is defined as:

$$\text{Sphericity } \psi = \frac{\text{surface area of sphere containing same vol. as particle}}{\text{surface area of the particle}}$$

These two surface areas are then solved

$$SA_p = 2TW + 2WL + 2TL$$

$$SA_p = 2(0.046875)(0.25) + 2(0.25)(0.75) + 2(0.046875)(0.75)$$

$$SA_p = 0.46875\ in^2$$

$$SA_{sphere} = \pi d^2$$

$$SA_{sphere} = \pi(0.256\ in)^2$$

$$SA_{sphere} = 0.206\ in^2$$

$$\psi = \frac{0.206 \text{ in}^2}{0.46875 \text{ in}^2}$$

$$\psi = 0.439$$

The consistency index K and flow behavior index n are then calculated. The below formulas were used and were from the Fluid Rheological Properties section of the notes.

$$n = 3.32 \log\left(\frac{\theta_{600}}{\theta_{300}}\right)$$

$$n = 3.32 \log\left(\frac{100}{60}\right)$$

$$n = 0.7365$$

$$K = \frac{510 \theta_{300}}{511^n}$$

$$K = \frac{510(60)}{511^{0.7365}}$$

$$K = 309.7 \text{ eq cP}$$

The apparent viscosity in the annulus is then calculated. The below formula assumes that the velocity is in ft/min.

$$\mu_a = \frac{K}{144}\left(\frac{d_a - d_p}{\frac{v_a}{60}}\right)^{1-n}\left(\frac{2 + \frac{1}{n}}{0.0208}\right)^n$$

$$\mu_a = \frac{309.7}{144}\left(\frac{6.5 \text{ in} - 3.5 \text{ in}}{\frac{90 \text{ ft/min}}{60}}\right)^{1-0.7365}\left(\frac{2 + \frac{1}{0.7365}}{0.0208}\right)^{0.7365}$$

$$\mu_a = 109.2 \text{ cP}$$

The slip velocity has three different equations depending on the Reynolds Number. For now assume that the Reynolds Number of the particle is between 1.0 and 2000. If this results in inaccurate numbers, change the equation used.

$$v_s = \frac{174 d_s (\rho_p - \rho_f)^{0.667}}{(\rho_f \mu_a)^{0.333}} \text{ (for } 1.0 \leq Re_p \leq 2000\text{)}$$

$$v_s = \frac{174(0.256 \text{ in})(20.8 - 14)^{0.667}}{[(14)(109.2)]^{0.333}}$$

$$v_s = 13.9 \text{ ft/min}$$

The Reynolds Number of the particle is then calculated to double check the previous assumption. When calculating the Reynolds Number, always use the equivalent spherical diameter of the cutting in inches. Also, the velocity should be in ft/sec.

$$Re_p = \frac{928 \rho_f v_s d_s}{\mu_a}$$

$$Re_p = \frac{928(14)(\dfrac{13.9 \text{ ft/min}}{60})(0.256 \text{ in})}{109.2 \text{ cP}}$$

$$Re_p = 7.07$$

The transport ratio can finally be calculated.

$$R_t = 1 - \frac{v_s}{v_a}$$

$$R_t = 1 - \frac{6.7}{90}$$

$$R_t = 0.845$$

Again, this same process was repeated for the maximum and minimum possible size of the cuttings to make sure that the slip velocity of the largest cuttings exceeded the annular fluid velocity. All the equations remained unchanged, and the results are shown below.

- Because the cuttings were flakes, and has a sphericity of less than 0.5, the Walker and Mayes correlation should produce more accurate results.
- The slip velocities calculated in both questions 1 and 2 were significantly less than Problem 378 because the equivalent spherical diameter was ranged from 0.19 to 0.31 in. compared to 0.375 in.
- Because slip velocity is directly proportional to the equivalent spherical diameter the equations used for the Moore Correlation, both the diameter and slip velocity have a value that is 68% of the original value if the mean cutting size were used.
- Although spherical diameter is not directly proportional to the slip velocity for Walker and Mayes, a smaller diameter and thickness of the disk will result in less shear stress and shear rate of the particle. The reduction of these two shear values in addition to the reduced diameter all contribute to a reduction in slip velocity.

Cuttings	Min	Max	Mean	Unit
Thickness	0.03125	0.0625	0.046875	in
Width	0.25	0.25	0.25	in
Length	0.5	1	0.75	in
n	0.736538	0.736538	0.736538	--
K	0.637113	0.637113	0.637113	lbf*sn/100 ft²
	309.6917	309.6917	309.6917	eq cP
μ_a	109.1785	109.1785	109.1785	cP
v_s	10.62656	16.86862	13.92474	ft/min
Re_p	4.118138	10.37706	7.071137	--
R_t	0.881927	0.812571	0.845281	--

380. Problem 10.36

An improved calculation would be to adjust the bulk density of the cuttings due to sphericity. Experiments have showed strong relationship between the sphericity, cuttings concentration and bulk density.

It can adjusted as follows

$\rho_{sb} = \rho_s \varphi^a$,

where

φ = sphericity calculated as above.

a = coefficient, usually 0.2.

A typical plot of Sphericity and bulk density with the sample flowrate is shown in Fig.10.2

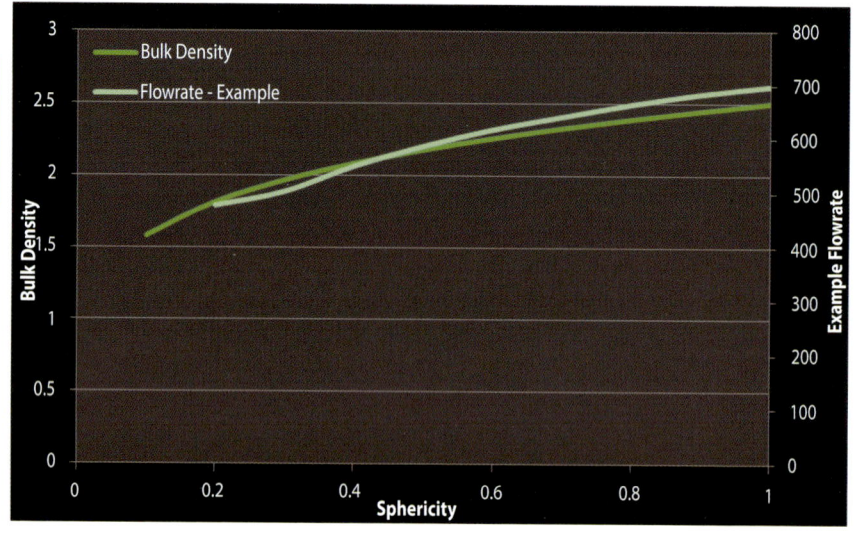

Figure 10.2 **Cuttings Sphericity**

CHAPTER 11

TUBULARS

This chapter focuses on the different basic calculations related to tubulars.

381. Problem 11.1
The transition between Drill Collars and Drill Pipes can be made using component/s known as _____
Solution:
Heavy weight drill pipe or HWDP.

382. Problem 11.2
Drill Collars are lighter than Drill Pipes.
   a. True
   b. False

Solution:
False – as drill collars are used to apply weight-on-bit.

383. Problem 11.3
Tubings may undergo wear and tear; they can be replaced if necessary.
   a. No
   b. Yes

Solution:
Yes – tubings are used as production string and can be replaced.

384. Problem 11.4
"Weight-on-Bit" is popularly expressed using _____ unit of measurement.
   a. Grams (g)
   b. Atomic mass unit (a.m.u.)
   c. Kilo-pounds (kips)

Solution:
Kilo-pounds.

385. Problem 11.5
A Drill Bit is approximately_____ long.
   a. 1 to 2 ft
   b. Between 50 ft and 100 ft
   c. Above 100 ft

Solution:
a. 1 to 2 ft

386. Problem 11.6
The Drill Bit, Drill Collar/s, Drill Pipe/s, and Heavy Weight Drill Pipe/s are constituents of the BHA (bottom hole assembly):
 a. True
 b. False

Solution:
b. False – as it does not include drill pipes.

387. Problem 11.7
In terms of Poisson's ratio (v) the ratio of Young's Modulus (E) to Shear Modulus (G) of isotropic elastic materials is
 a. $2(1-v)$
 b. $2(1+v)$
 c. $\dfrac{(1-v)}{2}$
 d. $\dfrac{(1+v)}{2}$

Solution:
Shear modulus is related to elastic modulus as $G = \dfrac{E}{2(1+v)}$.
So, solution is b.

388. Problem 11.8
Polar moment of inertia and axial moment of inertia are related by
 A. $J = 2I$
 B. $2J = I$
 C. $J = I$

Solution:
$J = 2I$
Polar Moment of Inertia is
$$= \frac{\pi}{32}\left(D_o^4 - D_i^4\right)$$

Axial Moment of Inertia is

$$= \frac{\pi}{64}\left(D_o^4 - D_i^4\right)$$

389. Problem 11.9

What can be the maximum load on the bottom of a column, if the cross-sectional area is 144 ft² and the compressive stress cannot exceed 200 lbf/ft²?

A. 20 kips
B. 22.4 kips
C. 28.8 kips
D. 30 kips

Solution:
$F = \sigma A = 200 \times 144 = 28{,}800$
$= 28.8$ kips

390. Problem 11.10

It is found that a tool stretches 1 in. when a force of F is applied to it.

The same force is applied to a tool of the same material but twice the diameter and twice the length. Calculate the stretch of the second tool.

Solution:
Using the equation $\dfrac{F}{F} = \dfrac{A_1}{A_2} \dfrac{E}{E} \dfrac{\Delta L_1}{\Delta L_2} \dfrac{L}{L}$

A = cross-sectional area
E = Young's model
L = length of the portion
ΔL = change in length
and substituting the values

$$\frac{F}{A} = E\frac{\Delta L}{L} = 0.5 \text{ in.}$$

391. Problem 11.11

For a pipe which has become partially plastic, upon removal of the load, the recovery is entirely elastic.

A. True
B. False

Solution:

A. True

392. Problem 11.12

Calculate the buoyancy factor and buoyed weight of 6000 ft of 6 ⅝" 27.7 ppf E grade Drill pipe in mud of density 10 ppg.

Solution:

Using steel density 65.4 ppg

$$\text{Buoyancy Factor} = \left(1 - \frac{\rho_m}{\rho_s}\right) = \left(1 - \frac{10}{65.4}\right) = 0.847$$

Buoyed Weight can be calculated as
Buoyed Weight = 0.847 × 27.7 × 6000 = 140,771.4 lbf = 140.8 kips

393. Problem 11.13

Buoyancy, Buoyed Weight, and Buoyancy Factor (BF) Calculations

The calculations are based on one fluid.

$$\text{Buoyancy} = \frac{\text{Weight of material in air}}{\text{Density of material}} \times \text{Fluid density}$$

$$\text{Buoyed Weight} = \left(\frac{\text{Density of material} - \text{Fluid density}}{\text{Density of material}}\right) \times \text{Weight of material in air}$$

$$\text{Buoyancy Factor} = \left(\frac{\text{Density of material} - \text{Fluid density}}{\text{Density of material}}\right)$$

$$\text{Buoyancy Factor} = \left(\frac{\rho_s - \rho_m}{\rho_s}\right) = \left(1 - \frac{\rho_m}{\rho_s}\right),$$

where
ρ_s = density of the steel/material.
ρ_m = density of the fluid/mud.

394. Problem 11.14
Buoyancy Calculations When the Fluid Densities Are Different

When the inside and outside fluid densities are different, the buoyancy factor can be given as follows:

$$\text{Buoyancy Factor (BF)} = \frac{A_o\left(1-\frac{\rho_o}{\rho_s}\right) - A_i\left(1-\frac{\rho_i}{\rho_s}\right)}{A_o - A_i},$$

where
A_o = external area of the component.
A_i = internal area of the component.

395. Problem 11.15

A drilling engineer was planning to calculate the maximum length of 9 ⅝" 47 ppf (air weight) casing that can be run into the hole. The mud weight of the drilling fluid in the hole is 12.5 ppg. The derrick capacity is 750 kips. Assume safety factor to be 1.1 for the load and the casing is run open ended.

Solution:

Using steel density 65.4 ppg

$$\text{Buoyancy Factor (BF)} = \left(1-\frac{\rho_m}{\rho_s}\right) = \left(1-\frac{12.5}{65.4}\right) = 0.809$$

Buoyed Weight for a length of the string L ft can be calculated as
Buoyed Weight = $0.809 \times 47 \times L = 38.023L$ lbf
Using the safety factor of 1.1 the allowed derrick load is = 750,000/1.1
= 681,818 lbf
So the length L that the casing can be run is = 681,818/38.023
= 17,931 ft

396. Problem 11.16

Calculate the buoyed weight of 5000 ft of 20" 106.5 ppf casing with drilling mud of density 9 ppg inside and 11 ppg cement outside the casing. Also estimate the buoyed weight of the casing with the same drilling fluid inside and outside before pumping cement. Neglect the tool joint effects.

Solution:

After pumping cement with full cement behind the casing
The inside diameter of the casing is 18.98 in.
Using steel density 65.4 ppg

$$\text{BF} = \frac{A_o\left(1-\frac{\rho_o}{\rho_s}\right) - A_i\left(1-\frac{\rho_i}{\rho_s}\right)}{A_o - A_i}$$

$$= \frac{0.7854 \times 20^2\left(1-\frac{11}{65.4}\right) - 0.7854 \times 19^2\left(1-\frac{9}{65.4}\right)}{0.7854 \times 20^2 - 0.7854 \times 18.98^2} = 0.5382$$

Buoyed Weight can be calculated as
Buoyed Weight = 0.5382 × 106.5 × 5000 = 286,618 lbf = 287 kips
Before pumping cement the buoyed weight can be estimated with steel density as 65.4 ppg.

$$\text{Buoyancy Factor} = \left(1-\frac{\rho_m}{\rho_s}\right) = \left(1-\frac{9}{65.4}\right) = 0.8623$$

Buoyed Weight can be calculated as
Buoyed Weight = 0.8623 × 106.5 × 5000 = 459,220.2 lbf = 459.2 kips

397. Problem 11.17

Effective Weight Calculations

Effective weight per unit length can be calculated using the following relation. Weight per foot in drilling mud is the weight per foot in air minus the weight per foot of the displaced drilling mud:

$$w_B = w_S + \rho_i A_i - \rho_o A_o$$

$$A_o = \frac{\pi}{4}\left(0.95 \times D_o^2 + 0.05 \times D_{oj}^2\right)$$

$$A_i = \frac{\pi}{4}\left(0.95 \times D_i^2 + 0.05 \times D_{ij}^2\right)$$

Without tool joints,
$$A_i = 0.7854 \times D_i^2 \text{ and } A_o = 0.7854 \times D_o^2$$

$$W_B = W_S + \rho_i A_i - \rho_o A_o$$

In the above equation, unit weight of the steel can be given as
$$W_S = \rho_S A_S$$

When the inside and outside fluid densities are the same
$$W_B = A_s(\rho_s - \rho_o) = A_s \rho_s \left(1 - \frac{\rho_o}{\rho_s}\right) = W_s \left(1 - \frac{\rho_o}{\rho_s}\right),$$

where
$$\left(1 - \frac{\rho_o}{\rho_s}\right) = \text{buoyancy factor.}$$

D_o = outside diameter of component body.
D_{oj} = outside diameter of tool joint.
D_i = inside diameter of component body.
D_{ij} = inside diameter of tool joint.
A_s = cross-sectional area of the steel/material.
ρ_o = annular mud weight at component depth in the wellbore.
ρ_i = internal mud weight at component depth inside the component.
ρ_s = density of the steel/material.

398. Problem 11.18

Starting from the effective equation prove that for different fluids inside and outside the drillstring, the buoyancy factor can be expressed as
$$BF = 1 - \frac{\rho_o r_o^2 - \rho_i r_i^2}{\rho_s \left(r_o^2 - r_i^2\right)},$$

where

p_o = density of the fluid outside the drillstring.
p_i = density of the fluid inside the drillstring.
p_s = density of the drillstring.
r_i = inside diameter of the drillstring.
r_o = inside diameter of the drillstring.

Solution:

Buoyancy Factor (BF) is

$$BF = \frac{W_b}{\rho_s} = \frac{\rho_s - \rho_o}{\rho_s} + \frac{A_i(\rho_i - \rho_o)}{A_s \rho_s} = 1 - \frac{A_s \rho_o - A_i \rho_i + A_i \rho_o}{A_s \rho_s},$$

where

p_o = density of the fluid outside the drillstring.
p_i = density of the fluid inside the drillstring.
p_s = density of the drillstring.
r_i = inside diameter of the drillstring.
r_o = inside diameter of the drillstring.

The above equation can further be written as

$$BF = \frac{\rho_s - \rho_o}{\rho_s} + \frac{A_i(\rho_i - \rho_o)}{A_s \rho_s} = 1 - \frac{A_s \rho_o - A_i \rho_i + A_i \rho_o}{A_s \rho_s}$$

Simplifying with area, results in

$$A_o = \frac{\pi}{4}\left(r_o^2\right)$$

$$A_i = \frac{\pi}{4}\left(r_i^2\right)$$

$$BF = 1 - \frac{\rho_o\left(r_o^2 - r_i^2\right) - \rho_i r_i^2 + r_i^2 \rho_o}{\rho_s\left(r_o^2 - r_i^2\right)}$$

$$BF = 1 - \frac{\rho_o r_o^2 - \rho_i r_i^2}{\rho_s\left(r_o^2 - r_i^2\right)}$$

399. Problem 11.19
Polar Moment of Inertia

If the total length of the tool joint is l and the total length of the pipe including tool joint is L, prove that the equivalent polar moment of inertia of the pipe can be calculated as:

$$J_{pipe} = \frac{J_{body} \times J_{jnt}}{\left(\frac{L-l}{L}\right)J_{jnt} \times \left(\frac{l}{L}\right)J_{body}},$$

where

J_{body} = polar moment of inertia of pipe body.
J_{jnt} = polar moment of inertia of tool joint.

When a rod is subjected to torque it undergoes twist given as

$$\theta = \frac{TL}{JG},$$

where
θ = angle of twist (radians) (can be $> 2\pi$).
L = length of section (ft).
T = torque (ft-lbf).
E = modulus of elasticity (psi).
μ = Poisson's ratio.
G = modulus of rigidity (psi) and is $G = \dfrac{E}{2(1+\mu)}$.
J = polar moment of inertia (in.^{4}).

400. Problem 11.20
Twist

The twist can be expressed for the combined pipe body and tool joint as

$$\theta = \frac{T}{G}\sum \frac{L}{J} = \theta = \frac{T}{G}\sum \left(\frac{l}{J_{jnt}} + \left(\frac{L-l}{J_{body}}\right)\right)$$

Tubulars

$$\frac{J_{pipe}}{L} = \frac{J_{body} \times J_{jnt}}{lJ_{jnt} \times (L-l)J_{body}}$$

$$J_{pipe} = \frac{J_{body} \times J_{jnt}}{lJ_{jnt} \times (L-l)J_{body}} \times L$$

Further simplifying it can be expressed as

$$J_{pipe} = \frac{J_{body} \times J_{jnt}}{\left(\frac{L-l}{L}\right)J_{jnt} \times \left(\frac{l}{L}\right)J_{body}}$$

If the exact length of the joint is not known the length can be approximated as 5% of the pipe length and the polar moment of inertia can be calculated as

$$J_{pipe} = \frac{J_{body} \times J_{jnt}}{0.95 J_{jnt} \times 0.05 J_{body}}$$

401. Problem 11.21

If the total length of the tool joint is l and the total length of the pipe including tool joint is L, prove that the equivalent area of the pipe when subjected to a uniform axial force can be calculated as:

$$A_{pipe} = \frac{A_{body} \times A_{jnt}}{\alpha A_{body} \times (1-\alpha) A_{jnt}},$$

where
  A_{body} = cross-sectional area of pipe body.
  A_{jnt} = cross-sectional area of tool joint.
  α = length factor for pipe body whereas 1 is for total pipe.

Solution:
  The total change in length is equal to the sum of the change in both sections

$$\Delta_{total} = \sum \frac{F_T L}{AE} = \frac{F_T L_{jnt}}{A_{jnt} E} + \frac{F_T L_{body}}{A_{body} E}$$

Using the length factors it can be written as

$$\Delta_{total} = \frac{F_T \alpha L}{A_{jnt} E} + \frac{F_T (1-\alpha) L}{A_{body} E}$$

$$\Delta_{total} = \frac{F_T L}{E} \left(\frac{\alpha}{A_{jnt}} + \frac{(1-\alpha)}{A_{body}} \right),$$

which can be further written as

$$\Delta_{total} = \frac{F_T L}{E} \left(\frac{\alpha A_{body} + (1-\alpha) A_{jnt}}{A_{jnt} A_{body}} \right)$$

So

$$A_{pipe} = \frac{A_{body} \times A_{jnt}}{\alpha A_{body} \times (1-\alpha) A_{jnt}}$$

402. Problem 11.22

If the total length of the tool joint is l and the total length of the pipe including tool joint is L, prove that the equivalent area of the pipe when volume is conserved can be calculated as:

$$A_{pipe} = \frac{\pi}{4}(1-\alpha)\left(D_{obody}^2 - D_{ibody}^2\right) + \alpha\left(D_{ojnt}^2 - D_{ijnt}^2\right),$$

where

D_{obody} = outside diameter of the pipe body.
D_{ibody} = inside diameter of the pipe body.
D_{ojoint} = outside diameter of the tool joint.
D_{ijoint} = inside diameter of the pipe body.
α = length factor for pipe body whereas the factor is 1 for total pipe length.

Tubulars

Solution:
Equivalent area volume can be given as

$$\frac{\pi}{4}\left(D_{pipe}^{2} - D_{ipipe}^{2}\right) = \frac{\pi}{4}\alpha\left(D_{obody}^{2} - D_{ibody}^{2}\right) + (1-\alpha)\left(D_{ojnt}^{2} - D_{ijnt}^{2}\right),$$

which can be further written as

$$\frac{\pi}{4}\left(D_{pipe}^{2} - D_{ipipe}^{2}\right) = \frac{\pi}{4}\left((1-\alpha)D_{obody}^{2} + \alpha D_{ojnt}^{2}\right) - \left((1-\alpha)D_{ibody}^{2} + \alpha D_{ijnt}^{2}\right)$$

Area inside volume

$$\frac{\pi}{4}D_{ipipe}^{2} = \frac{\pi}{4}\left((1-\alpha)D_{ibody}^{2} + \alpha D_{ijnt}^{2}\right)$$

Area outside volume

$$\frac{\pi}{4}D_{opipe}^{2} = \frac{\pi}{4}\left((1-\alpha)D_{obody}^{2} + \alpha D_{ojnt}^{2}\right)$$

Cross-sectional area = area outside volume − area inside volume
Substituting, above respective equations will result in

$$A_{pipe} = \frac{\pi}{4}(1-\alpha)\left(D_{obody}^{2} - D_{ibody}^{2}\right) + \alpha\left(D_{ojnt}^{2} - D_{ijnt}^{2}\right)$$

403. Problem 11.23

Calculate the air weight, buoyed weight in drilling fluid, buoyed weight when cement is inside and drilling fluid is in the annulus, buoyed weight when cement is outside and drilling fluid is inside. Casing outside diameter is 9 ⅝", casing inside diameter is 8.681", drilling fluid density is 10 ppg, cement slurry density is 12 ppg, and the depth of the well is 5000 ft.

Solution:
Air weight = 47 × 5000 = 235,000 lbf = 235 kips
Buoyed weight with drilling fluid

$$= \left(1 - \frac{10}{65.4}\right) \times 5000 \times 47 = 199{,}067 \, \text{lbf} = 199 \, \text{kips}$$

Buoyed weight with cement inside and drilling fluid outside is

$$BF = \frac{A_o\left(1-\dfrac{\rho_o}{\rho_s}\right) - A_i\left(1-\dfrac{\rho_i}{\rho_s}\right)}{A_o - A_i}$$

$$= \frac{0.7854 \times 9.625^2\left(1-\dfrac{10}{65.4}\right) - 0.7854 \times 8.681^2\left(1-\dfrac{12}{65.4}\right)}{0.7854 \times 9.625^2 - 0.7854 \times 8.681^2} = 0.98$$

$= 0.98 \times 5000 \times 47 = 230{,}406$ lbf $= 230$ kips

Buoyed weight with cement outside and drilling fluid inside is

$$BF = \frac{A_o\left(1-\dfrac{\rho_o}{\rho_s}\right) - A_i\left(1-\dfrac{\rho_i}{\rho_s}\right)}{A_o - A_i}$$

$$= \frac{0.7854 \times 9.625^2\left(1-\dfrac{12}{65.4}\right) - 0.7854 \times 8.681^2\left(1-\dfrac{10}{65.4}\right)}{0.7854 \times 9.625^2 - 0.7854 \times 8.681^2} = 0.6831$$

$= 0.6831 \times 5000 \times 47 = 160{,}541$ lbf $= 160$ kips

404. Problem 11.25

Calculate the air weight, buoyed weight in drilling fluid, buoyed weight when cement inside and drilling fluid in the annulus, buoyed weight when cement outside and drilling fluid inside.

Casing outside diameter = 9 ⅝"
Casing inside diameter = 8.681"
Drilling fluid density = 10 ppg
Cement slurry density 12 ppg
Depth of the well = 5000 ft

Solution:

Air weight = $47 \times 5000 = 235{,}000$ lbf $= 235$ kips
Buoyed weight with drilling fluid

$$= \left(1-\frac{10}{65.4}\right) \times 5000 \times 47 = 199{,}067 \text{ lbf} = 199 \text{ kips}$$

Tubulars

Buoyed weight with cement inside and drilling fluid outside

$$BF = \frac{A_o\left(1-\frac{\rho_o}{\rho_s}\right) - A_i\left(1-\frac{\rho_i}{\rho_s}\right)}{A_o - A_i}$$

$$= \frac{0.7854 \times 9.625^2 \left(1-\frac{10}{65.4}\right) - 0.7854 \times 8.681^2 \left(1-\frac{12}{65.4}\right)}{0.7854 \times 9.625^2 - 0.7854 \times 8.681^2} = 0.98$$

$= 0.98 \times 5000 \times 47 = 230{,}406 \text{ lbf} = 230 \text{ kips}$

Buoyed weight with cement outside and drilling fluid inside

$$BF = \frac{A_o\left(1-\frac{\rho_o}{\rho_s}\right) - A_i\left(1-\frac{\rho_i}{\rho_s}\right)}{A_o - A_i}$$

$$= \frac{0.7854 \times 9.625^2 \left(1-\frac{12}{65.4}\right) - 0.7854 \times 8.681^2 \left(1-\frac{10}{65.4}\right)}{0.7854 \times 9.625^2 - 0.7854 \times 8.681^2} = 0.6831$$

$= 0.6831 \times 5000 \times 47 = 160{,}541 \text{ lbf} = 160 \text{ kips}$

405. Problem 11.26

A drilling engineer is planning to calculate the weight of 50 drill collars in a drillstring with the following data:

OD = 8"; ID = 3"; Length = 30 ft; Density of the steel = 490 lbm/cu.ft; Density of mud = 10 ppg

Solution:

Weight of the drill collar in air

$$W_{dc} = \frac{\pi}{4} \times \left(\frac{8^2 - 3^2}{144}\right) \times 490 \times 30 = 4409.683 \text{ lb}$$

$$\text{Buoyancy Factor} = \left(1 - \frac{\rho_m}{\rho_s}\right) = \left(1 - \frac{10}{65.4}\right) = 0.847$$

Weight of 1 drill collar in mud = $4409 \times 0.847 = 3735$ lbm
So for 50 drill collars = 186 kips

406. Problem 11.27
Modulus of Elasticity Calculation

Modulus of elasticity is

$$E = \frac{\sigma}{\varepsilon} = \frac{F/A}{\Delta L/L} \text{ psi,}$$

where
  σ = unit stress, psi.
  ε = unit strain in inch per inch.
  F = axial force, lbf.
  A = cross-sectional area, in².
  ΔL = total strain or elongation, in.
  L = original length, in.

407. Problem 11.28

Explain the following stress-strain curves. Fig. 11.1

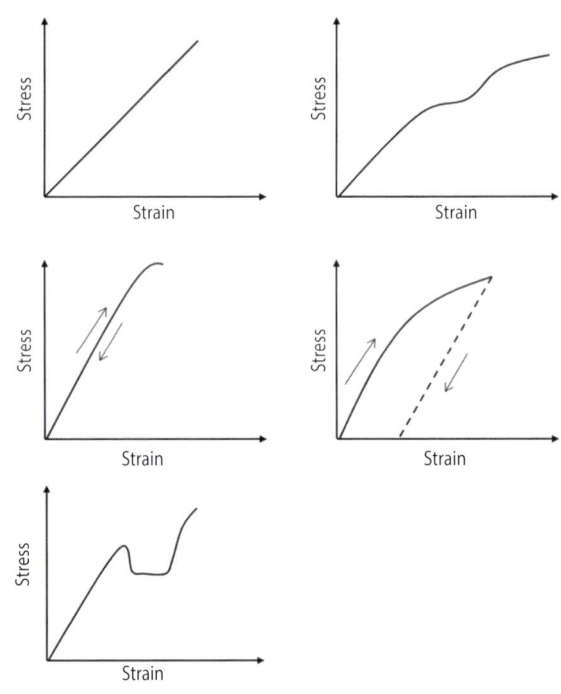

Figure 11.1   Stress Strain Curves

Solution:

    a. Linear material
    b. Non-liner material
    c. Elastic material returning to original state
    d. Plastic material remaining in deformed state
    e. Stress-strain behavior for some steels exhibiting the yield point (upper and lower yield points) phenomenon

408. Problem 11.29

Explain the points and paths in the following stress-strain curve (Fig. 11.2)

Solution:
Point A = Proportional limit of stress and strain
B = Elastic limit
C = Yield point
Path C-D = Yield plateaus
E = Ultimate strength
F = Rupture point

409. Problem 11.30

Consider a pipe with the following dimensions that carry an applied tensile load of 5000 lbs at the bottom. Calculate the maximum stress in the string.

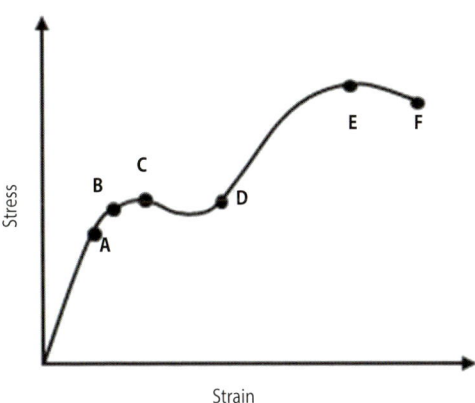

Figure 11.2   **Stress Strain Curves**

The pipe outside diameter = 5 in., the pipe inside diameter = 4 in., the pipe density = 490 lb/ft³, and the pipe length = 30 ft.

Solution:

Cross-sectional area of pipe $A = \dfrac{\pi}{4}(5^2 - 4^2) = 7.08$ in.²

Weight of the pipe $= \dfrac{\pi}{4}(5^2 - 4^2) \times 490 \times 30 \times 12 = 721.6$ lbf

Total force acting at the top of the pipe:
F = weight of the pipe + load applied
$F = 721.6 + 5000 = 5721.6$ lbf

Maximum stress at the top of the pipe

$$\sigma = \dfrac{F}{A} = \dfrac{5721.6}{7.08} = 809 \text{ psi}$$

410. Problem 11.31

Calculate the elongation of a cylindrical pipe of 5" in outside diameter, 4.0 in inside diameter and 10,000 ft long when a tensile load of 20,000 lbf is applied. Assume that the deformation is totally elastic and modulus of elasticity = 30×10^6 psi.

Solution:

From equation $E = \dfrac{F/A}{\Delta L/L}$, the elongation can be written as follows:

$$\Delta L = \dfrac{F/A}{E/L} = \dfrac{L \times F}{E \times A} = \dfrac{L \times F}{E \times \dfrac{\pi}{4}(D_o^2 - D_i^2)} = \dfrac{4L \times F}{E \times \pi \times (D_o^2 - D_i^2)}$$

Substituting the values

$$= \dfrac{4 \times 10,000 \times 12 \times 20,000}{30 \times 10^6 \times \pi \times (5^2 - 4^2)} = 11.32 \text{ in.}$$

411. Problem 11.32

A downhole tool with a length of 30 ft, an outside diameter of 5.5 in., and an inside diameter of 4.75 in. is compressed by an axial force of 30 kips.

The material has a modulus of elasticity 30,000 ksi and Poisson's ratio 0.3. Assume the tool is in the elastic range.

Calculate the shortening of tool.

Solution:

Using the Hook's law

$$\Delta L = \frac{F/A}{E/L} = \frac{L \times F}{E \times A} = \frac{30 \times 12 \times (-30,000)}{30 \times 10^6 \times \frac{\pi}{4}(5.5^2 - 4.75^2)} = -0.05962 \text{ in.}$$

Negative sign shows shortening of the tool.

412. Problem 11.33

A downhole tool with a length of 30 ft, an outside diameter of 5.5 in., and an inside diameter of 4.75 in. is compressed by an axial force of 30 kips. The material has a modulus of elasticity 30,000 ksi and Poisson's ratio 0.3. Assume the tool is in the elastic range.

Calculate the lateral strain

Solution:

Lateral strain can be obtained using the axial strain and Poisson's ratio:

Axial strain

$$= \frac{\Delta L}{L} = \frac{-0.05962}{360} = -0.0001656$$

Lateral strain

$$= v \times \frac{\Delta L}{L} = \frac{0.05962}{360} = 0.3 \times 0.0001656 = 4.968 \times 10^{-5}$$

413. Problem 11.34

A downhole tool with a length of 30 ft, an outside diameter of 5.5 in., and an inside diameter of 4.75 in. is compressed by an axial force of 30 kips. The material has a modulus of elasticity 30,000 ksi and Poisson's ratio 0.3. Assume the tool is in the elastic range.

Calculate the following:
- Increase in outer diameter
- Increase in inner diameter
- Increase in wall thickness

Solution:

Increase in outer diameter:

Increase in outer diameter is the lateral strain times the outer diameter:
$4.968 \times 10^{-5} \times 5.5 = 0.000273$ in.

Increase in inner diameter:

Increase in inner diameter is the lateral strain times the inner diameter:
$4.968 \times 10^{-5} \times 4.75 = 0.000236$ in.

Increase in wall thickness:

Increase in wall thickness can be estimated in a similar way as diameters:

$$4.968 \times 10^{-5} \times \left(\frac{5.5 - 4.75}{2}\right) = 1.8632 \times 10^{-5} \text{ in.}$$

414. Problem 11.35
Poisson's Ratio Calculations

$$\nu = \frac{\varepsilon_{lat}}{\varepsilon_{long}},$$

where

ε_{lat} = lateral strain in inches.
ε_{long} = longitudinal or axial strain in inches.
For most metals, Poisson's ratio varies from ¼ to ⅓.

Modulus of elasticity and shear modulus are related to Poisson's ratio as follows:

$$E = 2G(1+\nu)$$

Modulus of elasticity, shear modulus, and Poisson's ratio for common materials are given in Table 11.1.

Tubulars

Table 11.1 Modulus of elasticity, shear modulus, and Poisson's ratio at room temperature

Metal alloy	Modulus of elasticity		Shear modulus		Poisson's ratio
	Psi × 10⁶	MPa × 10⁶	Psi × 10⁶	MPa × 10⁶	
Aluminum	10	6.9	3.8	2.6	0.33
Copper	16	11	6.7	4.6	0.35
Steel	30	20.7	12	8.3	0.27
Titanium	15.5	10.7	6.5	4.5	0.36
Tungsten	59	40.7	23.2	16	0.28

415. Problem 11.36

An engineer pulls the casing string with 30,000 lbf above the buoyed weight of the casing. Casing OD = 7 in. with thickness 0.362 in. Calculate the new length of the casing. You may neglect the effect of temperature.

Solution:

$$A = \frac{\pi}{4} \times (7^2 - 6.276^2) = 7.549 \text{ in}^2$$

$$\sigma = \frac{F}{A} = \frac{30,000 \text{ lb}}{7.549 \text{ in}^2} = 3974 \text{ psi}$$

$$\varepsilon = \frac{\sigma}{E} = \frac{3974 \text{ psi}}{30 \times 10^6 \text{ psi}} = 0.00013247$$

$$\varepsilon = \frac{e}{L_o};$$

$$e = \varepsilon \times L_o = (0.00013247 \text{ in/in}) \times (10,000 \text{ ft} \times \frac{12 \text{ in}}{1 \text{ ft}}) = 15.89 \text{ in}$$

416. Problem 11.37

Minimum Yield Strength

Yield strength is defined as the stress that will result in specific permanent deformation in the material. The yield strength can be conveniently determined from the stress–strain diagram. Based on the test results, minimum and maximum yield strengths for the tubulars are specified Table 11.2.

417. Problem 11.38
Ultimate Tensile Strength

The ultimate tensile strength (UTS) of a material in tension, compression, or shear, respectively, is the maximum tensile, compressive, or shear stress resistance to fracture or rupture. It is equivalent to the maximum load that can be applied over the cross-sectional area on which the load is applied. The term can be modified as the ultimate tensile, compressive, or shearing strength.

418. Problem 11.39
Fatigue Endurance Limit

The endurance limit pertains to the property of a material and is defined as the highest stress or range of cyclic stress that a material can be subjected to indefinitely without causing failure or fracture. In other words, the endurance limit is the maximum stress reversal that can be

Table 11.2 API pipe properties

API Grade	Yield stress, psi Minimum	Yield stress, psi Maximum	Minimum ultimate Tensile, psi	Minimum Elongation (%)
H-40	40,000	80,000	60,000	29.5
J-55	55,000	80,000	75,000	24
K-55	55,000	80,000	95,000	19.5
N-80	80,000	110,000	100,000	18.5
L-80	80,000	95,000	95,000	19.5
C-90	90,000	105,000	100,000	18.5
C-95	95,000	110,000	105,000	18.5
T-95	95,000	110,000	105,000	18
P-110	110,000	140,000	125,000	15
Q-125	125,000	150,000	135,000	18

indefinitely subjected a large number of times without producing fracture. The magnitude of the endurance limit of a material is usually determined from a fatigue test that uses a sample piece of the material.

419. Problem 11.40
Twist Calculations

When a rod is subjected to torque it undergoes twist which is given as

$$\theta = \frac{TL}{GJ} \text{ rad,}$$

where
θ = angle of twist (radians) (can be > 2π).
L = length of section, ft.
T = torque, ft-lbf.
G = modulus of rigidity, psi.

$$G = \frac{E}{2(1+\nu)}$$

J = polar moment of inertia (in.^{4}) = $\frac{\pi}{32}(D_o^4 - D_i^4)$,

where
E = modulus of elasticity, psi.
ν = Poisson's ratio.

420. Problem 11.41

Calculate the angle of twist of a cylindrical pipe of 5" in outside diameter, 4.0 in inside diameter, and 10,000 ft long when a torque of 3000 ft-lbf is applied at the bottom of the pipe. Assume that the twisting is totally elastic and modulus of rigidity = 12×10^6 psi.

Solution:

Calculating the polar moment of inertial of the pipe

$$J = \frac{\pi}{32}(D_o^4 - D_i^4) = \frac{\pi}{32}(5^4 - 4^4) = 36.23 \text{ in}^4.$$

Using appropriate units

$$\theta = \frac{TL}{GJ} = \frac{3000 \times 10{,}000 \times 12 \times 12}{12 \times 10^6 \times 36.23} = 9.9374 \text{ rad} = 9.9374 \times \frac{180}{\pi} = 569.38°$$

421. Problem 11.42

Estimate the stiffness of a steel drill collar having an OD 6 ¼" and ID of 2 3/16".

Solution:

Bending Stiffness

$$EI = 30 \times 10^6 \times \frac{\pi}{64}\left(6.25^4 - \left(2\frac{3}{16}\right)^4\right) = 2{,}213{,}332{,}493 \text{ lbf-in}^2$$

Torsional Stiffness

$$GJ = 12 \times 10^6 \times \frac{\pi}{32}\left(6.25^4 - 2\frac{3}{16}^4\right) = 1{,}770{,}657{,}995 \text{ lbf-in}^2$$

Similarly, for Tungsten using $E = 16.5 \times 10^6$ psi

422. Problem 11.43

A downhole tool has a composite shaft made of a tight aluminum sleeve inside a steel drill collar jacket. A torque of magnitude 5162 lbf-ft is applied at the free end of the composite drill collar whereas the other end is fixed. Geometrical dimensions are of the tools are steel (OD = 8 in., ID = 4 in.), Aluminum (ID = 3.5 in.), length 8.202 ft. Assume a modulus of rigidity of 11,167.905 kpsi for steel and 3916.018 kpsi for the aluminum.

Calculate the following:
a. maximum shearing stress experienced in the inner aluminum sleeve
b. maximum shearing stress experienced in the steel jacket
c. the angle of twist at the end where torque is applied

Solution:

For steel

$$J = \frac{\pi}{32}\left(8^4 - 4^4\right) = 376.8 \text{ in.}^4$$

Torsional stiffness

$$GJ = 11{,}167{,}905 \times 376.8 = 4{,}207{,}984{,}677 \text{ in.}^4$$

For inner aluminum sleeve

$$J = \frac{\pi}{32}\left(4^4 - 3.5^4\right) = 10.6 \text{ in.}^4$$

Torsional stiffness

$$GJ = 3916.018 \times 10.6 = 41{,}490{,}060 \text{ in.}^4$$

$$\text{Angle change per unit length} = \frac{T}{(GJ)_{Al} + (GJ)_{steel}}$$

$$= \frac{5162 \times 144}{41{,}490{,}060 + 4{,}207{,}984{,}677}$$

$$= 0.000174954 \text{ rad/ft}$$

$$\text{Angle of twist} = 0.000174954 \times \frac{180}{\pi} = 0.082°$$

Shearing stress in steel

$$\left(\frac{OD}{2}\right)_{steel} \times \frac{T}{(GJ)_{Al} + (GJ)_{steel}} = 0.000174954 \times \frac{8}{2 \times 12} = 651 \text{ psi}$$

Shearing stress in aluminum

$$\left(\frac{OD}{2}\right)_{Al} \times \frac{T}{(GJ)_{Al} + (GJ)_{steel}} = 0.000174954 \times \frac{4}{2 \times 12} = 114 \text{ psi}$$

423. Problem 11.44

Using the data from Problem 11.43 if the materials are switched in such a way that aluminum as made as jacket and steel as sleeve determine the following and draw conclusions against the results obtained in Problem 11.43:

(a) maximum shearing stress experienced in the inner steel sleeve
(b) maximum shearing stress experienced in the aluminum jacket
(c) the angle of twist at the end where torque is applied

Solution:
For Aluminum
$$J = \frac{\pi}{32}(8^4 - 4^4) = 376.8 \text{ in.}^4$$

Torsional stiffness
$$GJ = 3,916,018 \times 376.8 = 1,475,527,094 \text{ in.}^4$$

For inner steel sleeve
$$J = \frac{\pi}{32}(4^4 - 3.5^4) = 10.6 \text{ in.}^4$$

Torsional stiffness
$$GJ = 11,167,905 \times 10.6 = 118,323,505 \text{ in.}^4$$

$$\text{Angle change per unit length} = \frac{T}{(GJ)_{Al} + (GJ)_{steel}}$$

$$= \frac{5162 \times 144}{1475,527,094 + 118,323,505}$$

$$= 0.000466457 \text{ rad/ft}$$

$$\text{Angle of twist} = 0.000466457 \times \frac{180}{\pi} = 0.2192°$$

Shearing stress in steel
$$\left(\frac{OD}{2}\right)_{steel} \times \frac{T}{(GJ)_{Al} + (GJ)_{steel}} = 0.000466457 \times \frac{8}{2 \times 12} = 608 \text{ psi}$$

Shearing stress in aluminum
$$\left(\frac{OD}{2}\right)_{Al} \times \frac{T}{(GJ)_{Al} + (GJ)_{steel}} = 0.000466457 \times \frac{4}{2 \times 12} = 869 \text{ psi}$$

424. Problem 11.45

Calculate the total angle change at the bottom of the well of depth 12,000 ft. The drillstring assembly consists of only drill pipe of 5" OD (outside diameter) × 4.26" ID (inside diameter) and 1200 ft of 8" OD × 2 13/16" ID drill collar.

Solution:

Twist angle can be given as

$$\theta = \frac{TL}{GJ} \text{ rad,}$$

where

$$J = \frac{\pi}{32}\left(D_o^4 - D_i^4\right).$$

$$J_{body} = \frac{\pi}{32}\left(5^4 - 4.4276^4\right) = 28.54.$$

$$J_{collar} = \frac{\pi}{32}\left(8^4 - 2.8125^4\right) = 396 \text{ in.}^4$$

$$\theta_M = \frac{650 \times 12}{11.5 \times 10^6}\left(\frac{10,800}{28.54} + \frac{1200}{396}\right) = 3.13 \text{ rad} = 79°.$$

425. Problem 11.46

Calculate the length of 5" – 19.5 ppf E-grade drill pipe for a total angle change of 180° when used with 1200 ft of 8" × 2 13/16" drill collar. The bit generated torque is 650 ft-lbf. Use shear modulus for steel (11.4 × 10⁶ psi).

Solution:

Twist angle can be given as

$$\theta = \frac{TL}{GJ} \text{ rad,}$$

where

$$J = \frac{\pi}{32}\left(D_o^4 - D_i^4\right).$$

Inside diameter of 5" drill pipe for 19.5 ppf is 4.4276"
Polar moment of inertia of pipe is

$$J_{body} = \frac{\pi}{32}\left(5^4 - 4.4276^4\right) = 28.54$$

Polar moment of inertia of drill collar

$$J_{collar} = \frac{\pi}{32}\left(8^4 - 2.8125^4\right) = 396 \text{ in.}^4$$

Length of drill collar is 1200 ft and the length of drill pipe is L ft.

Angle change is 182° which is equal to $\dfrac{\pi}{180} \times 182 = 3.176$ rad

$$\theta_M = \frac{650 \times 12 \times 12}{11.5 \times 10^6}\left(\frac{L}{28.54} + \frac{1200}{396}\right) = 3.176$$

$$L = 11{,}130 \text{ ft}$$

426. Problem 11.47

Calculate the angle change at a well depth 12,200 ft with the Drillstring comprising of drill pipe of 5" – 19.5 ppf E-grade and 1200 ft of 8" × 2 13/16" drill collar. The bit generated torque is 650 ft-lbf. Use shear modulus for steel (11.4×10^6 psi).

Solution:

Inside diameter of 5" drill pipe for 19.5 ppf is 4.4276"
Polar moment of inertia of pipe is

$$J_{body} = \frac{\pi}{32}\left(5^4 - 4.4276^4\right) = 28.54$$

Polar moment of inertia of drill collar

$$J_{collar} = \frac{\pi}{32}\left(8^4 - 2.8125^4\right) = 396 \text{ in}^4$$

Length of drill collar = 1200 ft
Length of drill pipe = (12,200 – 1200) = 11,000 ft

$$\theta_M = \frac{650 \times 12 \times 12}{11.5 \times 10^6}\left(\frac{11{,}000}{28.54} + \frac{1200}{396}\right) = 3.1616$$

Angle change is 3.1616 which is equal to $\frac{180}{\pi} \times 3.1616 = 181.2°$

427. Problem 11.48
Friction Calculations
Coefficient of friction

The coefficient of friction (COF) is defined as the ratio of the frictional force to the normal force acting at the point of contact. It is given as

$$\mu = \frac{F_f}{F_n},$$

where
F_f = friction force, lbf.
F_n = normal force, lbf.

The COF is a scalar dimensionless value that depends on the surface but is independent of the surface area. Table 11.3 and Table 11.4 shows typical COFs for various materials.

Table 11.3 Typical coefficient of friction (Rabbat, 1985)

Material 1	Material 2	Dry		Lubricated	
		Static	Sliding	Static	Sliding
Steel	Steel	0.78	0.42	0.05–.11	0.29 – 0.12
Aluminium	Aluminum	1.05 – 1.35	1.4	0.3	
Aluminum	Mild Steel	0.61	0.47		
Copper-Lead	Steel	0.22			
Diamond	Diamond	0.1			
Diamond	Metal	0.15			
Steel	Concrete	0.57 – 0.75	0.45		
Steel	EmbeddedSand		0.7		

Table 11.4 Range of friction factors

Fluid type	Friction factors	
	Cased hole	Open hole
Oil-based	0.16–0.20	0.17–0.25
Water-based	0.25–0.35	0.25–0.40
Brine	0.30–0.4	0.3–0.4
Polymer-based	0.15–0.22	0.2–0.3
Synthetic-based	0.12–0.18	0.15–0.25
Foam	0.30–0.4	0.35–0.55
Air	0.35–0.55	0.40–0.60

428. Problem 11.49

Types of Friction

Static friction is

$$\mu_s = \frac{F_{sf}}{F_n}$$

Kinetic friction is

$$\mu_k = \frac{F_{kf}}{F_n}$$

A typical static and kinetic coefficient plot is shown in Figure 11.3. Rolling friction is

$$\mu_r = \frac{F_{rf}}{F_n}$$

Angle of Friction

Angle of friction is

$\varphi = \tan^{-1} \mu_s$

Slide/roll friction is

$\varphi = \tan^{-1} \mu_s$

The kinetic friction and the friction angle are related as follows:

$$\mu_k = \tan \varphi - \frac{a_x}{g \sin \varphi}$$

Tubulars

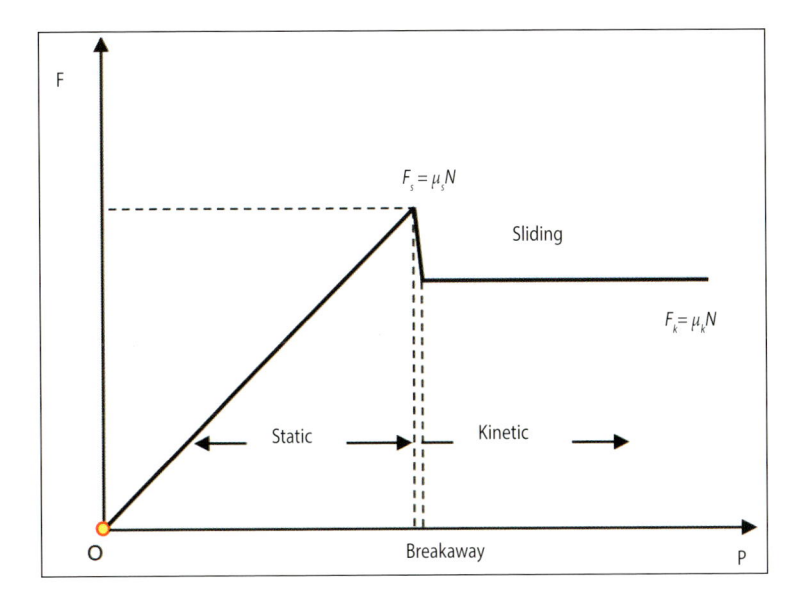

Figure 11.3 Friction force as a function of pulling force

429. Problem 11.50

Friction and Rotational Speed

The following empirical equation provides a good representation and coupling of the friction effects and drillstring rotating speed as well as tripping speed:

$$\mu_v = \mu_s \times e^{-k|V_{rs}|}$$

The resultant velocity, V_{rs}, of a contact point on the drillstring is the vector sum of two components: circumferential velocity V_c (caused by rotation) and axial velocity V_{ts} (affected by drilling rate or tripping speed).

The friction factor, which has the dependency on the side force, kinematics, temperature, and geometrical parameters of the contacting surfaces, is given by

$$\mu_v = \frac{\mu_s}{1 + \left(\dfrac{\mu_s \sigma_n}{k\Delta t}\right)|V_{rs}|}$$

where

σ_n = normal stress at the contact.
Δt = average contact temperature.
$|V_{ts}|$ = trip speed.
$|V_{rs}|$ = resultant speed = $\sqrt{(V_{ts}^2 + \omega^2)}$.
$|\omega|$ = angular speed = diameter × π × $\dfrac{N}{60}$.
N = pipe rotational speed, rpm.

430. Problem 11.51

What is the kinetic energy of the hammer inside a mechanical jar whose weight is 10 lbm travelling at 60 ft/sec?

Solution:

Mass = 10/32.2 = 0.31055 slugs
Velocity = 60 ft/sec
Kinetic energy = $\dfrac{1}{2}mv^2 = \dfrac{1}{2} 0.31055 \times 60^2$ = 559 ft-lbf

431. Problem 11.52

The side force at the bit in anisotropic formation can be calculated using the equation

$$F_s = p\sqrt{EI}\left[\frac{W}{24}\left(\frac{24r}{EIp\sin\theta}\right)^{\frac{3}{4}} - \left(\frac{3}{2}\frac{r}{EIp\sin\theta}\right)^{\frac{1}{4}}\right],$$

where
p = unit weight of the drill collar in mud.
EI = stiffness.
r = radial clearance between the wellbore and drill collar.
W = weight-on-bit.
θ = inclination angle.

Derive an equation for the side force to drill a horizontal section of the well.

Solution:

The general equation to calculate the side force at the bit is given as

$$F_s = p\sqrt{EI}\left[\frac{W}{24}\left(\frac{24r}{EIp\sin\theta}\right)^{\frac{3}{4}} - \left(\frac{3}{2}\frac{r}{EIp\sin\theta}\right)^{\frac{1}{4}}\right],$$

where

I = moment of inertia of drill collar above the bit, in.^{4}
r = radial clearance between the drill collar and hole diameter, ft.
W = weight-on-bit in lbf.
E = Modulus of elasticity.
θ = inclination angle, deg.
p = buoyed weight of the drill collar per unit length (ppf).

For horizontal wells $\theta = 90°$ and substituting in the above equation, it reduces to

$$F_s = p\sqrt{EI}\left[\frac{W}{24}\left(\frac{24r}{EIp}\right)^{\frac{3}{4}} - \left(\frac{3}{2}\frac{r}{EIp}\right)^{\frac{1}{4}}\right]$$

432. Problem 11.53

Calculate the side force required to maintain a horizontal section while drilling in an anisotropic formation. The weight-on-bit applied is 5 kips. Drillstring contains drill collar 6 ¼" × 2 ½" 88 ppf; bit diameter 8 ½", PDC bit; mud density, 9 ppg. Estimate the side torque for a bit friction factor of 0.1.

Solution:

Buoyancy factor, BF = 1 − 9/65.4 = 0.86
Unit weight, P = 88 × 0.86 = 75.88
Stiffness, EI = 15,204,985 lbf

Substituting the respective values in the above equation

$F_s = 75.88 \times 15{,}204{,}985$

$$\left[\frac{5000}{24}\left(\frac{24 \times 0.09375}{15{,}204{,}985 \times 75.8}\right)^{\frac{3}{4}} - \left(\frac{3}{2}\frac{0.09375}{15{,}204{,}985 \times 75.8}\right)^{\frac{1}{4}}\right] = -965 \text{ lbf}$$

433. Problem 11.54

A drilling engineer was planning to calculate the maximum length of the 9 ⅝" 47 ppf (air weight) casing that can be run into the hole. The mud weight of the drilling fluid in the hole is 12.5 ppg. The derrick capacity is 750 kips. Assume a safety factor to be used is 1.1 for the load and the casing is run open ended. Assume vertical well and calculate the maximum depth the casing can be run.

Solution:

BF = (1−12/65.5) = 0.816
Maximum load that can be handled = 750/1.1 = 681 kips
Maximum depth it can be run = 681,000/47 × 0.816 = 17,756 ft

434. Problem 11.55

Maximum Weight-on-Bit

Maximum weight on bit that can be applied for a given length of the drill collar is given as

$$WOB = \frac{L_{dc} \times w_{dc} \times BF \times \cos\alpha}{SF}$$

Alternatively the length of the drill collar required can be calculated as

$$L_{dc} = \frac{WOB \times SF}{w_{dc} \times BF \times \cos\alpha},$$

where
WOB = weight-on-bit in lbs.
  SF = design factor ranges from 1 to 2.
  w_{dc} = unit weight of the collar in lbf/ft in air.
  BF = buoyancy factor.
  α = wellbore inclination in degrees.

435. Problem 11.56

The ultimate strength of a downhole tool measured from five samples is given in the range as 130–150 kpsi. Calculate the mean value, standard deviation, and the coefficient of variation.

Tubulars

Solution:

The mean value is given as

$$\text{Mean value, } x = x_1 + x_2 + x_3 + \ldots X_n = \frac{1}{n}\sum_{i=1}^{n} x_i,$$

where

x_1, x_2, x_3 = values of different measurements of the same downhole tool material.

n = number of measurement.

Mean value can be taken as 140 kpsi

Standard deviation is

$$\text{Std deviation, } \sigma = \sqrt{\left(\frac{1}{n-1}\sum_{i=1}^{n}(x_i - x)^2\right)}$$

So the standard deviation = (150 − 130)/5 = 4 kpsi
Coefficient of variation = 4/140 = 0.0285

Smaller the standard deviation and coefficient of variation, more the homogenous material.

436. Problem 11.57

Estimate the size and length of the drill collars for the following data:
Hole size = 8 ½"
Well inclination = 20°
Casing size 7", 38 ppf, P-110 – BTC casing
Mud weight = 12 ppg
Weight-on-bit desired = 20 kips
Design factor = 2

Solution:

Coupling OD of the 7" 38 ppf BTC is 7.656 in.
Diameter of the drill collar is
D_{dc} = 2 × 7.656 − 8.5 = 6.812".
The size of the drillcollar = 6.5"

The buoyancy factor is

$$BF = 1 - \frac{12}{65.39} = 0.816$$

The length of the drill collar is

$$L_{dc} = \frac{20{,}000 \times 2}{99 \times 0.82 \times \cos 20} = 525 \text{ ft}$$

437. Problem 11.58

Composite Materials Calculations

For longitudinal directional ply and longitudinal tension, modulus can be given as

$$E = V_m E_m + V_f E_f,$$

where

E_m = elastic modulus of base pipe, psi.
E_f = elastic modulus of rubber attachment.
V_m = volume fraction of matrix.
V_f = volume fraction of fiber attachment.
Also, $V_m + V_f = 1$, the same way Poisson's ratio can be calculated:
$v = V_m v_m + V_f v_f$

When the stress is applied perpendicular to the fiber orientation, the modulus of the elasticity of the composite material can be given as

$$\frac{1}{E} = \frac{V_m}{E_m} + \frac{V_f}{E_f}$$

438. Problem 11.59

Estimate the modulus of the composite shaft with 25 % of the total volume with fibers. Assume the modulus of elasticity for the fiber is 50×10^6 psi and the matrix 600 psi and the load is applied longitudinally as well as perpendicular to the fibers.

Solution:

When the load is applied longitudinally to the fibers

$E = V_m E_m + V_f E_f = 500 \times 0.75 + 25{,}000{,}000 \times 0.25 = 6{,}250{,}450$ psi

When the load is applied perpendicular to the fibers

$$\frac{1}{E} = \frac{V_m}{E_m} + \frac{V_f}{E_f} = \frac{.25}{500} + \frac{0.75}{25,000,000} = 0.00125$$

Thus, $E = 800$ psi

439. Problem 11.60
What are the four modes of vibration?
Solution:
  Axial, torsional, lateral, and coupled. (Fig.11.4)

440. Problem 11.61
What are the different types of vibration?
Solution:
  Free vibration – usually it happens when drillstring is rotating off-bottom
  Forced Vibration – usually it happens during drilling
  Self-excited vibration – usually it happens due to mud motor operation
  Steady vibration – usually it happens during drilling
  Harmonic excitation – usually it happens during drilling

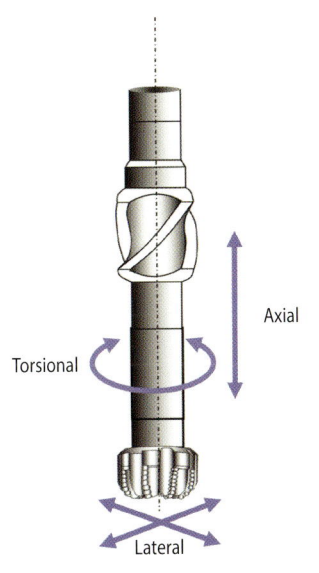

Figure 11.4   Vibration Types

Transient vibration – occurs in all operations

Parametric vibration – due to multiple sources such as stabilizers, under-reamer

Non-Linear vibration – occurs in all operations

Non-damped vibration – happens usually in underbalanced drilling operations

441. Problem 11.62

Explain why an increase in a rotary torque yields a decrease in load capacity of drill pipe.

Solution:

From Equation of Effective Stress for Axial and Torsional Criteria

$$\delta_e^2 = \delta_n^2 + 3t^2,$$

where

$$\delta_n = \frac{P}{A} \text{ and } t = \frac{T}{Z}.$$

$$\delta_e^2 = \left(\frac{P}{A}\right)^2 + 3\left(\frac{T}{Z}\right)^2.$$

$$P = A\sqrt{\delta_e^2 - 3\left(\frac{T}{Z}\right)^2}.$$

From the above expression, it is seen that an increase in Rotary Torque (T) yields a decrease in axial load capacity of the drill pipe.

442. Problem 11.63

A 15,000 ft string of 7 in., 26 lb/ft N80 casing is run in a hole filled with 10 ppg mud. The string is equipped with a float shoe. Calculate the following:
  a. Hook load of an empty closed end casing string in mud (air weight = 15,000 ft × 26 lb/ft = 390,000 lb)
  b. Hook load of a mud filled string in mud (same mud weights inside and outside)

c. Hook load of 16 ppg cement-filled string in mud

Solution:

a. The displaced volume of the mud is
$(\pi/4)d^2L = (\pi/4)(7/12)^2(15,000) = 4007 \text{ ft}^3$
and the displaced weight of the mud is
$4007 \times 12 \times 7.48 = 359{,}668 \text{ lbs}$
The displaced weight of the mud is a force acting upward on the casing, trying to float it out of the well. It has the effect of trying to lighten the total load.
Thus, hook load
$= 390{,}000 - 359{,}668 = 30{,}332 \text{ lbs}$ (this ignores the weight of the air)

b. Since the only action is on the cross-section of the steel body, the buoyancy factor difference formula is adequate.
$$b_f = 1 - \frac{\rho_f}{\rho_s} = 1 - (10 \times 7.48/489.5) = 0.847$$
$W_b = W_a b_f = (390{,}000 \text{ lbs})(0.847) = 330{,}330 \text{ lbs}$

c. Total buoyant weight = weight of casing in air + weight of cement − buoyancy of displaced mud.

Weight of cement is

$(\pi/4)(6.276 \text{ in.})^2 (16 \text{ ppg})(0.052)(15{,}000 \text{ ft}) = 385{,}878 \text{ lbs}.$

From part (a), the buoyant force is 359,668 lbs. Thus, the total buoyant weight is

$390{,}000 + 385{,}878 - 359{,}668 = 416{,}210 \text{ lbs}.$

It should be noticed that the hook load with 16 ppg cement inside the casing is significantly higher than the air weight of the casing string. Hence, proper hook load design parameters should be selected in accordance with the well plan.

443. Problem 11.64

Calculate the True Tension and Effective Tension Calculation at the surface and at 2750 ft for a tripping out case:

Given data:
Wellbore depth 5500 ft – 10" OD, vertical
Drillstring Data:
Drill pipe: 5" × 3" – 19.5 ppf with flush tool joints
Mud Density – 9.2 ppg
Also determine both tensions when a surface pressure of 1000 psi is applied on the drillstring side.

Solution:

Tension without Surface Pressure

True Tension Calculation:
Weight component = 5500 × 19.5 × cos(0) = 107.25 kips
For the Tripping Out Case:
WOB = 0 kips as the case is tripping out
DF = 0 as the wellbore is vertical
ΔF_{area} = 0 as the cross-sectional area is same
F_{bot} = the compressive force due to fluid pressure applied at the bottom of the pipe

$F_{bot} = P_e(A_e - A_i) = 0.052 \times 5500 \times 9.2 \left(\frac{\pi}{4}5^2 - \frac{\pi}{4}3^2\right) = -33$ kips
(force is compressive)

Effective Tension at Surface:
F_{bs} = 0 due to the fact that $P_i = P_e = 0$, pertaining to this case.
So the true tension and effective tension at the surface:
= 107.25 – 0 + 0 – 33 + 0 = 74.2 kips

True Tension at 2750 ft:
Weight component = 2750 × 19.5 × cos(0) = 53.6 kips

$F_{bot} = P_e(A_e - A_i) = 0.052 \times 5500 \times 9.2 \left(\frac{\pi}{4}5^2 - \frac{\pi}{4}3^2\right) = -33$ kips
(force is compressive)

So the true tension at 2750 ft:
= 53.6 – 0 + 0 – 33 = 20.6 kips
$F_{bs} \neq 0$ as P_i, $P_e \neq 0$ at this depth

Tubulars

$$F_{bs} = P_e A_e - P_i A_i = 0.052 \times 2750 \times 9.2 \left[\left(\frac{\pi}{4} 5^2\right) - \left(\frac{\pi}{4} 3^2\right)\right] = 16.5 \text{ kips}$$

Effective tension at this depth
$= T_t + F_{bs} = 20.6 + 16.5 = 37.1$ kips

444. Problem 11.65

Tension with Surface Pressure (Drillstring Side)

True Tension at surface when a surface pressure of 1000 psi is applied at the string side:

Weight component = $5500 \times 19.5 \times \cos(0) = 107.25$ kips

$F_{bs} \neq 0$ as $P_e = 0$, $P_i \neq 0$ at the surface due to the pressure at the string side.

$$F_{bs} = -P_i A_i = -1000 \left[\left(\frac{\pi}{4} 3^2\right)\right] = 7.07 \text{ kips}$$

True tension at surface:
= $107.25 - 0 + 0 - 33 + 7.07 = 81.32$ kips

Effective tension at the surface:
= $81.32 - 7.07 = 74.2$ kips

True Tension at 2750 ft:

Weight component = $2750 \times 19.5 \times \cos(0) = 53.6$ kips

$$F_{bot} = P_e(A_e - A_i) = 0.052 \times 5500 \times 9.2 \left(\frac{\pi}{4} 5^2 - \frac{\pi}{4} 3^2\right) = -33 \text{ kips}$$
(force is compressive)

So the true tension at 2750 ft:
= $53.6 - 0 + 0 - 33 + 7.07 = 27.67$ kips

$F_{bs} \neq 0$ as P_i, $P_e \neq 0$ at this depth
$F_{bs} = P_e A_e - P_i A_i$

$$= \left(0.052 \times 2750 \times 9.2 \left(\frac{\pi}{4} 5^2\right)\right) - \left((0.052 \times 2750 \times 9.2 + 1000)\left(\frac{\pi}{4} 3^2\right)\right)$$

= 9.46 kips

Effective tension at this depth = $T_t + F_{bs} = 27.67 + 9.46 = 37.1$ kips

445. Problem 11.66
Tension with Surface Pressure (Annulus Side)

True Tension at Surface when a surface pressure of 1000 psi is applied on the annulus side:

Weight component = 5500 × 19.5 × cos(0) = 107.25 kips

$F_{bs} \neq 0$ as $P_e \neq 0$, $P_i = 0$ at the surface due to the pressure at the string side.

$$F_{bs} = P_e A_e = 1000\left[\left(\frac{\pi}{4}5^2\right)\right] = 19.64 \text{ kips}$$

$$F_{bot} = P_e(A_e - A_i) = 0.052 \times 5500 \times 9.2\left(\frac{\pi}{4}5^2 - \frac{\pi}{4}3^2\right) = -33 \text{ kips}$$
(force is compressive)

True tension at surface:
= 107.25 − 0 + 0 − 33 − 19.64 = 54.61 kips

Effective tension at the surface:
= 54.61 + 19.64 = 74.2 kips

True Tension at 2750 ft:
Weight component = 2750 × 19.5 × cos(0) = 53.6 kips

$$F_{bot} = P_e(A_e - A_i) = 0.052 \times 5500 \times 9.2\left(\frac{\pi}{4}5^2 - \frac{\pi}{4}3^2\right) = -33 \text{ kips}$$
(force is compressive)

So the true tension at 2750 ft:
= 53.6 − 0 + 0 − 33 − 19.64 = 1 kip

$F_{bs} \neq 0$ as P_i, $P_e \neq 0$ at this depth.

$F_{bs} = P_e A_e - P_i A_i$

$$= \left((0.052 \times 2750 \times 9.2 + 1000)\left(\frac{\pi}{4}5^2\right)\right) - \left((0.052 \times 2750 \times 9.2)\left(\frac{\pi}{4}3^2\right)\right)$$

= 36.18 kips

Effective tension at this depth = $T_t + F_{bs}$ = 1 + 36.18 = 37.1 kips

446. Problem 11.67
Tension with Surface Pressures (String and Annulus sides)
True Tension at Surface when a surface pressure of 1000 psi is applied on the annulus side:

Weight component = $5500 \times 19.5 \times \cos(0) = 107.25$ kips

$$F_{bot} = P_e(A_e - A_i) = (0.052 \times 5500 \times 9.2 + 1000)\left(\frac{\pi}{4}5^2 - \frac{\pi}{4}3^2\right) = -45.6$$

kips (force is compressive)

$F_{bs} \neq 0$ as $P_e \neq 0$, $P_i \neq 0$ at the surface due to the pressure at the string side

$$F_{bs} = P_e A_e - P_i A_i = (1000)\left[\left(\frac{\pi}{4}5^2\right) - \left(\frac{\pi}{4}3^2\right)\right] = 12.57 \text{ kips}$$

True tension at surface:
= 107.25 − 0 + 0 − 45.6 = 61.65 kips

Effective tension at the surface:
= 61.65 + 12.57 = 74.2 kips

True Tension at 2750 ft:

Weight component = $2750 \times 19.5 \times \cos(0) = 53.6$ kips

$$F_{bot} = P_e(A_e - A_i) = (0.052 \times 5500 \times 9.2 + 1000)\left(\frac{\pi}{4}5^2 - \frac{\pi}{4}3^2\right) = -45.6$$

kips (force is compressive)

So the true tension at 2750 ft:
= 53.6 − 0 + 0 − 45.6 = 8 kip

$F_{bs} \neq 0$ as P_i, $P_e \neq 0$ at this depth

$$F_{bs} = P_e A_e - P_i A_i$$

$$= \left[(0.052 \times 2750 \times 9.2 + 1000)\left(\frac{\pi}{4}5^2\right)\right] - \left[(0.052 \times 2750 \times 9.2 + 1000)\left(\frac{\pi}{4}3^2\right)\right]$$

= 29.11 kips

Effective tension at this depth = $T_t + F_{bs}$ = 8 + 29.11 = 37.1 kips

Summary

	At surface				At 2750 ft			
	No surface pressures	Drillstring surface pressure	Annulus surface pressure	Both sides	No surface pressures	Drillstring surface pressure	Annulus surface pressure	Both sides
True Tension	74.2	81.32	54.6	61.5	20.6	27.67	1	8
Effective tension	74.2	74.2	74.2	74.2	37.1	37.1	37.1	37.1

447. Problem 11.68

Calculate the True Tension and Effective Tension at the surface for the following data:

Wellbore depth 7500 ft – 10" OD, Drillstring Data: Drillpipe: 5" – 19.5 ppf E grade Class pipe. Tool joint effects may be neglected.

Mud Density – 9.5 ppg. Assume a wellbore inclination of 30°. Azimuth = 0

Friction factor: 0.25

a. Compute the tensions when the tripping out speed is 30 fpm.
b. Estimate the tensions when tripping in speed is 30 fpm with no rotation and with simultaneous rotation of 20 rpm.

Solution:

a. Tension without Surface Pressure – Tripping Out Operation
True Tension Calculation:
Average weight = 19.5 ppf since tool joint effect is neglected. Else average weight will be 20.89 ppf
Weight component = 7500 × 19.5 × cos(30) = 126.66 kips
WOB = 0 kips as the case is tripping out
Buoyancy factor = (1 – 71.065/489.024) = 0.85468
Side force = 0.85468 × 19.5 × Sin (30) = 8.31(lb/ft)
DF = 8.31 × 7500 × 0.25 = 15.58 kips
ΔF_{area} = 0, as the cross-sectional area is same

Tubulars

F_{bot} = compressive force due to fluid pressure applied at the bottom of the pipe

The bottom force should be equal to the stability force at the bottom of the pipe. This bottom force is applied throughout the pipe uniformly whereas the buckling stability force is calculated using the same equation but with different depths.

$$F_{bot} = P_e(A_e - A_i) = 0.052 \times 7500 \times \cos 30 \times 9.5\left(\frac{\pi}{4}5^2 - \frac{\pi}{4}4.276^2\right)$$

= −16.92 kips (force is compressive)
Effective Tension at Surface:
F_{bs} = 0 due to the fact that $P_i = P_e$ = 0, pertaining to this case.
So the true tension and effective tension at the surface for tripping out operation:
= 126.66 + 15.58 − 16.9 = 125.34 kips

448. Problem 11.69
Tension without Surface Pressure – Tripping in Operation

True Tension Calculation:
With Tripping Speed = 30 fpm and RPM = 0
Weight component = 7500 × 19.5 × cos(30) = 126.66 kips
WOB = 0 kips as the case is tripping in
ΔF_{area} = 0 as the cross-sectional area is same
F_{bot} = − 16.92 kips (force is compressive)
Effective Tension at Surface:
F_{bs} = 0 due to the fact that $P_i = P_e$ = 0, pertaining to this case.
So the true tension and effective tension at the surface for tripping out operation:
= 126.66 − 15.58 − 16.9 = 94.18 kips
Linear Velocity = 30/60 = 0.5 ft/sec
Tangential Velocity = 0.0523 × 20 × 5/16 = 0.436 ft/sec
Resultant Velocity = $\sqrt{0.5^2 + 0.436^2}$ = 0.66 ft/sec

Linear Velocity Ratio = $\sqrt{\dfrac{0.5}{0.66}}$ = 0.753

DF = 8.3 × 7500 × 0.25 × 0.753 = 11.74 kips

Effective Tension at Surface:

F_{bs} = 0 due to the fact that $P_i = P_e$ = 0, pertaining to this case

So the true tension and effective tension at the surface for tripping in operation:

= 126.66 − 11.74 − 16.9 = 98 kips

449. Problem 11.70

Example Problem for Riserless Drilling

Calculate the True Tension and Effective Tension Calculation at the surface and at 5000 ft for a tripping out case:

Given data:

Wellbore depth 6000 ft – 20" OD, vertical

Drillstring Data:

Drill pipe: 5" × 4.276" – 19.5 ppf with flush tool joints

Mud Density – 9.5 ppg

1. When inside and outside are with uniform density of 9.5 ppg.
2. Density of 8.4 ppg up to a depth of 5000 ft in the annulus and rest with the fluid of 9.5 ppg outside and 9.5 ppg inside.

Also determine both tensions when a surface pressure of 1000 psi is applied on the drillstring side.

Solution:

a. Tension without Surface Pressure

True Tension Calculation:

Weight component = 6000 × 19.5 × cos(0) = 117 kips

WOB = 0 kips as the case is tripping out

DF = 0 as the wellbore is vertical

ΔF_{area} = 0 as the cross-sectional area is same

F_{bot} = the compressive force due to fluid pressure applied at the bottom of the pipe

The bottom force should be equal to the stability force at the bottom of the pipe. This bottom force is applied throughout the pipe uniformly whereas the buckling stability force is calculated using the same equation but with different depths.

$$F_{bot} = P_e(A_e - A_i) = 0.052 \times 6000 \times 9.5\left(\frac{\pi}{4}5^2 - \frac{\pi}{4}4.276^2\right) = -15.63$$

kips (force is compressive)

Effective Tension at Surface:

$F_{bs} = 0$ due to the fact that $P_i = P_e = 0$, pertaining to this case

So the true tension and effective tension at the surface:

= 117 − 0 + 0 − 15.63 + 0 = 101.3 kips

True Tension at 5000 ft:

Weight component = (6000 − 5000) × 19.5 × cos(0) = 19.5 kips

$$F_{bot} = P_e(A_e - A_i) = 0.052 \times 6000 \times 9.5\left(\frac{\pi}{4}5^2 - \frac{\pi}{4}4.276^2\right) = -15.63$$

kips (force is compressive)

So the true tension at 5000 ft:

= 19.5 − 0 + 0 − 15.63 = −3.9 kips

$F_{bs} \neq 0$ as P_i, $P_e \neq 0$ at this depth.

$$F_{bs} = P_e A_e - P_i A_i = 0.052 \times 5000 \times 9.5\left[\left(\frac{\pi}{4}5^2\right) - \left(\frac{\pi}{4}4.276^2\right)\right] = 13.02 \text{ kips}$$

Effective tension at this depth = $T_t + F_{bs}$ = − 3.9 −13.02 = −16.92 kips

b. Tension with different density to a certain depth in the annulus – no surface pressures

True Tension Calculation:

Weight component = 6000 × 19.5 × cos(0) = 117 kips

$F_{bs} = 0$ as $P_e = P_i = 0$, at the surface due to different fluid densities

$$F_{bot} = P_e(A_e - A_i) = 0.052 \times 5000 \times 8.4\left(\frac{\pi}{4}5^2 - \frac{\pi}{4}4.276^2\right)$$
$$+ 0.052 \times 1000 \times 9.5\left(\frac{\pi}{4}5^2 - \frac{\pi}{4}4.276^2\right)$$
= − 10 kips (force is compressive)

True Tension at Surface:

= 117 − 0 + 0 − 0 − 10 = 107 kips

Effective tension at the surface:
= 107 + 0 = 107 kips

True Tension at 5000 ft:
Weight component = (6000 − 5000) × 19.5 × cos(0) = 19.5 kips

$$F_{bot} = P_e A_e - P_i A_i = 0.052 \times 5000 \times 8.4 \left(\frac{\pi}{4} 5^2\right)$$

$$- 0.052 \times 5000 \times 8.4 \left(\frac{\pi}{4} 5^2\right) = 7.14 \text{ kips}$$

So the true tension at 5000 ft:
= 19.5 − 0 + 0 − 7.14 = 9.5 kips
$F_{bs} \neq 0$ as P_i, $P_e \neq 0$ at this depth
Effective tension at this depth = $T_t + F_{bs}$ = 9.5 + 7.4 = 16.9 kips

CHAPTER 12

TUBULAR WEAR

Tubular wear includes drillpipe wear, casing wear, riser wear, and downhole tools wear. This chapter focuses on the different wear models, basic calculations, and solved problems related to casing, riser wear.

450. Problem 12.1

What are the different wear estimation models?

Solution:
- specific energy model
- linear wear-efficiency model
- nonlinear casing wear model
- Hertzian model
- impact wear model
- wellbore energy model

451. Problem 12.2

What are wear factors?

Solution:

Wear factor, f_w, controls wear efficiency and is defined as

$$f_w = \frac{f}{e} \text{ in}^2/\text{lb or } 1/\text{Pa}.$$

e = specific energy, in-lb/in.^{3} = lb/in.^{2}

$$f_{Nf} = \frac{f_w}{W}.$$

f_W = casing wear condition coefficient, (m/Pa).
W = half of the contact width, (m).
f_{NF} = casing wear factor of nonlinear model, (1/Pa).

452. Problem 12.3

List the typical wear factors.

Solution:

Typical wear factors are listed in Table 12.1.

453. Problem 12.4

Calculate the amount of N-80 worn material based on the linear wear-efficiency model.

Table 12.1 Wear factors for the listed cases

Casing material	Mud type with additives	Tool joint	Rang of wear factor (e-10/psi)
N-80	Water	S-S	28.8–42.3
N-80	Water C. DrilBeads	S-S	0.12–4.7
N-80	Water EP-Lube	S-S	0.09–8.80
N-80	WB 2% Sand 3% Barite	S-S	0.8–1.7
N-80	WB 4% Ironox	S-S	3.70–5.50
N-80	WB 7% Sand	ARMACOR-M (Smooth)	0.6–0.64
N-80	WB 7% Sand	HB-S	10–1835
N-80	WB 7% Sand C. DrilBeads	S-S	3.5–5.2
N-80	WB 7% Sand EP Lube	S-S	0.9
N-80	Brine (Supersaturated)	S-S	0.43
N3-80	WB 7% Sand	S-S	1.2–4.6
N4-80	WB 7% Sand	ARMACOR M	1.1–2.8
L-80	ASP-POL/K-37SPHERES	S-S	0.55–1.17
P-110	WB 7% Sand	S-S	8.50–10.8
P-110 (BP-1)	WB 7% Sand	S-S	7.26
K-55	WB 7% Sand	S-S	10.3–13.6
K-55	WB 7% Sand 1% DL100 Lube	S-S	1.8
C-75 (PB-4)	WB 7% Sand	S-S	6.39
C-110 (BP-4)	WB 7% Sand	S-S	6.56
C-90 (MOBIL 1)	WB 7% Sand	S-S	7.47
C-95(ORYX)	WB 7% Sand	S-S	10.91
Q-125	WB 7% Sand 0.6% Ep Lube	S-S	6.31–8.20

Dogleg length = 100 ft
Tooljoint diameter = 6.625"
Sideforce = 5000 lbf
Assume a friction coefficient of 0.25
Total expected revolutions of the pipe = 1 million
The average values of the wear efficiency and $\frac{\eta}{H}$ is given in Table 12.2.

Solution:
Circumference rubbed = $\pi \times 6.25$
Total sliding length = $\pi \times 6.25 \times 1 \times 10^6 = 208.13 \times 10^6$ in.
Friction force = $0.25 \times 5000 = 1250$ lbf
$\frac{\eta}{H}$ from table is 0.81×10^{-9}
So, wear volume = $0.81 \times 10^{-9} \times 208.13 \times 10^6 = 210.73$ in³

If the volume is uniformly distributed, the wear volume will be 210.73/100 = 2.107 in³/ft

454. Problem 12.5

Calculate the casing wear for the following data:

Operation: Drilling from 5000 to 9000 ft. Rotary speed is 120 rpm. Rate of penetration (ROP) is 100 ft/hr.

Drill Pipe: use E-75 IEU XH drill pipe. Its length is 9000 ft. Its outer diameter is 5 in. and inner diameter is 4.276 in. Its weight is 20.9 lb/ft. The outer diameter of its tool-joint is 6.375 in.

Table 12.2   Average values of wear efficiency and η/H

Mud type	Casing grade	η	η/H
			psi-1×10⁻⁹
Water	K-55	0.0001	0.36
	N-80	0.00025	0.81
	P-110	0.00063	1.4
Oil	K-55	0.0006	2.2
	N-80	0.0012	3.9
	P-110	0.0017	4.2

Courtesy SPE Drilling Engineering—White and Dawson

Tubular Wear

Casing: and the inner diameter is 3.75 in. Tooljoint contact length is 14 in. and length of drill pipe joint is 30 ft.
use 9.625", L80 and DAMS new casing. Shoe depth is 5000 ft. Its outer diameter is 9.625 in. Its inner diameter is 8.681 in., its density is 490 lb/ft³.

Wellbore: Inclination at 3000 ft is 20° and dogleg is 2.0°.

Wear factor: 1.0 E-10/psi

Assumption: Drag force in drill pipe at 3000 ft is 103,440 lbf. Normal force on drill pipe at 3000 ft is 30 lbf/ft during drilling from 5000 to 9000 ft. It is to be noted that the normal force and drag force change as the well is drilled.

Correction factor is shown in Figure 12.1.

Solution:

Contact time between the tool-joint and casing, t, is

$$t = \frac{\left[9000(\text{ft}) - 5000(\text{ft})\right]\left(\frac{14(\text{in})}{12}\right)}{\frac{100\left(\frac{\text{ft}}{\text{hr}}\right) \times 30(\text{ft})}{60}} = 93.33 \text{ min}$$

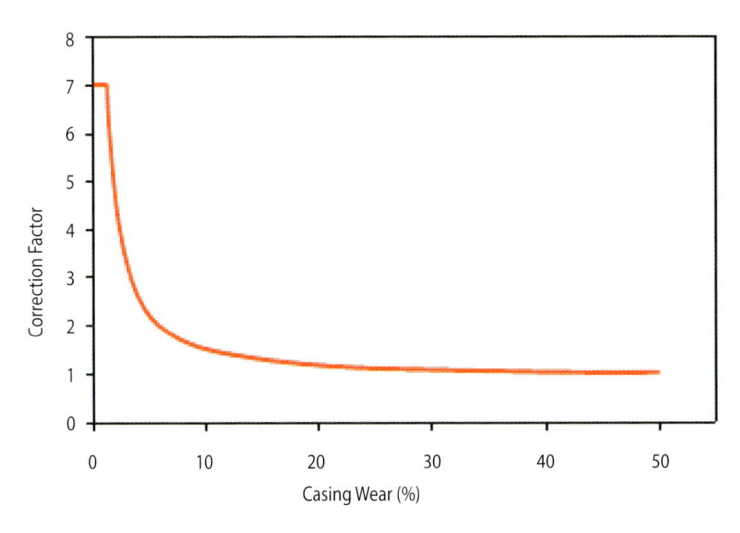

Figure 12.1 Correction factor for problem 12.5

Total sliding distance, $D_{sliding}$, between the tooljoint and casing is given by

$D_{sliding} = 3.14159 \times 120(\text{rpm}) \times 6.375(\text{in}) \times 93.33(\text{min}) = 2{,}24{,}309.5$ in.

Tool-joint side force per foot, F_s, is given by

$$F_s = \frac{30\left(\frac{\text{lbf}}{\text{ft}}\right) \times 30(\text{ft})}{14(\text{in})\cdot 12\left(\frac{\text{in}}{\text{ft}}\right)} = 771.43 \text{ lbf/ft}$$

Volume of material worn away in the crescent wear groove of a unit length, V_w, is calculated as

$V_w = 1.0 \times 0.0000000001 \times 771.43 \times 224{,}309.5 = 0.0173 \text{ in}^3/\text{ft}$

Solving nonlinear equations, depth of penetration into the casing/riser wall, h, is given by

$h = 0.004$ (in)

Casing wear percentage, CWP, is

$$\text{CWP} = \frac{0.004}{0.472} \times 100 = 0.847\%$$

The nonlinear correction factor can be obtained from Figure 12.1. The correction factor, f_{cf}, is 7.0, and then the modified CWP_{final} is given by

$\text{CWP}_{final} = \text{CWP} \times f_{cf} = 0.847 \times 7 = 5.932\%$

Remainder thickness of casing, RTC, is given by

$$\text{RTC} = \left(1 - \frac{5.932}{100}\right) \times 0.472 = 0.444\%$$

455. Problem 12.6

Calculate the riser wear for the following data:
Operation: Drilling from 8100 to 15,000 ft.
Rotary speed is 60 rpm.

Rate of penetration (ROP) is 60 ft/hr.
Hook load = 450,000 lbs.

Drill Pipe: Length is 15,000 ft. Its outer diameter is 5.975 in. and inner diameter is 5.045 in. Its weight is 28.4 lb/ft. The outer diameter of its tool-joint is 7.0 in. Tooljoint contact length is 10 in. and drill pipe joint length is 31.5 ft.

Riser: Riser length is 8011 ft. Outer diameter is 17.25 in. Inner diameter is 15.41 in, density is 489.00 lb/ft,^{3} and yield strength is 70,000 psi.

Riser Shape: the offset of rig is 10 ft. Inclination at 708 ft is 0.031° and dogleg is 0.0407°/100 ft.

Wear factor: 8.0 E-10/psi.

Assumption: Hook load (not include traveling assembly weight) is 450,000 lbf. Drag force in drill pipe at 708.0 ft is 448,305 lbf. Normal force on drill pipe at 708.0 ft is 2.8 lbf/ft during drilling from 8100 to 15,000 ft. It is to be noted that the normal force and drag force change as the well is drilled.

Correction factor for riser wear is given in Figure 12.2.

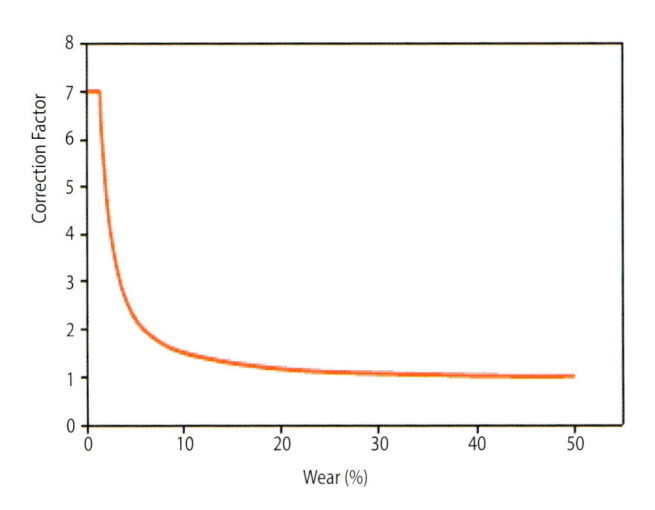

Figure 12.2 Correction factor for problem 12.6

Solution:

Contact time between the tool-joint and riser, t, is given by

$$t = \frac{[15{,}000(\text{ft}) - 8100(\text{ft})]\left(\dfrac{10(\text{in})}{12}\right)}{\dfrac{60\left(\dfrac{\text{ft}}{\text{hr}}\right) \times 31.5(\text{ft})}{60}} = 182.6 \text{ min}$$

Total sliding distance, $D_{sliding}$, between the tooljoint and riser is given by
$D_{sliding} = 3.14159 \times 60(\text{rpm}) \times 7(\text{in}) \times 182.6(\text{min}) = 240{,}934.8 \text{ in.}$

Tool-joint side force per foot, F_s, is calculated as

$$F_s = \frac{2.8\left(\dfrac{\text{lbf}}{\text{ft}}\right) \times 31.5(\text{ft})}{\dfrac{10(\text{in})}{12\left(\dfrac{\text{in}}{\text{ft}}\right)}} = 105.8 \text{ lbf/ft}$$

Volume of material worn away in the crescent wear groove of a unit length, V_w, is calculated as

$V_w = 8.0 \times 0.0000000001 \times 105.8 \times 240{,}934.8 = 0.02039 \text{ in}^3/\text{ft}.$

Solving nonlinear equation, the depth of penetration into the riser wall, h, is given by

$h = 0.006 \text{ (in)}$

Riser wear percentage, CWP, is

$$\text{CWP} = \frac{0.006}{0.92} \times 100.0 = 0.65(\%)$$

The nonlinear correction factor can be obtained from Figure 12.2. The correction factor, f_{cf}, is 7.0, and then the modified CWP_{final} is given by

$$\text{CWP}_{final} = \text{CWP} \times f_{cf} = 0.65 \times 7 = 4.55\%$$

Remainder thickness of riser, RTC, is given by

$$\text{RTC} = \left(1 - \frac{4.55}{100}\right) \times 0.92 = 0.878\%$$

456. Problem 12.7

Calculate the Stress Concentration Factor (SCF) and the derated burst pressure with the following data:

Casing size: 7" N-80, 32 ppf
Groove depth: 0.10", Groove diameter: 0.21"
Groove depth: 0.15", Groove diameter: 0.20"

Solution:

First Case:

Diameter of the groove, $d = 0.21$"
Depth of the groove, $h = 0.10$"
Since $2h < d$, the equation to calculate SCF is

$$SCF = \frac{b^2\gamma - (a-h)\sqrt{b^2 - (a-h)^2}}{b^2\gamma - a^2\beta - (a-h)^2(\tan\gamma - \tan\beta) + \dfrac{a^3(4m - 5m\mu + 3a^2n)}{(7 - 5\mu)}} K,$$

where

$$K = \frac{27 - 15\mu}{14 - 10\mu}.$$

$$a = \frac{d^2 + 4h^2}{8h}.$$

$$b = a - h + t.$$

$$\beta = \cos^{-1}\left(\frac{a-h}{a}\right).$$

$$\gamma = \cos^{-1}\left(\frac{a-h}{b}\right).$$

$$m = \frac{\sin\gamma - \sin\beta}{a - h} + \frac{\beta}{a} - \frac{\gamma}{b}.$$

$$n = \frac{3\sin\gamma - 3\sin\beta - \sin^3\gamma + \sin^3\beta}{3(a-h)^3} + \frac{\beta}{a^3} - \frac{\gamma}{b^3}$$

For the condition when the depth of the groove h equals half the groove diameter d,
$a = h = d/2$:
ID of the casing = 6.094"
$t = 0.906$"
Poisson's ratio $\mu = 0.3$

$$a = \frac{0.21^2 + 4 \times 0.1^2}{8 \times 0.1} = 0.1051 \text{ and } b = 0.1051 - 0.1 + 0.906 = 0.91$$

$$\gamma = \cos^{-1}\left(\frac{0.1051 - 0.1}{0.91}\right) = 1.5652 \text{ and } \beta = \cos^{-1}\left(\frac{0.1051 - 0.1}{0.1051}\right) = 1.52$$

Calculating the other constants
$m = 12.99$ and $n = 1318.53$

$$K = \frac{27 - 15 \times 0.3}{14 - 10 \times 0.3} = 2.045$$

$$SCF = \frac{b^2\gamma - (a-h)\sqrt{b^2 - (a-h)^2}}{b^2\gamma - a^2\beta - (a-h)^2(\tan\gamma - \tan\beta) + \frac{a^3(4m - 5m\mu + 3a^2n)}{(7 - 5\mu)}} K$$

$$SCF = \frac{0.91^2 \times 1.565 - (0.105 - 0.1)\sqrt{0.91^2 - (0.105 - 0.1)^2}}{0.91^2 \times 1.565 - 0.1051^2 \times 1.52 - (0.1051 - 0.1)^2(\tan 1.565 - \tan 1.52)}$$
$$+ \frac{0.105^3(4 \times 9.734 - 5 \times 9.734 \times 0.3 + 3 \times 105^2 \times 666.2)}{(7 - 5 \times 0.3)} \times 2.045$$

$= 2.05$

The corrected burst pressure is:

$$P_b = 0.875 \times 2 \times 80{,}000 \left(\frac{1}{7/0.906}\right)\left(\frac{1}{2.05}\right) = 4429 \text{ psi}$$

Second Case:
Diameter of the groove, $d = 0.15$"
Depth of the groove, $h = 0.20$"

Since $2h > d$, the equation used to calculate the SCF is
For the condition when the depth of the groove h equals half the groove diameter d, $a = h = d/2$:

$$SCF = \frac{\dfrac{27-15\mu}{14-10\mu}}{1-k_1\dfrac{a^3}{t^3}-k_2\dfrac{a^5}{t^5}}$$

where

$$k_1 = -\left(\frac{27-15\mu}{14-10\mu}\right)\left[\frac{5-4\mu^2}{(6-4\mu)(1+\mu)}\right]+2.5.$$

$$k_2 = \left(\frac{27-15\mu}{14-10\mu}\right)\left[\frac{5-4\mu^2}{(6-4\mu)(1+\mu)}\right]-1.5.$$

μ = Poisson's ratio
a = radius of a spherical corrosion groove, in.
r = distance to the center of the groove, in.
t = casing/tubing wall thickness, in.
ID of the casing = 6.094"
t = 0.906"
The constants are
a = 0.1083 and b = 0.86
Calculating the dimensionless parameters for this case

$$\psi = \cos^{-1}\left(\frac{0.2-0.1083}{0.86}\right) = 1.981 \text{ and } \omega = \cos^{-1}\left(\frac{.15-0.1083}{0.1083}\right) = 1.18$$

$$w = \pi - 1.18 - \frac{0.1083(\sin 1.117 - \sin 1.18)}{0.2-0.1083} - \frac{0.1083(\pi - 1.117)}{0.86}$$
$$+ \frac{3(1.117-1.18)}{4-5\times 0.3} = 1.9815$$

Other parameters are
u = 1191.21

For this condition, the SCF is

$$SCF = \frac{\pi\left[b^2 - b^2 \cdot \psi + (h-a)\sqrt{b^2 - (h-a)^2}\right]K}{(b^2 - a^2)(\pi - \psi) - a^2(\psi - \omega) + (h-a)^2(\tan\psi - \tan\omega) + K_1}$$

$$SCF = \frac{\pi\left[0.86^2 - 0.86^2 \times 1.18 + (0.2 - 0.1083)\sqrt{0.86^2 - (0.2 - 0.1083)^2}\right] \times 2.045}{(0.86^2 - 0.1083^2)(\pi - 1.18) - 0.1083^2(1.18 - 1.498)}$$

$$\frac{}{+(0.2 - 0.1083)^2(\tan 1.18 - \tan 1.498) + 0.03348} = 2.06$$

The corrected burst pressure is

$$P_b = 0.875 \times 2 \times 80,000 \left(\frac{1}{7/0.906}\right)\left(\frac{1}{2.06}\right) = 4405 \text{ psi}$$

457. Problem 12.8

Using the following data, calculate the Holmquist-Nadai collapse reduction factor and estimate the derated collapse strength of the pipe.

If the casing is worn out 5%, calculate the reduction in the collapse strength of the pipe.

7500 ft of 7.625 in K-55 26.4 ppf casing.
Mud weight: 9.50 ppg
Inclination of the well: 19.08°
Inside diameter of casing: 6.969 in.
Body collapse rating: 2890 psi
Body tensile yield rating: 414,000 lb
Connection tensile yield rating: 342,000 lb

Solutions:
Cross sectional area is given as

$$A = \frac{\pi}{4}(OD^2 - ID^2) = \frac{\pi}{4}(7.625^2 - 6.969^2) = 7.519 \text{ in}^2$$

$$T = 7500 \times 26.4 - 9.5 \times 0.052 \times 7087.6 = 171,700 \text{ lb}$$

$$\sigma_z = \frac{T}{A} = \frac{171,700}{7.52} = 22,835 \text{ psi}$$

$$\sigma_v = \frac{T_v}{A} = \frac{342,000}{7.52} = 45,484 \text{ psi}$$

$$T_r = \frac{\sigma_z}{\sigma_v} = \frac{22,835}{45,484} = 0.502$$

Using the Equations $\sigma_{yeq} = \sigma_y \left(\sqrt{1 - 0.75 t_1^2} - 0.5 t_1 \right)$

$\sigma_{y1} = \sigma_{yeq}$
$\sigma_{y2} = \sigma_{y1}^2$
$\sigma_{y3} = \sigma_{y1} \times \sigma_{y2}^2$

$$C_r = \frac{1}{2} \left[\sqrt{4 - 30.502^2} - 0.502 \right]$$

$C_r = 0.65$

Therefore, the collapse rating will be derated to 2890 × 0.65 = 1879 psi
If the casing is worn out to 5%, the new casing inside diameter is 6.969 × 10/100 = 7.31745 in.
Cross-sectional area is given as
A = 3.609 in²
$C_r = 0.6119$
Therefore the collapse rating will be derated to 2890 × 0.6119 = 1768 psi

The collapse strength reduction will be $\frac{1879 - 1768}{1879} \times 100 = 5.9\%$

458. Problem 12.9

Using the following data, calculate the collapse reduction factor using the equations provided in Table 12.3 and estimate the derated collapse strength of the pipe when the inside of the pipe is worn out to 2%. 7500 ft of 7.625 in K-55 26.4 ppf casing.

Mud weight: 9.50 ppg
Inclination of the well: 19.08°
Inside diameter of casing: 6.969 in.
Body collapse rating: 2890 psi
Body tensile yield rating: 414,000 lb
Connection tensile yield rating: 342,000 lb

Table 12.3 Collapse calculations

Failure mode	Equation	Range	Additional formula
Yield Collapse	$P_{cry} = 2\sigma_{ye}\left(\dfrac{d_{ot}-1}{d_{ot}^2}\right)$	$d_{ot} \leq (d_{ot})_{\text{limit 1}}$	$(d_{ot})_{\text{limit 1}} = \dfrac{\sqrt{t_1^2 + 8t_2} + t_1}{2t_2}$ $t_1 = \dfrac{\sigma_z}{\sigma_y},\ t_1 = a-2;\ t_2 = b + \dfrac{c}{\sigma_{y1}}$
Plastic Collapse	$P_{crp} = \sigma_{y1}\left(\dfrac{a}{d_{ot}} - b\right) - c$	$(d_{ot})_{\text{limit 1}} \leq d_{ot} \leq (d_{ot})_{\text{limit 3}}$	$(d_{ot})_{\text{limit 1}} = \dfrac{\sqrt{t_1^2 + 8t_2} + t_1}{2t_2}$ $(d_{ot})_{\text{limit 2}} = \dfrac{\sigma_{y1}(a-f)}{(c+\sigma_{y1}(b-g))}$ $(d_{ot})_{\text{limit 3}} = \dfrac{2+t_2}{3t_2}$ $t_1 = 46.95 \times 10^6$ $t_2 = \dfrac{b}{a},\ t_3 = \dfrac{3t_2}{2+t_2}$ $f = \dfrac{t_1 t_3^3}{\sigma_{y1}(t_3 - t_2)(1-t_3)^2}$ $g = f \times t_2$

Transition Collapse	$P_{crt} = \sigma_{y1}\left(\dfrac{f}{d_{ot}} - g\right)$	$(d_{ot})_{\text{limit 2}} \leq d_{ot} \leq (d_{ot})_{\text{limit 3}}$	$(d_{ot})_{\text{limit 2}} = \dfrac{\sigma_{y1}(a-f)}{(c+\sigma_{y1}(b-g))}$
			$(d_{ot})_{\text{limit 3}} = \dfrac{2+t_2}{3t_2},\ t_2 = \dfrac{b}{a}$
Elastic Collapse	$P_{cre} = \dfrac{t_1}{d_{ot}(d_{ot}-1)^2}$	$d_{ot} \geq (d_{ot})_{\text{limit 3}}$	$(d_{ot})_{\text{limit 3}} = \dfrac{2+t_2}{3t_2}$
			$t_1 = 46.95 \times 10^6$

$$a = 2.8762 + 0.10679 \times 10^{-5} \sigma_{y1} + 0.21301 \times 10^{-10} \sigma_{y2} - 0.53132 \times 10^{-16} \sigma_{y3}$$

$$b = 0.026233 + 0.50609 \times 10^{-6} \sigma_{y1}$$

$$c = -465.93 + 0.030867 \sigma_{y1} - 0.10483 \times 10^{-7} \sigma_{y2} + 0.36989 \times 10^{-13} \sigma_{y3}$$

$$\sigma_{yeq} = \sigma_y \left(\sqrt{1 - 0.75 t_1^2} - 0.5 t_1\right)$$

$$\sigma_{y1} = \sigma_{yeq}$$

$$\sigma_{y2} = \sigma_{y1}^2$$

$$\sigma_{y3} = \sigma_{y1} \times \sigma_{y2}^2$$

Solutions:
a. Under Zero Tension:
Cross-sectional area is given as

$$A = \frac{\pi}{4}(OD^2 - ID^2) = \frac{\pi}{4}(7.625^2 - 6.969^2) = 7.519 \text{ in.}^2$$

$$d_{ot} = \frac{2 \times 7.625}{7.625 - 6.969} = 23.25$$

The values of a, b, c, f, and g are further calculated using the values from Table 12.3 where

$t_1 = \dfrac{\sigma_z}{\sigma_y} = 0$ since there is no tensile load for the first part of the question.

$$\sigma_{yeq} = \sigma_y = \frac{414{,}000}{7.52} = 55{,}059.5 \text{ psi}$$

Therefore, $\sigma_{y1} = 55{,}059.5$, $\sigma_{y2} = 3{,}031{,}549{,}094$ and $\sigma_{y3} = 1.669 \times 10^{14}$
$a = 2.8762 + 0.10679 \times 10^{-5} \times 55059.5 + 0.21301 \times 10^{-10} \times 3{,}031{,}549{,}094$
$- 0.53132 \times 10^{-16} \times 1.669 \times 10^{14} = 2.99$
$b = 0.026233 + 0.50609 \times 10^{-6} \times 55{,}059.5 = 0.054$
$c = -465.93 + 0.030867 \times 55059.5 - 0.10483 \times 10^{-7} \times 3{,}031{,}549{,}094 +$
$0.36989 \times 10^{-13} \times 1.669 \times 10^{14}$
$c = 1207.98$
$f = 1.9893$
$g = 0.03598$

Then determine the different $(d_{ot})_{\text{limit}}$ limits to establish the collapse type:

Limit 1:

$$(d_{ot})_{\text{limit 1}} = \frac{\sqrt{t_1^2 + 8t_2} + t_1}{2t_2} = 14.80$$

Limit 2:

$$(d_{ot})_{\text{limit 2}} = \frac{55059.5(2.99 - 1.9893)}{(1207.98 + 55059.5(0.054 - 0.03598))} = 25$$

Limit 3:

$$(d_{ot})_{limit\,3} = \frac{2 + 0.01808}{3 \times 0.01808} = 37.186$$

For the above calculated limits, it can be inferred that d/t falls between $(d_{ot})_{limit\,1} \leq d_{ot} \leq (d_{ot})_{limit\,3}$. From Table 12.3 this range represents the plastic collapse and hence the reduced collapse resistance is

$$P_{crp} = 55{,}059.5\left(\frac{2.99}{23.247} - 0.054\right) - 1207.986 = 2890 \text{ psi}$$

When the casing is worn to 2%, the new limits need to be established.
Limit 1:
$d_{ot} = 29.52$
$(d_{ot})_{limit\,1} = 12.33$
Limit 2:
$(d_{ot})_{limit\,2} = 20.14$
Limit 3:
$(d_{ot})_{limit\,3} = 25.63$
For the limits calculated above, it can be inferred that d/t falls between $d_{ot} \geq (d_{ot})_{limit\,3}$ which is elastic collapse.
Hence the reduced collapse resistance is

$$P_{crp} = \frac{4{,}695{,}000}{29.51(29.51-1)^2} = 1955 \text{ psi}$$

It is to be noted that $469{,}500 = \frac{2E}{(1-v^2)}$,
where
E = Young's Modulus for Steel (30×10^6).
v = Poisson's ratio (0.3).

When the casing is worn to 5%, the new casing collapse strength would be 2017 psi which would result in a reduction of 32%.

b. For an axial load of 171,700 lbf

$$\sigma_{yeq} = 55{,}059.5\left(\sqrt{1 - 0.75 \times 0.4147^2} - 0.5 \times 0.4147\right) = 39{,}968 \text{ psi}$$

Equivalent stress is

$$\sigma_{yeq} = \sigma_y = \frac{414,000}{7.52} = 55,059.5 \text{ psi}$$

The other calculated values are
$a = 2.9495$, $b = 0.04646$, $c = 753.376$, $f = 2.0633$, $g = 0.03250$

Limit 1:
$(d_{ot})_{\text{limit 1}} = 16.40$

Limit 2:
$(d_{ot})_{\text{limit 2}} = 27.009$

Limit 3:
$(d_{ot})_{\text{limit 3}} = 42.656$

Again d/t falls between limit 1 and limit 2 and the applicable collapse criteria is transition collapse.

The reduce collapse rating is

$$P_{crp} = 39,968 \left(\frac{2.949}{23.247} - 0.046 \right) - 753.37 = 2461 \text{ psi}$$

If the reduced connection tensile rating of 342,000 lbs is considered as a limiting load, then the reduced collapse resistance will be 2062 psi.

CHAPTER 13

DRILLING OPERATIONS

In this chapter basic calculations and problems related to drilling operations are presented.

459. Problem 13.1

The tricone bits pulled out from a well were graded as below. Explain the dull bit grading:

T_3B_8I
$T_8B_8O_{1/2}$
T_3B_8I
$T_1B_1O_{1/4}CP$
$T_8B_8O_{1/4}LIH$

Solution:

T_3B_8I
⅜ of the teeth worn out, life of bearing left is zero and bit is in gage

$T_8B_8O_{1/2}$
Teeth and bearing are complete worn out
O – out of gage 1/2 an inch

$T_1B_1O_{1/4}CP$
⅛ of the teeth worn out, ⅛ of life of bearing is used and bit is out of gage ¼" and the reason for pulling out is core point reached

$T_8B_8O_{1/4}LIH$
Life of the bit is completely used up and cone left in the hole.

460. Problem 13.2

What is Mechanical Specific Energy?

It is generally called MSE which is the amount of energy required to remove unit volume of rock or formation. It is only the mechanical energy contributed by the mechanical parameters weight-on-bit (WOB) and bit rotational speed. It is given as

$$\text{MSE} = \frac{\text{WOB}}{A_b} + \frac{120\pi NT}{A_b ROP},$$

where
  WOB = weight-on-bit.
  A_b = area of bit.

N = rotational speed of the bit.
T = torque at bit.
ROP = rate of penetration.

461. Problem 13.3
Hydro-Mechanical Specific Energy (HMSE)

A more inclusive version of the Mechanical Specific Energy is the Hydro-Mechanical Specific Energy that takes into account the Hydraulic component of the energy which cannot be neglected. The hydro-mechanical specific energy is the total amount of energy expended to remove unit volume of rock and is given as

$$\text{HMSE} = \frac{\text{WOB}_e}{A_b} + \frac{120\pi NT + 1154\eta \Delta P_b Q}{A_b ROP},$$

where
$\text{WOB}_e = \text{WOB} - \eta F_j$.
WOB_e = effective weight-on-bit.
η = dummy factor for energy reduction.
F_j = force due to nozzle jets.
ΔP_b = pressure drop across the bit.
Q = flow rate.

The mechanical specific energy or the hydro-mechanical specific energy can be combined with the laser specific energy to determine the overall efficiency of the drilling operation.

462. Problem 13.4
Laser Specific Energy (LSE)

It has been defined and used in literature and is given by:

$$\text{LSE} = \frac{\text{Power Intensity} \times \text{Time}}{\text{Thermal Penetration Depth}}$$

To determine the overall drilling efficiency by using the hybrid bit and hole-openers, we can use either one of the following equations or combinations to determine the total energy input to the drilling system and try to minimize the total input energy.

Total Input Energy = MSE + LSE
Total Input Energy = HMSE + LSE

463. Problem 13.5

Initial wellbore inclination 68.01°, azimuth 224.83°.
BHA model results in the following results:
Bit tilt in the inclination direction = 0.12°
Bit tilt in the azimuth direction = 0°
Rate of penetration = 10 ft/hr
Bit side force in the inclination direction = 2651 lbf
Bit side force in the azimuth direction = 0 lbf
Calculate the new inclination and azimuth.

$$S.E = \frac{\text{Power Intensity} \times \text{Time}}{\text{Thermal Penetration Depth}}$$

Solution:

Classic Drillahead model generally combines the BHA response with the bit's side cutting ability.

With that the new inclination angle as well as new azimuth can be calculated using the following equations:

$$\alpha_n = \alpha_o + \varphi_i + \tan^{-1}\left(\frac{S_x}{S_z}\right)$$

$$\phi_n = \phi_o + \varphi_a + \tan^{-1}\left(\frac{S_y}{S_z}\right),$$

where

α_o = old inclination angle, deg.
φ_i = bit tilt in the inclination direction, deg.
S_x, S_y, S_z = lateral penetration rates in x, y directions and axial penetration rates in x direction.

ϕ_o = old azimuth angle, deg.
φ_a = bit tilt in the azimuth direction, deg.

Since there is no side cutting the lateral penetration in the inclination and azimuth direction is zero, i.e., $S_x, S_y = 0$

Using the given equations for new inclination and azimuth

$\alpha_n = 66.85 + 0.12 + 0 = 66.97°$

$\phi_n = 224.82 + 0 + 0 = 224.82°$

464. Problem 13.6

The bit tilt in the inclination and azimuth direction change when the weight-on-bit is changed prior to drilling.

True or False

Solution:

True. When the weight-on-bit is applied, the compressive force changes the position of the string resulting in the change in the bit tilt.

465. Problem 13.7

If the bit is cutting sideways and if the ratio of the lateral penetration to axial penetration can be related to the ratio of bit side force to axial force (WOB) calculate the new inclination and azimuth for the following condition:

Initial wellbore inclination 60°, azimuth 254°
BHA model results in the following results:
Bit tilt in the inclination direction = 0.71°
Bit tilt in the azimuth direction = 0.04°
Rate of penetration = 10 ft/hr
Bit side force in the inclination direction is = 599 lbf
Bit side force in the azimuth direction is = 31 lbf
Calculate the new inclination and azimuth.

Solution:

New inclination is $\alpha_n = \alpha_o + \varphi_i + \tan^{-1}\left(\dfrac{F_x}{WOB}\right)$

Substituting the values

$$\alpha_n = 60 + 0.71 + \tan^{-1}\left(\frac{-599}{10{,}000}\right) = 57.28°$$

New azimuth angle

$$\phi_n = \phi_o + \varphi_a + \tan^{-1}\left(\frac{F_y}{\text{WOB}}\right)$$

Substituting the values

$$\phi_n = 254 - 0.04 + \tan^{-1}\left(\frac{-31}{10{,}000}\right) = 253.78°$$

466. Problem 13.8

A hammer inside a jar weighing 10 lbm hits the head of the anvil which makes the bottom of the jar which is stuck move 6 in. The velocity of the hammer at the time of impact is 60 ft/sec. What is the resistive force at the stuck point?

Solution:

$$Fx = \frac{1}{2}mv^2$$

$$\text{so } F = \frac{\frac{1}{2}mv^2}{x}$$

x = 6/12 = 0.5 in.
Mass = 10/32.2 = 0.31055 slugs
Velocity = 60 ft/sec

$$F = \frac{\frac{1}{2}0.31055 \times 60^2}{\frac{6}{12}} = 1118 \text{ lbf}$$

467. Problem 13.9

A hammer inside a jar weighing 10 lbm hits the head of the anvil which makes the stuck portion close to the jar move 6 in. Portion below the stuck

Drilling Operations

point is modeled as a spring as shown in Figure 13.1. The spring constant is 250 kips/in. What is the velocity of the impact? Neglect the mass of the spring and other losses.

Solution:

Kinetic energy of the hammer before hitting is equal to the potential energy of the anvil and spring system when compressed to maximum.

$$\frac{1}{2}kx^2 = \frac{1}{2}mv^2 \text{ and so}$$

$$v = \sqrt{\frac{kx^2}{m}}$$

$x = 6/12 = 0.5$ in.
Mass $= 10/32.2 = 0.31055$ slugs
Velocity $= 60$ ft/sec

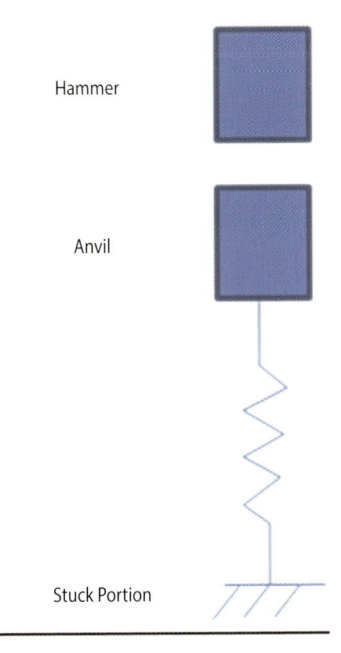

Figure 13.1 Problem 13.1

$$v = \sqrt{\frac{250{,}000 \times 6^2 \times 12}{0.31055}} \times \frac{1}{12} = 4914 \text{ ft/sec}$$

468. Problem 13.10

Calculate the acoustic velocity in steel, tungsten, and aluminum. Assume ambient temperature.

Solution:

$$c = \sqrt{\frac{E}{\rho}}$$

For Steel 30×10^6 psi or 2×10^{11} N/m² and density 7800 kg/m³

$$c = \sqrt{\frac{E}{\rho}} = \sqrt{\frac{2 \times 10^{11}}{7.8 \times 10^3}} = 5.06 \times 10^3 \text{ m/sec}$$

For Tungsten 58×10^6 psi 4×10^{11} N/m² and density 19,250 kg/m³

$$c = \sqrt{\frac{E}{\rho}} = \sqrt{\frac{4 \times 10^{12}}{19.25 \times 10^3}} = 4.6 \times 10^3 \text{ m/sec}$$

For Aluminum 10×10^6 psi or 6.9×10^{10} N/m² and density 2700 kg/m³

$$c = \sqrt{\frac{E}{\rho}} = \sqrt{\frac{6.9 \times 10^{10}}{2.7 \times 10^3}} = 5.05 \times 10^3 \text{ m/sec}$$

469. Problem 13.11

Explain briefly the Slack-off test.

Solution:

The evaluation of axial force transfer depends on the sinusoidal or helical modes of the string and the contact force thereby increased drag forces. It can be seen that the additional force due to the helical mode of the string results in increase in side force. Furthermore it can also be seen that the additional side force increase is square of the axial load which in turn results in extremely increased drag force which further reduces the axial load or increased axial compressive load. Further effect of this side force results in cascading effect on the final axial load and thereby decreased efficiency of the axial force transfer. Increase in compressive load may result in yield failure of the string (permanent corkscrewing) and also may result in axial transfer to the bottom of the string to zero (Lockup).

Lockup is a post-helical condition and is more pronounced in horizontal and extended reach wells preventing reaching the desired depth.

This necessitates establishing a method to estimate the efficiency of the axial force transfer so that the string is "locked up" or goes into permanent cork screwing. This will help to delineate the safe operating envelope so that maximum rate of penetration and operating life of downhole tools such as downhole motor is achieved. The axial force transfer efficiency of the string can be estimated in the field by conducting a simple test, called the "**BLF-slack-off-WOB**" test by observing the hook load and the weight-on-bit at the bottom. For this test measurement drilling tool should be needed to monitor the weight-on-bit at the bottom.

With the bit off bottom, the reference hook load is noted. After slacking off the desired amount of hook load the weight-on-bit at the bottom is noted. This test is repeated with different slack of loads so that the change in the slope is observed. The tests can be extended to lock up or close to lock up condition or no more weight-on-bit can be applied to the bit by slacking off the string at the surface $\left(\eta_{slackoff} = \dfrac{\Delta WOB}{\Delta HL} = 0 \right)$,

where
ΔHL = change in slack off load.
ΔWOB = corresponding change in downhole weight-on-bit.

Also the test can be performed with rotation as well with or without flow rate.

A. Real-time data (required)

1. Time-depth information (high sample-rate)
2. Local magnetic parameters (whatever is used with surveys of record)

3. Surveys of record (detailed) with all corrections applied
4. RPM
5. Weight-on-bit
6. Torque on bit
7. Bending moments (*x* and *y*) (DrillDoc)
8. Mud weight
9. BHA configuration used with distances from bit
10. Borehole diameter in all sections

B. Test conditions for non-rotating/rotating cases
1. Change in weight-on-bit/trend
2. Rotational checkshot surveys at known depths
3. Slack-off in 5 to 10 kips increments with no rotation
4. Slack-off in 5 to 10 kips increments with 10 rpm increments of rotation
5. Increase hookload in 5 to 10 kips increments with no rotation
6. Increase hookload in 5 to 10 kips increments with 10 rpm increments of rotation

470. Problem 13.12
Buckling Limit Factor

What is Buckling Limit Factor (BLF)? Explain with details.

Solution:

Buckling limits commonly used is based on the theory that as the pipe when compressed inside the wellbore the string goes initially into snaking or lateral buckling mode. It is also called sinusoidal buckling mode. This condition still allows the pipe to be compressed and after exceeding the threshold constant calculated by the researchers, snaps into helical mode. This causes the wall force to increase that will result in lockup state of the pipe. Usually lock up is defined as the ratio between the changes in the downhole weight to the change in surface slack off weight less than 2%.

In the past, various limits were published to define the regions of no buckling, sinusoidal buckling, and helical buckling.

Based on the work by Lubinski Dawson and Paslay and Paslay and Bogy the compression force to induce onset of sinusoidal buckling is given as

$$F_s = 2 \times \sqrt{\frac{EIW \sin\theta}{r}},$$

where
I = moment of inertial for component.
E = Young's modulus of elasticity.
W = tubular weight in mud.
θ = wellbore inclination.
r = radial clearance between wellbore and component.

Using the curvilinear model the side force can be given as

$$F_s = 2\sqrt{\frac{EIw_c}{r}}$$

In which the contact force, w_c between the pipe and wellbore is given as

$$w_c = \sqrt{\left(w_{bp}\sin\theta + F_b\theta'\right)^2 + \left(F_b \sin\theta\phi'\right)^2},$$

where
ϕ = azimuth angle.
' is the derivative with respect to measured depth.

For constant curvature wellbore, the contact force can be expressed as

$$w_c = \sqrt{\left(w_{bp}n_z - F_b\kappa\right)^2 + \left(w_{bp}b_z\right)^2},$$

where
n_z = vertical component of the normal to the curve.
b_z = vertical component of the binormal to the curve.

Compression force to induce onset of helical buckling is given as

$F_s = F \times F_s,$

where
F = constant.

Various buckling constants used for the onset of helical buckling by various authors are listed in Table 13.1.

There is no consensus among the authors and the validations carried out with either 50 ft or less than 100 ft acrylic pipe with aluminum or steel rods. Acrylic pipe used in the lab for testing whether straight or slightly undulated do not translate close to downhole conditions. Also the models have been developed, in tested, in a discrete fashion. The model does not take into effect of the pipe condition as a whole which may be completely different when analyzed piecewise. For example, the buckling condition may be different when the pipe is in J-type well as opposed to S-type well.

There were very limited field data available and with the data available it has been found that the downhole weight change does not follow the theoretical prediction. In most of the cases it very well aligns up to the onset of the sinusoidal buckling and quickly goes into lockup state without the onset of helical buckling mode.

For the compressive axial loads between 1.4 and 2.8 times the sinusoidal buckling force, there is enough strain energy in the pipe to sustain helical buckling, but not enough energy to spontaneously change from

Table 13.1 Buckling constants

Model	F
Chen and Cheatham, 1990	−2.83
He and Kyllingstad, 1995	−2.83
Lubinski and Woods, 1953	−2.85
Lubinski and Logan, 1962	−2.4
Qui, Miska and Volk, 1998	−5.66
Qui, Miska and Volk, 1998	−3.75
Wu and Juvkam Wold, 1993	−3.66
Wu and Juvkam Wold, 1995	−4.24

sinusoidal buckling to helical buckling. That is, if you could reach in and lift the pipe up into a helix, it would stay in the helix when you let go. This means that in an ideal situation, without external disturbances, the pipe would stay in a sinusoidal buckling mode until the axial force reached 2.8 times the sinusoidal buckling force. At this point, the pipe would transition to the helical buckling mode. This is the "loading" scenario.

Once the pipe is in the helical buckling mode, the axial force can be reduced to 1.4 times the sinusoidal buckling force, and the helical mode will be maintained. If the axial force falls below 1.4 times the sinusoidal buckling force, the pipe will fall out of the helix into a sinusoidal buckling mode. This is the "unloading" scenario.

Loading Model

$$F_h = 2.828427 \times F_s$$

Unloading Model

$$F_h = 1.414213 \times F_s,$$

where
F_h = compression force to induce onset of sinusoidal buckling.
F_s = compression force to induce onset of helical buckling.

471. Problem 13.13

Key Factors Influencing Buckling

The key factors influencing buckling are as follows:

1) Lateral Clearance – Hole Wash Out
2) Localized Pipe Heating - Flows Behind Pipe
3) Temperature Increase – Drilling, Production
4) Formation Sticking – Axial Restraints
5) Incremental Compressive Load
6) Wellbore Interaction – Friction and Side Loading
7) Wellbore Trajectory and tortuosity

472. Problem 13.14
Values of Buckling Limit Factor

The *Buckling Limit Factor* in short called as BLF is a multiplying factor used to adjust the constants used in the buckling equation. This helps to calibrate the model and adjust the buckling limits lines based on the wellbore tortuosity, borehole quality, or shape.

The buckling force $F_{s(modified)}$ is modified as below:

$$F_{s(modified)} = BLF \times 2\sqrt{\frac{EIw_c}{r}}$$

Compression force to induce onset of helical buckling is given as
$F_s = F \times F_{s(modified)}$
Loading Model
$F_h = 2.828427 \times F_{s(modified)}$
Unloading Model
$F_h = 1.414213 \times F_{s(modified)}$

A zero or null value in the field will be assumed to be a factor of 1. This will help the previous version to default to the original calculations so that buckling limits are not changed. A higher buckling limit factor will provide a higher buckling limit, and a smaller factor will reduce the buckling limit.

Advantages:

1. It helps the user to manipulate the buckling equation and adjust the limit lines.
2. It helps to calibrate and use a standard value depending on company policies.
3. It helps to adjust the both buckling lines simultaneously.

Suggested BLF with respective to the models are given in Table 13.2. The reference is based on the Wellplan model (He and Kyllingstad).

Table 13.2 Buckling limit factor

Model	F	BLF
Chen and Cheatham, 1990	−2.83	1
He and Kyllingstad, 1995	−2.83	1
Lubinski and Woods, 1953	−2.85	1.007
Lubinski and Logan, 1962	−2.4	0.848
Qui, Miska and Volk, 1998	−5.66	2
Qui, Miska and Volk, 1998	−3.75	1.326
Wu and Juvkam Wold, 1993	−3.66	1.295
Wu and Juvkam Wold, 1995	−4.24	1.498

473. Problem 13.15

The drilling rate is given by $\frac{dF}{dt} = (15)10^{-0.0004F}$ where F is the footage drilled in ft and time t in hours. Calculate the time in hours to drill 200 ft.

Solution:

Integrating between 0 to F and 0 to T

$$t = \frac{1}{15}\int_0^F 10^{kF}\, dF = \frac{1}{15}\frac{10^{kF}}{k\ln 10}\bigg|_0^F$$

$$= \frac{1}{15}\frac{(10^{0.0004F} - 1)}{0.0004 \times \ln 10} = 14.6 \text{ hrs}$$

CHAPTER 14

CEMENTING

In this chapter basic cementing equations and related calculations will be presented.

474. Problem 14.1

Steps Involved in Cement Manufacturing

Solution:

- Quarrying
- Transporting
- Crushing and grinding
- Blending
- Calcinated (1100°C)
- Convert to Clinker
- Clinker grinding and testing
- Final blending
- Storage

475. Problem 14.2

What is Heat of Hydration in Cementing?

Solution:

    Heat is released when the cement is mixed with the water and during the hydration process due to various components. The hydration reaction is an exothermic reaction and generates considerable quantity of heat.

476. Problem 14.3

    What are the main types of Cementing?

Solution:

    Primary Cementing
      - Main cementing after casing is run
    Secondary cementing
      - Additional cementing operations after the main cementing job
        - plug back
        - squeeze cementing
        - balanced plug

477. Problem 14.4
Temperature Calculations

Geothermal gradient is given in °F/100 ft. For normal wells, the gradient usually is 1.5° F/100 ft. The conversions used are as follows:

$$T\ ^\circ F = \frac{9}{5}(T\ ^\circ C) + 32$$

$$T\ ^\circ C = \frac{5}{9}(T\ ^\circ F - 32)$$

Other conversions to Kelvin and Rankine are as follows:

$$T\ ^\circ K = T\ ^\circ C + 273.15$$

$$T\ ^\circ K = \frac{5}{9}(T\ ^\circ F + 459.67)$$

$$T\ ^\circ K = \frac{5}{9} \times\ ^\circ R$$

$$T\ ^\circ R = (T\ ^\circ F + 459.67)$$

For offshore wells, the temperature profile inside the seawater is different depending on the depth, place, and time of the year. Usually below the thermocline the temperature remains fairly constant, and the profile is similar to Figure 14.1 Approximately below the depth of 5000 ft, the temperature of the seawater reaches close to 40–45° F and drops slowly as the depth increases. Deeper depths' temperature varies between 34°F and 40°F.

478. Problem 14.5

The temperature profile of seawater is given empirically as
$T = 68.016 - 7.896 \times ln(3.28084 \times D_w)\ ^\circ C\ D_w \leq 914\ m$
and
$T = 5.3966 - 0.0002063\ (3.28084 \times D_w)\ ^\circ C\ D_w > 914\ m,$

![Figure 14.1 Seawater temperature profile showing temperature vs depth curves for High Altitudes, Mid Altitudes - Winter, Mid Altitudes - Summer, and Low Altitudes]

Figure 14.1 Seawater temperature profile

where
D_w = depth of the seawater, in feet.
Calculate the temperature for the depths 762 m and 2500 m.

Solution:

For water depth 762 m
$T = 68.016 - 7.896 \times ln(3.28084 \times 762) = 6.23\ °C$
For water depth 2500 m
$T = 5.3966 - 0.0002063(3.28084 \times 2500) = 3.7\ °C$

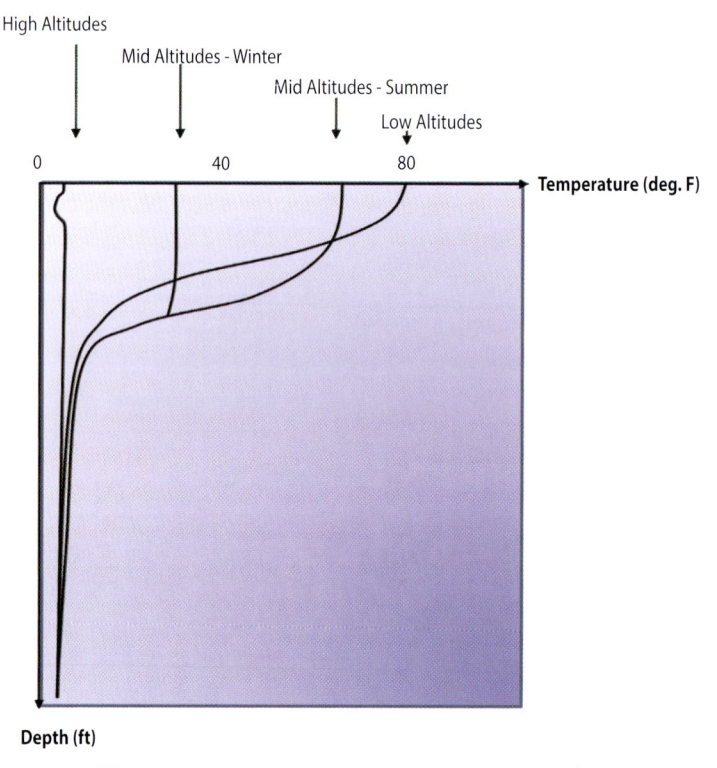

479. Problem 14.6

Convert $80°$ F to Centigrade, Kelvin, and Rankine.

Solution:

Fahrenheit to Centigrade:

$$T\ ^\circ C = \frac{5}{9}(T\ ^\circ F - 32) = \frac{5}{9}(80 - 32) = 26.67\ ^\circ C$$

Fahrenheit to Kelvin:

$$T\ ^\circ K = \frac{5}{9}(T\ ^\circ F + 459.67) = \frac{5}{9}(80 + 459.67) = 299.82\ ^\circ K$$

Fahrenheit to Rankine:

$$T\ ^\circ R = (T\ ^\circ F + 459.67) = (80 + 459.67) = 539.7\ ^\circ R$$

480. Problem 14.7

The temperature profile of seawater is given empirically as
$T = 154.43 - 14.214 \times \ln(D_w)\ ^\circ F\ D_w \leq 3000$ ft
and
$T = 41.714 - 0.0003714\ (D_w)\ ^\circ F\ D_w > 3000$ ft,
where
D_w = depth of the seawater in feet.
Calculate the temperature for the depths 2500 ft and 5500 ft.

Solution:

For water depth 2500 ft
$T = 154.43 - 14.214 \times \ln(2500) = 43.22\ ^\circ F$
For water depth 5500 ft
$T = 41.714 - 0.0003714\ (5500) = 39.67\ ^\circ F$

481. Problem 14.8

Basic Cement Slurry Requirement Calculations

Three basic requirements are

- Density
- Yield
- Water requirement

The density of the slurry is given as

$$\rho_c = \frac{W_c}{V_c},$$

where
V_c = cement slurry volume, gal.
W_c = total weight of slurry, lbm.
The number of sacks of cement is calculated as

$$N_c = \frac{V_{sl}}{Y} \text{ sacks},$$

where
V_{sl} = slurry volume, cubic ft.
Y = yield of cement, cubic ft/sack.

The volume of slurry obtained per sack of cement used is called the yield of cement. It should not be confused with the yield point of a fluid.
It is expressed in ft³/sack.
Yield of the slurry and is given by

$$Y = \frac{V_c}{7.48} \text{ ft}^3/\text{sack}$$

The mix water requirement is calculated as

$$V_w = V_i \times N_c \text{ gal},$$

where
V_i = mix water per sack.
Percent mix can be expressed in terms of Weight percent.

$$\text{percent mix} = \frac{\text{water weight}}{\text{cement weight}} \times 100$$

482. Problem 14.9
Procedure
What is a liner and what are the different types of liner?
Solution:
A liner is a casing string which does not go all the way to the surface. It is hung inside the previous casing by drillpipe. They eliminate complete

Cementing

casing strings up to the surface. Liner hangers are used to run the liner to set against the prior casing. The section where the liner runs inside another string is the overlap section. An overlap will be maintained as shown.

Different types of liners
- Drilling liner
- Production liner
- Stub liner or tieback. The tieback consists of a downhole mechanical sealing assembly in a hanger, into which a linear string or the tieback string is stabbed, to complete the seal. A cement job seals the liner into place in the casing and prevents leakage from the formation into the casing.
- Scab liner

483. Problem 14.10

Procedure

What are the steps involved in single stage cementing (Fig. 14.2)?

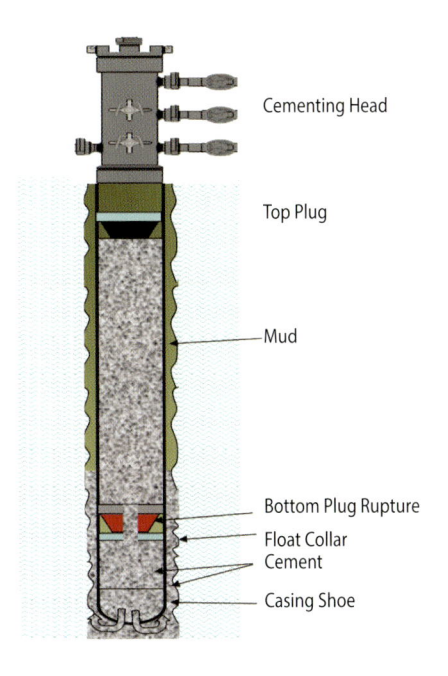

Figure 14.2   Single stage cementing

Solution:

- Circulate and condition mud
- Pump pre flush
- Drop bottom plug
- Pump cement slurry
- Drop top plug
- Pump post flush
- Displace cement with mud
- Bump the plug

484. Problem 14.11

Procedure

What are the steps involved in balanced plug cementing?

Solution:

It is the same procedure as explained in Problem 14.10 for horizontal drilling but may need more stabilization in the lateral section with centralizers as shown in Figure 14.3.

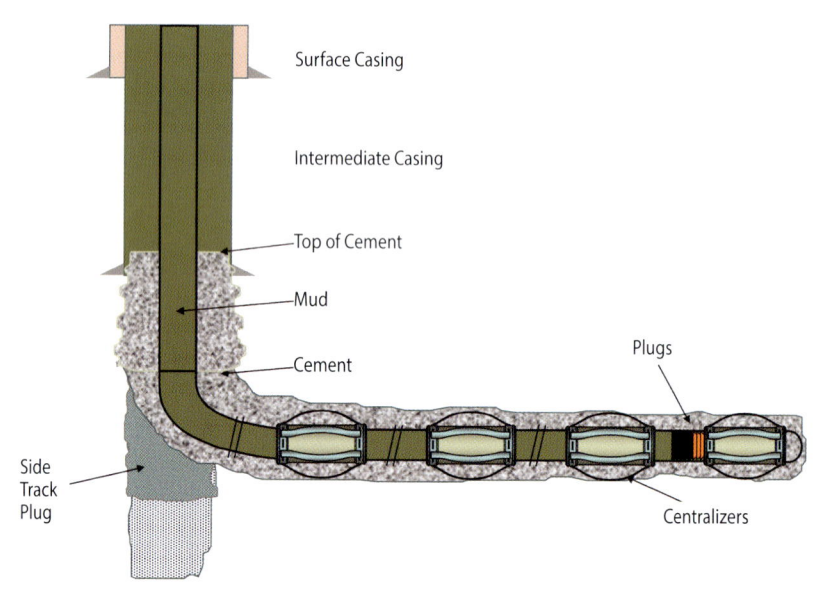

Figure 14.3   Horizontal well cementing

485. Problem 14.12
Procedure

What are the steps involved in balanced plug cementing?

Solution:

- Circulate and condition mud
- Pump water spacer
- Pump cement slurry
- Displace cement with mud
- Pull the pipe

486. Problem 14.13
Procedure

What is inner string cementing?

Solution:

- It is also called stab-in cementing
- Large-diameter casing strings are cemented using this method. With this method drill pipe string is run to the bottom of the casing and stabbed into the float collar or float shoe. The cement is pumped through the drillpipe
- Reduces rig time required for displacement
- Allows continuous mixing and pumping of cement
- It is used in multilateral system
- Results in improved displacement efficiency
- Reduces the cement volume and displacement time
- Avoids the use of large plugs

487. Problem 14.14
Procedure

What is stage cementing?

Solution:

Cementing is carried out in two or more stages.

It helps to isolate the weak formation causing cement loses and/or contamination of the producing zone.

It is used in wells with narrow pore and fracture gradients.

488. Problem 14.15
Procedure

What is the procedure involved in stage cementing?

Solution:

- Circulate and condition mud
- Pump water spacer
- Pump first stage cement slurry
- Drop first stage top plug
- Displace first stage cement with mud
- Drop opening bomb and open stage collar
- Pump second stage water spacer
- Pump second stage cement slurry
- Drop second stage top plug
- Displace second stage cement with mud
- Bump the plug and close stage collar

489. Problem 14.16
Procedure

What is conditioning mud?

Solution:

- Conditioning mud is important for proper cementation.
- Mud circulation is carried out once the casing is at bottom.
- Fluids: Drilling mud inside the pipe and in the annulus

The details are given in Figure 14.4.

490. Problem 14.17
Procedure

What is excess volume in cementing calculations?

Solution:

- An excess factor is used to account the irregularities of the wellbore to achieve proper cement top and placement behind the casing.
- The excess factor is applied to the calculated theoretical cement volume.

Cementing

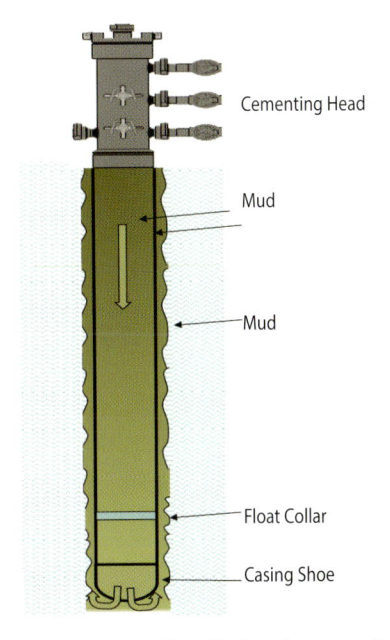

Figure 14.4 Conditioning mud

- Accuracy of the excess factor depends on experience and wellbore size data availability.

491. Problem 14.18
Procedure
What is shoe track and the condition in cementing?
Solution:
Shoe track is the cement between the float collar and shoe after the completion of the primary cement job.

This will act as barrier. Sometimes wet shoe track is observed which may be due unset or contaminated or no cement due to various reasons.

Figure 14.5 shows the top plug when it lands on the float collar. The top plug is solid and will not rupture. This will cause pressure to increase at the surface. The bumping of the plug by of increase in surface pressure will confirm the completion of the displacement of cement. The cement will also be present inside the casing between the float collar and float shoe (casing shoe).

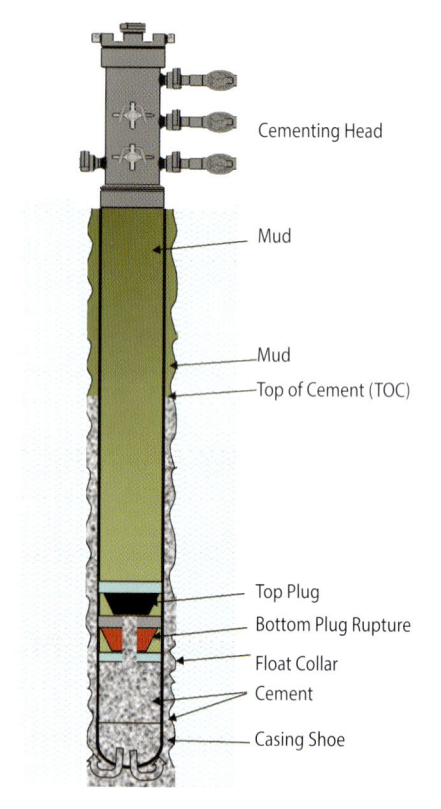

Figure 14.5   Shoe track

492. Problem 14.19
Procedure

What is complex fluid pumping in cementing?
Solution:

Complex fluid pumping includes

Pre flush/spacer – a spacer or flush fluid that will be placed in the annulus or pumped out. Sometimes to break the gelled mud, make the displacement easy and minimize the cement contamination.

Lead slurry – top section with usually lower strength

Tail slurry – bottom section with higher strength

Post flush

493. Problem 14.20
Procedure

What are the steps involved in liner cementing?

Solution:

- Landing collar or Float collar and Float shoe (Fig. 14.6)
- Usually Float collar is placed above 2 joints of casing from the bottom.
- Conditioning Mud:
  - Conditioning mud is important for proper cementation. Sometimes last two joints are circulated to remove the cuttings/fill-ups at the bottom of the wellbore.
  - Mud circulation is carried out once the liner is at bottom.
  - Sometimes liner is reciprocated at least with one joint stroke length.
  - Fluids: Drilling mud inside the pipe and in the annulus
- Liner hanger and setting tool is connected to the top of the casing in the liner string. Above the liner hanger the drillpipe is connected. Liner hanger can be of
- Mechanical
- Hydraulic

Figure 14.6 Float collar and float shoe/casing shoe *(Courtesy: Davis Lynch)*

- Hydro-Mechanical
- Expandable (VersaFlex)
    - Mechanical set hanger: There are different types and the simple type is with a J-Slot. The drillpipe is rotated at the surface which disengages a J-slot. The drillstring is slacked to set the slips against the casing wall.
    - Hydraulic set hanger: There are different types and simple type force slips upwards using the differential pressure between the circulation pressure and actuating piston.
    - Hydro-Mechanical set hanger: This can be set either mechanically or hydraulically.
    - Expandable: Multiple elastomeric elements are used in the solid hanger to seal the annulus. No slips are used.
    - There are different methods:
- Liner is hung before pumping of cement which reduces the annular clearance between the hanger and the previous casing pipe.
- Liner is hung after cement operation.
    - Detaching the drill pipe from the liner before cementing minimizes the risk of being unable to detach from the liner once the cement is in place. For the first condition, liner is hung by setting the slips at the desired depth and the liner weight is released.
    - If the liner is hung off prior to circulating cement, the small bypass area around the liner offers a greater restriction to fluid flow and thus may result in loss circulation because of the back-pressure of the flowing cement.
    - If there are cement channels, as well as a large hydrostatic pressure difference between the inside and the outside of the running tools, the cups or seals can give way before cementing of the liner is complete.
    - The displacement efficiency of cement around the tubulars when pipe is not moved is lessened.
    - The liner setting tool is released and ensured it is free to move and then lowered to make a tight seal with the lower part of the liner

Cementing

hanger assembly. The circulation is reestablished to ensure free circulation. The circulation is through the annular area between the liner and the previous casing. This is a very narrow clearance.

- After pumping the calculated volume of cement slurry, the drillpipe wiper plug or wiper dart (Fig. 14.7) or pump down plug is released from the cementing head. This wipes the cement inside the drillpipe. Then displacement fluid is pumped which further moves the cement into the annulus.
- When the drillpipe plug reaches the liner setting tool it latches to the liner wiper plug. The shear pin holding the liner wiper plug is sheared from the running tool and further both the plugs move in unison. This wipes the inside of the liner.
- This will cause pressure to increase at the surface. The bumping of the plug by increase in surface pressure will confirm the completion of the displacement of cement. The cement will also be present inside the casing between the float collar and float shoe (casing shoe).
    - Shoe track is the cement between the float collar and shoe after the completion of the primary cement job.
    - This will act as barrier. Sometimes wet shoe track is observed which may be due unset or contaminated or no cement due to various reasons.
- Once the plug bumps the pressure is increased to the desired values to test the integrity of the casing and shoe.

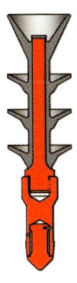

Figure 14.7 Wiper plug

- Release the setting tool from the liner.
- Few stands, usually 10 stands, of drillpipe are pulled out or above the top of the cement and reversed out to clean the drillpipe.
- Pull out the drillstring completely.

494. Problem 14.21

The required slurry density = 20 ppg
Iron oxide to be used = 0.42 gal per sack of cement
Water requirement = 5 gal/sack of cement
Calculate the iron oxide required to achieve the slurry density.
Assume
1 sack = 94 lbm
Density of iron oxide = 40 ppg
Density of cement = 26 ppg

Solution:

The total water requirement of the slurry is given by $5 + 0.0042\,x$.
Where
x lbm is the iron oxide per 100 lbm cement.
Slurry density is given as

$$\rho_{sl} = \frac{\text{total mass of slurry in (lbm)}}{\text{total volume of slurry in (gal)}}$$

Total mass of slurry

$$94 + x + 8.33(5 + 0.0042x)$$

Total volume of slurry

$$\frac{94}{26} + \frac{x}{40} + (5 + 0.0042x)$$

Substituting in the slurry density

$$18 = \frac{94 + x + 8.33(5 + 0.0042x)}{\frac{94}{26} + \frac{x}{40} + (5 + 0.0042x)},$$

$$\left(\frac{x}{40}+(0.0042x)\right)-x-0.0042x = 94-\left(\frac{94}{26}\times 18+5\right)+8.33\times 5$$

The iron oxide required is 64.5 lbm /per sack of cement.

495. Problem 14.22

A cementing engineer is planning to prepare a cement slurry using class H cement with the following content:
sand = 35%
water = 40%
retarder = 5%
Determine the following:
slurry density, slurry yield, and mixing water volume.
Assume the density of cement = 26 ppg
Density of sand = 21 ppg
Density of retarder = 8.5 ppg
Weight of cement per sack = 94 lbm

Solution:

The calculation can be done for a single sack of cement basis i.e., 94 lbm.

Weight of cement = 94 lbm/sack.
Weight of sand (35%) = 94 × 35/100 = 32.9 lbm/sack.
Weight of water (40%) = 94 × 40/100 = 43.24 lbm/sack.
Weight of retarder (5%) = 94 × 5/100 = 4.7 lbm/sack.
Total weight of the components = 32.9 = 43.24 + 4.7 = 174.84 lbm.
Volume of cement is

$$=\frac{94}{26}=3.61\text{ gal}$$

Volume of sand is calculated as

$$=\frac{32.9}{21}=1.56\text{ gal}$$

Volume of water is calculated as

$$= \frac{43.24}{8.33} = 5.19 \text{ gal}$$

Volume of retarder is calculated as

$$= \frac{4.7}{8.5} = 0.553 \text{ gal}$$

Total volume of the components = 3.6 + 1.5 + 5.2 + 0.55 = 10.93 gal
Slurry density = Total weight/total volume
= 174.84 / 10.93 = 16.41 ppg
The yield of the slurry:

$$= \frac{\text{total volume}}{7.48} = \frac{10.93}{7.48} = 1.46 \text{ ft}^3/\text{sack}$$

Mix water volume = 5.19 gal/sack

496. Problem 14.23

It was desired to obtain a class H cement slurry weighing 17 ppg by adding sand and water with sand = 35%.
Determine the amount of water needed.
Assume the density of cement = 26 ppg
Density of sand = 21 ppg
Density of retarder = 8.5 ppg
Weight of cement per sack = 94 lbm

Solution:

Calculation can be done for a single sack of cement basis i.e., 94 lbm.
Weight of cement = 94 lbm/sack.
Weight of sand (35%) = 94 × 35/100 = 32.9 lbm/sack.
Weight of water (x%) = 94 x × /100 = 94x/100 lbm/sack.
Total weight of the components = 32.9 + 0.94x lbm
Volume of cement is

$$= \frac{94}{26} = 3.61 \text{ gal}$$

Volume of sand is calculated as

$$= \frac{32.9}{21} = 1.56 \text{ gal}$$

Volume of water is calculated as

$$= \frac{94x}{8.33} \text{ gal}$$

Total volume of the components = 3.6 + 1.56 + 11.29x gal
Slurry density = Total weight/total volume
= 32.9 + 0.94x / 5.16 + 11.29x ppg
Slurry density is given as 17 ppg and so it can be written as
17 = 32.9 + 0.94x / 5.16 + 11.29x
Solving will yield 52% water requirement.

CHAPTER 15

OFFSHORE

In this chapter basic offshore-related problems and calculations will be presented.

497. Problem 15.1

What are the six offshore vessel motions?

Solution:

Surge: Translation fore and aft (X-axis)
Sway: Translation port and starboard (Y-axis)
Yaw: Rotation about the Z-axis (rotation about the moonpool)
Heave: Translation up and down (Z-axis)
Roll: Rotation about the X-axis
Pitch: Rotation about the Y-axis

498. Problem 15.2

Which offshore vessel motions are translations?

Solution:

Surge: Translation fore and aft (X-axis)
Sway: Translation port and starboard (Y-axis)
Heave: Translation up and down (Z-axis)

499. Problem 15.3

Which offshore vessel motions are rotational?

Solution:

Yaw: Rotation about the Z-axis (rotation about the moonpool)
Roll: Rotation about the X-axis
Pitch: Rotation about the Y-axis

500. Problem 15.4

What do the following Acronyms represent?
MSL
BML
TVDBML
FTD
MODU
BOEMRE
LMRP

Solution:
    MSL – Mean Sea Level
    BML – Below Mud Line
    TVDBML – True Vertical Depth below Mud Line
    FTD – Final Total Depth
    MODU – Mobil Offshore Drilling Unit
    BOEMRE - Bureau of Ocean Energy Management, Regulation and Enforcement; formerly MMS.
    LMRP - Lower Marine Riser Package

501. Problem 15.5

With the aid of figures, show the details of the offshore vessels sides as well the types of vessel motions commonly used. Also show the draft and freeboard in the figure.

Solution:
Offshore vessels sides as well the types of vessel motions encountered are shown in Figures 15.1 and 15.2, respectively.

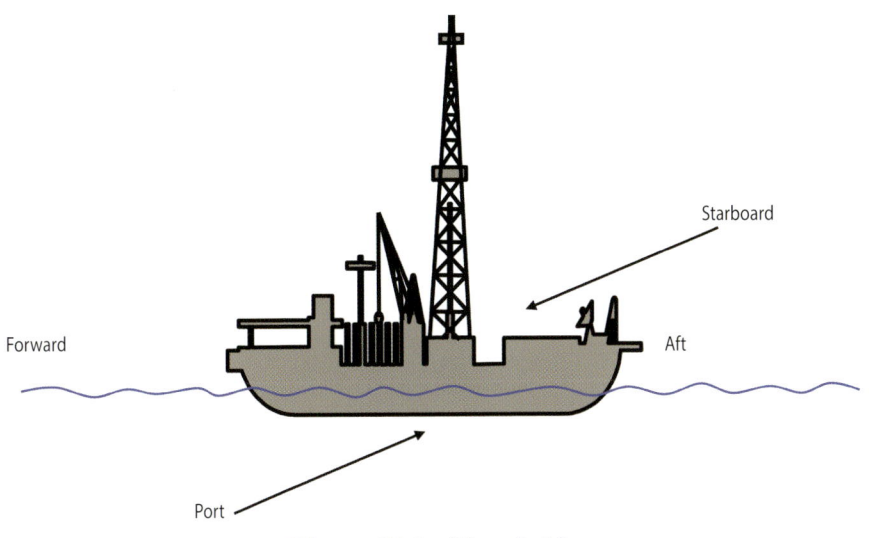

Figure 15.1   Vessel sides

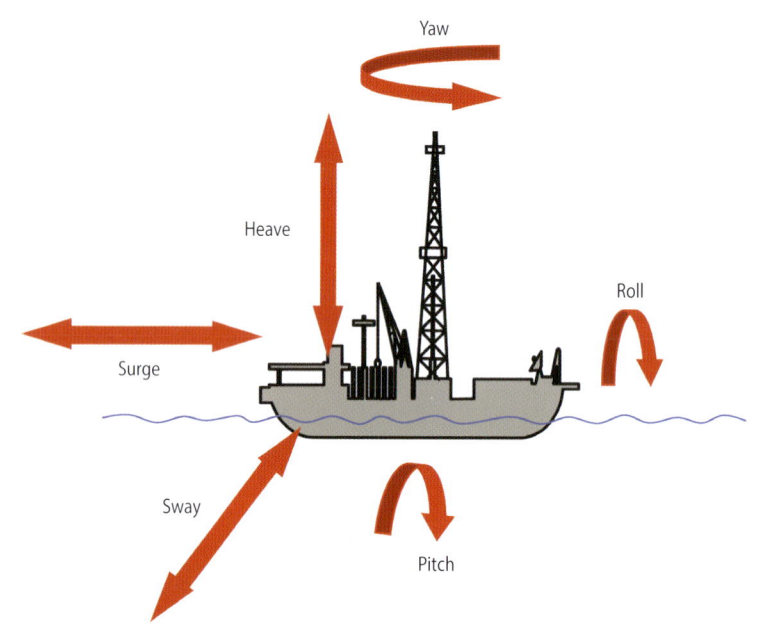

Figure 15.2 Vessel motion

Draft is the depth of the vessel in the water whereas the freeboard is the distance above the water. They are shown in Figure 15.3.

502. Problem 15.6

A drilling engineer standing on an offshore rig looking towards the stern of the rig observes a problem on the port side of the rig. Which direction is he observing the incident?

a. Left
b. Right
c. Backside

Solution:

a. Right side as shown in Figure 15.4

Offshore

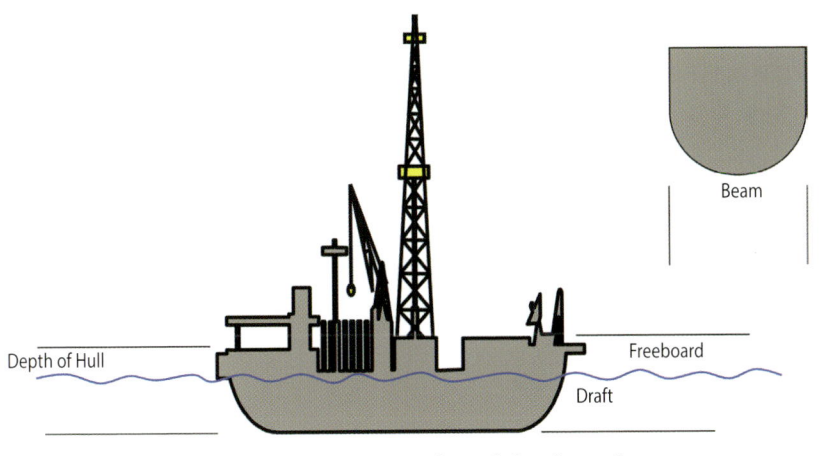

Figure 15.3 Draft and freeboard

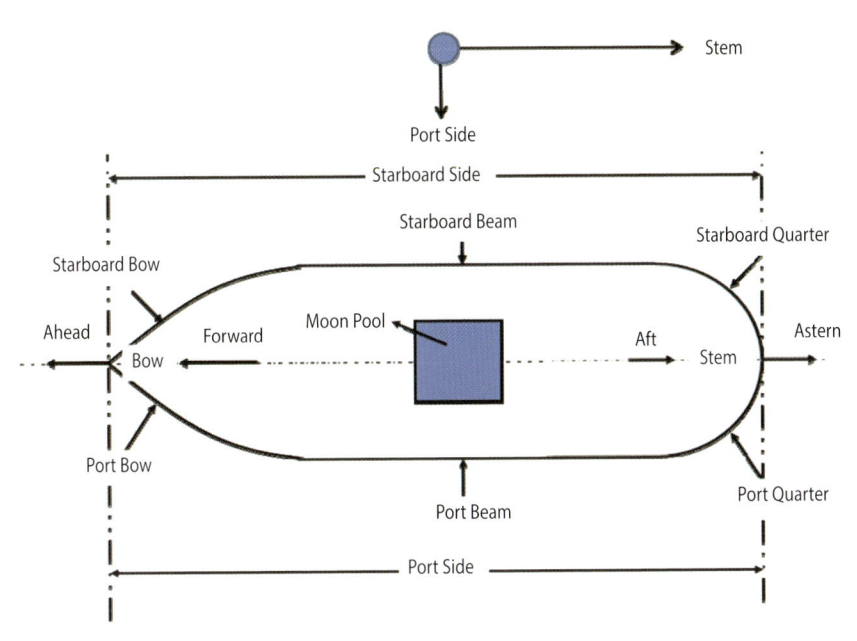

Figure 15.4 Vessel position

503. Problem 15.7

List three important environmental forces that affect the offshore rigs and show it using a figure.

Solution:

1. Wind Force
2. Wave Force
3. Current Force

These forces are shown in Figure 15.5.

504. Problem 15.8

Provide an equation to calculate the wind force that is encountered in an offshore rig.

Solution:

Wind force is given by

$$F_w = 0.00338 \, V_w^2 \, C_h \, C_s \, A,$$

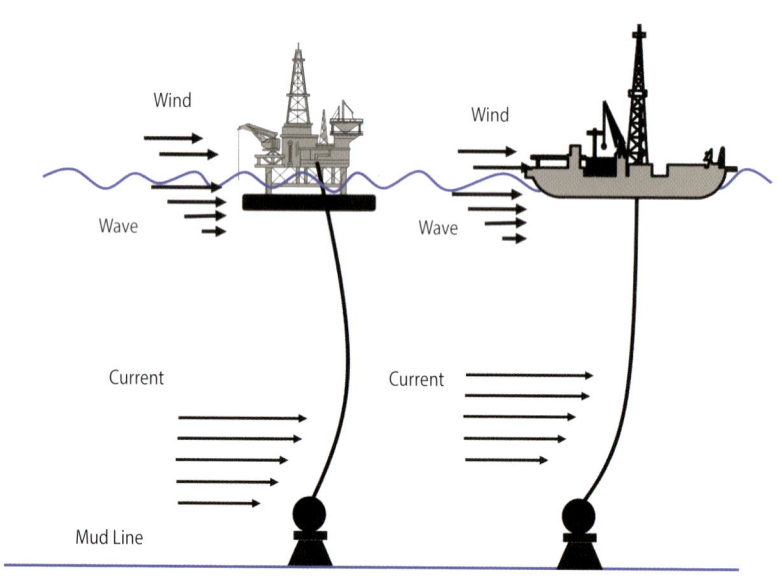

Figure 15.5 **Environmental forces**

where
F_A = wind force, lbf.
V_A = wind velocity, knots.
C_s = shape coefficient.
C_h = height coefficient.
A = projected area of all exposed surfaces, sq.ft.
Shape coefficients can be estimated from Table 15.1.
Height Coefficients is estimated from Table 15.2.

Table 15.1 Shape coefficients

Shape	Cs
Cylindrical shapes	0.5
Hull (surface type)	1.0
Deck House	1.0
Isolated structural shapes (cranes, beams..etc)	1.0
Under Deck areas (smooth surfaces)	1.0
Under Deck areas (exposed beams, girders)	1.3
Rig derrick (each face)	1.25

Table 15.2 Height coefficients

From to (ft)	Ch
0–50	1.0
50–100	1.1
100–150	1.2
150–200	1.3
200–250	1.37
250–300	1.43
300–350	1.48
350–400	1.52
400–450	1.56
450–500	1.6

505. Problem 15.9

Provide an equation to calculate the current force that is encountered in an offshore rig.
Solution:
Current force is calculated from

$$F_c = g_c V_c^2 C_s A,$$

where
F_c = current drag force, lbf.
V_c = current velocity, ft/sec.
C_s = drag coefficient same as wind coefficient.
A = projected area of all exposed surfaces, sq.ft.

506. Problem 15.10

Provide an equation to calculate the wave force that is encountered in an offshore rig.
Solution:
Wave force is calculated for various conditions.
Bow forces are under different conditions depending on the wave period.

If $A > 0.332\sqrt{L}$

$$F_{bow} = \frac{0.273 H^2 B^2 L}{T^4}$$

If $A < 0.332\sqrt{L}$

$$F_{bow} = \frac{0.273 H^2 B^2 L}{\left(0.664\sqrt{L} - A\right)^4},$$

where
F_{bow} = bow force, lbf.
A = wave period, sec.
L = vessel length, ft.

H = significant wave height, ft.
B = vessel beam length, ft.

Beam Forces are

If $T > 0.642\sqrt{B+2D}$

$$F_{beam} = \frac{2.10H^2B^2L}{A^4}$$

If $A < 0.642\sqrt{B+2D}$

$$F_{beam} = \frac{2.10H^2B^2L}{\left(1.28\sqrt{B+2D} - A\right)^4},$$

where

F_{beam} = bow force, lbf.
A = wave period, sec.
L = vessel length, ft.
H = significant wave height, ft.
B = vessel beam length, ft.
D = vessel draft, ft.

507. Problem 15.11

List few types of marine risers and sketch them.

Solution:

- Steel Catenary (SCR)
- Lazy Wave
- Lazy S
- Steep S
- Lazy Wave
- Steep wave
- Hybrid

Different marine risers are shown in Fig. 15.6 and Fig. 15.7.

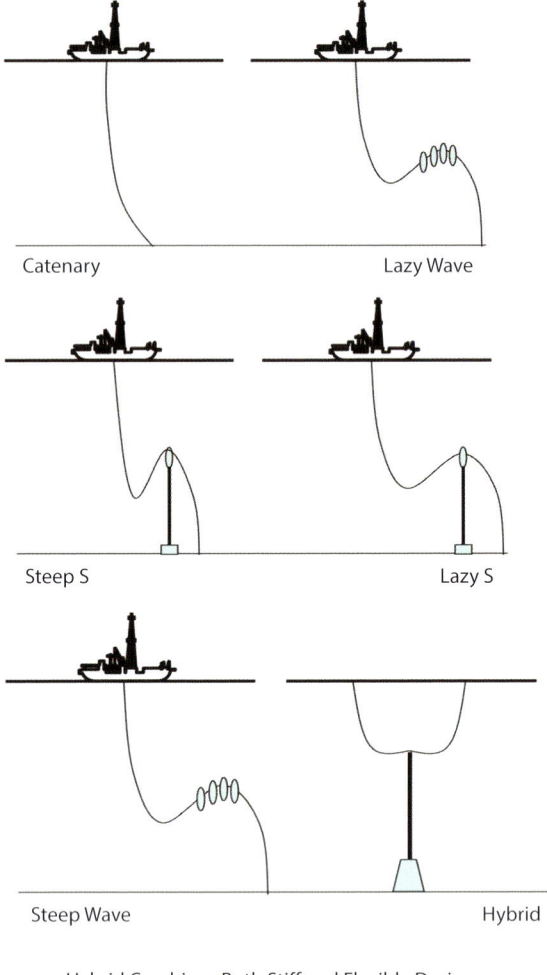

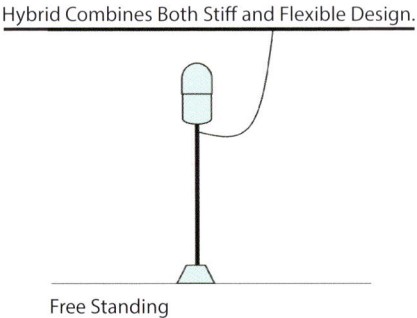

Figure 15.6 Marine risers

Offshore

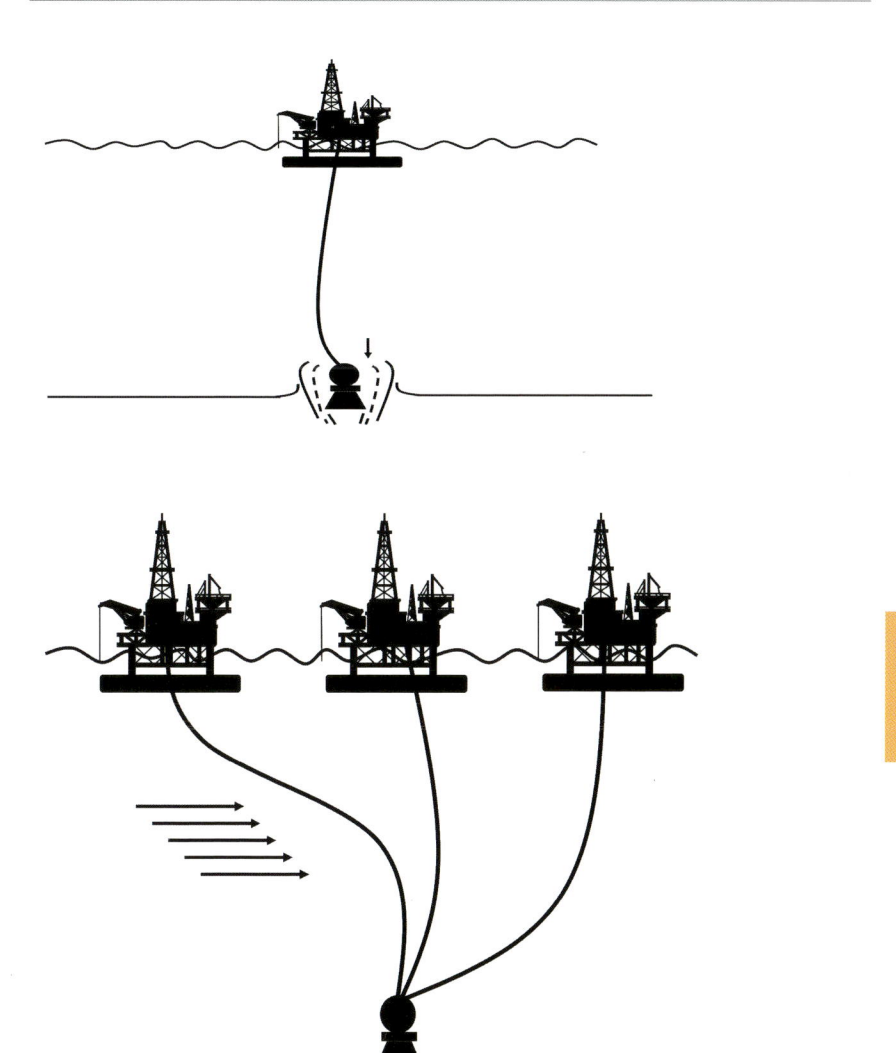

Figure 15.7 **Problems with marine risers**

508. Problem 15.12

List the functional loads of a riser system.
Solution:

1. Functional Loads
    - Weight of riser

- Buoyancy
- Top tension
- Running tools
- Pressure – internal/external
- Thermal effects
- Tension/compression
- Drillstring loads

509. Problem 15.13

List the environmental loads of a riser system.
Solution:

Environmental Loads
- Vessel motion
- Waves
    Regular waves
    Irregular Waves
- Current
- Snow and ice

510. Problem 15.14

List the problems encountered in a riser system.
Solution:

- Fatigue
- Disconnect and hang-off
- Drift-off due to vessel black-out
- Riser recoil
- Vortex induced vibration
- Flex-joint rotation
- Ball joint damage by drill bit, downhole tools, drillcollars, etc.
- Emergency disconnect failure
- Excessive stresses in riser connections
- BOP stack collapse to seabed
- Riser tensioner failure

511. Problem 15.15
Flex Joint Rotation

What is fleet angle and show the ball joint and fleet angle in the riser system.

Solution:

Fleet angle is the angle between the vertical axis and the wire line. Large fixed end moments are avoided by using ball joints at the riser ends. Ball joints are normally designed for rotations up to 8–10° in any direction. However, the drill string has to pass freely through the joints which limit the drilling operations to take place at ball joint angles of maximum 4–5°.

- To allow angular misalignment between riser and BOP
- Top and bottom and may be at intermediate levels

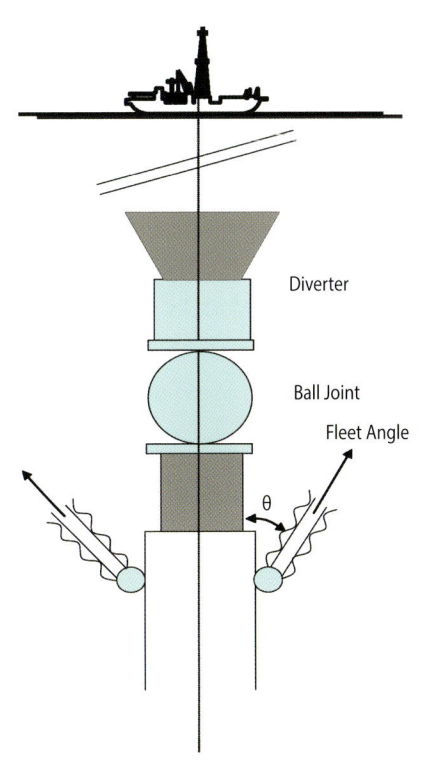

Figure 15.8 Fleet angle

512. Problem 15.16

Provide an equation to calculate the riser angle that is encountered in an offshore rig.

Solution:

It measures the riser angle relative to vertical. Riser angle are measured with respect to x and y axis and the resultant riser angle is given as

Exact equation

$$\theta = \tan^{-1}\sqrt{\tan^2\theta_x + \tan^2\theta_y}$$

Approximate equation

$$\theta \cong \sqrt{\theta_x^2 + \theta_y^2},$$

where
θ_x = riser angle in x- direction, deg.
θ_y = riser angle in y- direction, deg.
θ = resultant angle, deg.

513. Problem 15.17

Calculate the percentage error in using the approximate equation to calculate the resultant riser angle with the riser angle in X and Y directions are 4^0 and 5^0, respectively.

Solution:

Exact equation

$$\theta = \tan^{-1}\sqrt{\tan^2\theta_x + \tan^2\theta_y} = \tan^{-1}\sqrt{\tan^2 4 + \tan^2 5} = 6.39^0$$

Approximate equation

$$\theta \cong \sqrt{\theta_x^2 + \theta_y^2} = \sqrt{4^2 + 5^2} = 6.40^0$$

$$\text{Percentage error} = \frac{6.39^0 - 6.40^0}{6.39^0} = 0.2\%$$

514. Problem 15.18

With the following information regarding the floater, calculate the wind and current forces.

Floater details: Draft – 45 ft
                Freeboard – 50 ft
                Length – 450 ft
                Width – 80 ft
Substructure: Substructure 40 ft × 25 ft
                Rig Derrick:
                bottom section = 40 ft × 90 ft height and 20 ft width at the top
                top section = 20 ft × 20 ft height and 10 ft width at the top
Heliport Truss: 200 ft² area of wind path
Wind velocity – 50 mph (towards port side perpendicular to floater)
Current velocity – 3 ft/sec (towards port side perpendicular to floater)

Solution:

Rig Wind Force = $0.00338 \times 43.45^2 \times 1.1 \times 1.25$
$$\times \left[\frac{1}{2}(20+10)20 + \frac{1}{2}(40+20)90\right] = 26,322 \text{ lbf}$$

Helipad = $0.00338 \times 43.45^2 \times 1.1 \times 1.25 \times 200 = 1754$ lbf
Hull = $0.00338 \times 43.45^2 \times 1 \times 1 \times 450 \times 50 = 143,575$ lbf
Total Wind Force = 180 kips
Current Force = $1 \times 1 \times 3^2 \times 450 \times 45 = 182,250$ lbf smooth surface assumed

515. Problem 15.19

Provide the various relationships of nautical mile.

Solution:

- One nautical mile (NM) is one minute of latitude
- A speed of one nautical mile per hour is termed the Knot
- Determined by latitude (not longitude)
- 1 min of latitude = 1 nautical mile – 6076 ft
- 1° of latitude = 60 NM

516. Problem 15.20

Using the effective weight method calculate the riser tension for a riser of 146 ppf with drilling fluid of 12 ppg mud inside. Water depth is 1000 ft. Riser diameter 22" and thickness 0.5 in. Assume seawater density = 8.44 ppg. Also estimate the number of tensioners needed if each tensioner can handle 80,000 lbs.

Solution:

Effective weight per unit length can be calculated using the relation: Weight per foot in drilling mud is the weight per foot in air – the weight per foot of the displaced drilling mud.

$$w_B = w_s + \rho_i A_i - \rho_o A_o$$

$$\text{Total weight} = (w_B \times L) \times SF,$$

where
L = length of the riser.
SF is the safety factor, usually 1.2.
Without additional weight:

$$A_i = \frac{\pi}{4}(ID_p)^2$$

$$A_o = \frac{\pi}{4}(D_p)^2$$

Substituting the values

$$w_B = 147 + \frac{12 \times \frac{\pi}{4} \times 21^2}{12^2} - \frac{8.44 \times \frac{\pi}{4} \times 22^2}{12^2}$$

$$\text{Total weight} = \left(147 + \frac{12 \times \frac{\pi}{4} \times 21^2}{12^2} - \frac{8.44 \times \frac{\pi}{4} \times 22^2}{12^2}\right) \times 1000 \times 1.2$$

$$= 384,295 \text{ lbs}$$

Number of tensioner needed = 3,84,295/80,000 = 5

517. Problem 15.21

Calculate the response time of the tensioner based on sine wave when the heave height is 20 ft and the period is 20 sec.

Solution:

Maximum velocity can be estimated as

$$V_{avg} = \frac{\pi \times H}{P} \text{ ft/sec}$$

Substituting the values

$$V_{avg} = \frac{\pi \times 20}{20} = 3.14 \text{ ft/sec}$$

518. Problem 15.22

List the common components that are used in a drilling riser system from the LMRP.

Solution:

1. LMRP
2. Lower flex joint
3. Ball Joint
4. Riser joints/connectors
5. Slip joint
6. Upper flex joint
7. Riser tensioners

Additional attachments

1. Buoyancy modules
2. Choke and Kill lines
3. Hydraulic lines
4. Booster lines
5. Diverter assembly

519. Problem 15.23

Calculate the riser length for a semisubmersible rig with the following data:

Water depth: 2500 ft

Solution:

Riser length is given as

$$L_{R_{min}} = D - (L_{wh} + L_{BOP} + L_{LMRP} + L_{TL} + L_{Div}),$$

where

L_{wh} = length of wellhead.
L_{Tl} = telescopic joint length (mid stroke).
L_{div} = RKB to diverter bottom.
D = RKB to mudline.

CHAPTER 16

WELL COST

This chapter focuses on the different basic calculations involved in well cost due to various drilling and fishing operations.

520. Problem 16.1

What is AFE?

Solution:

AFE (Authorization for Expenditure) is an effective method of accounting the expenditures involved in drilling and completing the well. This gives the detailed cost estimate for a well.

521. Problem 16.2

What is Drilling Time Curve?

Solution:

Drilling time curve is days vs depth drilled also it can be overlaid as depth vs cost. This helps to compare the actual progress against the planned. When the differences occur between the predicted and actual corrective measures can be taken. (Fig. 16.1)

522. Problem 16.3

What is Cost per foot?

Solution:

Cost per foot during drilling is given as

$$C_d = \frac{\text{Bit cost} + \text{Tools cost} + \text{Mud cost} + (T_d + T_t + T_c)(C_r + C_s + C_t)}{F \times T_d},$$

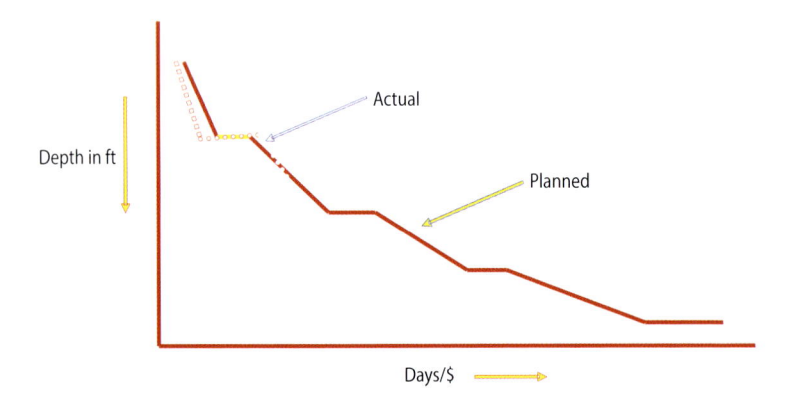

Figure 16.1 Drilling time curve

where
T_d = drilling time, hr.
T_t = trip time, hr.
T_c = connection time and other time chargeable to non-drilling task, hr.
C_r = rig rate, $/hr.
C_s = support rate, $/hr.
C_t = tool rental rate, $/hr.
F = footage, ft/hr.

523. Problem 16.4

The cost per foot equation for coring is given as

$$C = \frac{C_b + C_r(T_d + T_t + T_{rc})}{F \times R_c},$$

where
C_b, C_r = core bit and rig cost, respectively.
F = footage cored (ft).
T_t, T_{rc} = trip and core recovery times, respectively (hrs).
T_d = total coring time (hrs).
R_c = core recovery percentage.

The coring rate is given as

$$\frac{dF}{dt} = ke^{-\alpha t},$$

where
t = coring time (hrs).
k, α = constants.
Calculate the cost per foot if $T_d + T_t + T_{rc}$ = 24 hrs, T_d = 12 hrs, daily rig rental cost = $24,000, core bit cost = $10,000,

$$R_c = \frac{30{,}000\alpha}{k},$$

$$\alpha = 0.00035$$

Solution:

$$\frac{dF}{dT_d} = ke^{-\alpha T_d} \text{ can be written as}$$

$$dF = ke^{-\alpha T_d} dT_d$$

Integrating between the limits

$$\int_0^F dF = \int_0^{T_d} ke^{-\alpha t} dt$$

$$F = \frac{k}{\alpha}\left(1 - e^{-\alpha T_d}\right)$$

Substituting in the cost equation

$$C = \frac{C_b + C_r\left(T_d + T_t + T_c\right)}{\frac{k}{\alpha}\left(1 - e^{-\alpha T_d}\right)} = \frac{\alpha}{k}\left[\frac{C_b + C_r\left(T_d + T_t + T_c\right)}{\left(1 - e^{-\alpha T_d}\right)}\right]$$

Daily rig rental cost = $24,000 so C_r = 24,000/24 = 1000
$T_d + T_t + T_{rc}$ = 24 hrs
The cost per foot is = 270 $/ft

524. Problem 16.5

The cost per foot equation for coring is given as

$$C = \frac{C_b + C_r\left(T_d + T_t + T_{rc}\right)}{F \times R_c},$$

where

C_b, C_r = core bit and rig cost, respectively.
F = footage cored (ft).
T_t, T_{rc} = trip and core recovery times, respectively (hrs).
T_d = total coring time (hrs).
R_c = core recovery percentage.

The coring rate is given as

$$\frac{dF}{dt} = ke^{-\alpha t},$$

where

t = coring time (hrs).

k, α = constants.

Calculate the cost per foot if $T_d + T_t + T_{rc}$ = 24 hrs. T_d = 12 hrs, daily rig rental cost = \$24,000. core bit cost = \$10,000.

$$R_c = \frac{30{,}000\alpha}{k}.$$

$\alpha = 0.00035.$

Solution:

$\dfrac{dF}{dT_d} = ke^{-\alpha T_d}$ can be written as

$$dF = ke^{-\alpha T_d} dT_d$$

Integrating between the limits

$$\int_0^F dF = \int_0^{T_d} ke^{-\alpha t} dt$$

$$F = \frac{k}{\alpha}\left(1 - e^{-\alpha T_d}\right)$$

Substituting in the cost equation

$$C = \frac{C_b + C_r(T_d + T_t + T_c)}{\frac{k}{\alpha}(1 - e^{-\alpha T_d})} = \frac{\alpha}{k}\left[\frac{C_b + C_r(T_d + T_t + T_c)}{(1 - e^{-\alpha T_d})}\right]$$

525. Problem 16.6

A drilling engineer is planning for a core job was interested to calculate the cost involved in the trip. 30 ft of core is requested by the geologist in a soft formation where the recovery is expected not to be more than 50%. Depth the of the well for coring is 21,100 ft. Calculate the coring cost per foot if a 60-ft core is planned to offset loss of core due at 50% recovery.

- Trip time = 12 hrs
- Coring time = 12 hrs

- Circulation and ball drop time = 1 hr
- Rig cost = 50,000 $/day
- Core bit cost = $10,000
- Core barrel rental rate = $1000/day

Solution:

Cost per foot during drilling is given as

$$C_d = \frac{1}{\text{Recovery Percentage}} \left(\frac{\text{Core Bit cost + Tools cost + Mud cost} + (T_d + T_t + T_c)(C_r + C_s + C_t)}{F \times T_d} \right),$$

where

T_d = coring time, hrs.
T_t = trip time, hrs.
T_c = connection time and other time chargeable to non-drilling task, hr.
C_r = rig rate, $/hr.
C_s = support rate, $/hr.
C_t = tool rental rate, $/hr.
F = coring footage, ft/hr.

Substituting the respective values the coring cost is

$$C_d = \frac{1}{0.5} \left(\frac{10{,}000 + 15{,}000 + \text{Mud cost} + \dfrac{50{,}000 + 1000}{24}(12 + 12 + 1)}{60} \right)$$

$= 2675\,$/hr$

526. Problem 16.7
Economic Fishing Time Calculations

Economic fishing time (EFT) is the maximum time allowed for a fishing operation to be concluded without resulting in a cost that is higher than the expected costs of not fishing, which are the costs incurred when the fish is abandoned and the well sidetracked. This time should be determined probabilistically rather than deterministically.

$$E = \text{Probablility of success} \times \frac{\text{Total cost to abandon and sidetrack}}{\text{Daily cost of fishing operation}}$$

According to this concept, a fishing job completed in a time equal or less than EFT will be considered successful.

527. Problem 16.8
Decision Tree Analysis

A decision analysis method provides a way to take decision when to continue or discontinue an operation. It can be given in the form of events, decision points and control points or nodes are explained below.

Nodes

Decision Node = decision is made, shown as a square

Event Node = uncertainty is resolved and occurrences is known, shown as a circle

Terminal Node = combination of decision and events and are the end points, shown as a vertical bar.

Branches

Action branches:

Branches extending from decision nodes are possible course of actions

Event Branches:

Branches extending from event nodes are possible event occurrences and are assigned probability. Sum of the probabilities must be equal to one.

528. Problem 16.9
Procedure for Fishing

Decision Tree Sketching
Steps for sketching decision tree

1. Draw the starting point where the analysis is to be carried out by a square and write down the cost either inside the square or above the square.
2. Branch out with the decision and for fishing to fish or sidetrack with the circle at the end.

3. On the to fish side the events are branched out one above as success and one below failure. The success branch ends and the regular operation continues.
4. On the contrary, the failure branch will result in control circle or node which branches into two either to sidetrack and complete the operation or continue fishing operation.
5. Control nodes will draw success or failure and the inferior or less profitable action is pruned off
6. Repeat this for n number of days.

529. Problem 16.10
Procedure for Fishing – Decision Tree Calculations

Steps for calculating the values in the decision tree:

- Calculation starts from the terminal node and steps backwards sequentially.
- Fill all the calculated values at the control nodes with respective probabilities.
- Compare the values with decision nodes.
- If the sidetrack value is less than the control values discontinue the fishing operation and vice versa.

530. Problem 16.11
After 2 days of mobilizing services and initial attempt has resulted in $100,000, the cost of sidetracking the well is estimated to be $200,000. The daily fishing operation cost is $25,000. The probability of success on the first day is 75% whereas on the second day it is 50%. Draw the decision tree and find out whether it is profitable to continue fishing after second day. (Fig. 16.2)
Solution:

- Start with a square depicting an amount of 100K.
- Bottom branch will terminate with a cost of (initial cost + side track cost = 300 K).

Well Cost

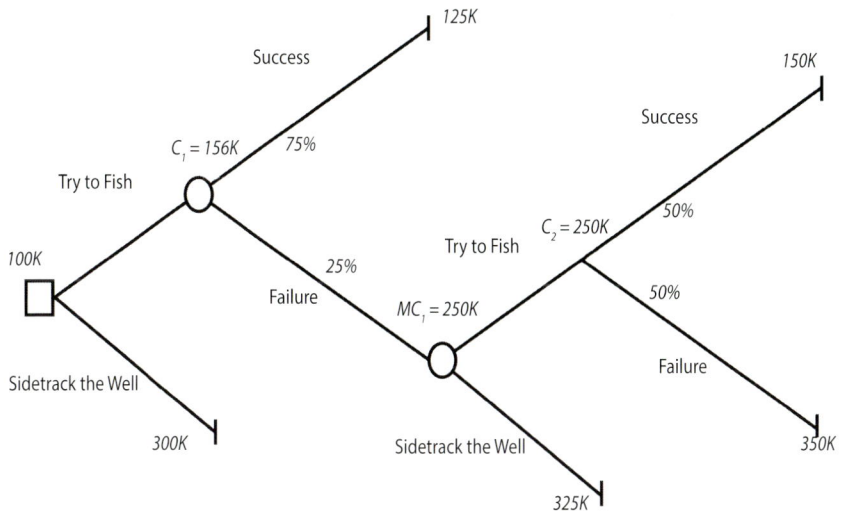

```
                              125K
                       Success                                         150K
            C₁ = 156K   75%                                    Success
  Try to Fish                                                   50%
100K                          C₂ = 250K
                              Try to Fish
                       25%                                      50%
                       Failure                                 Failure
  Sidetrack the Well          MC₁ = 250K
                                                                       350K
              300K
                       Sidetrack the Well
                              325K
```

Figure 16.2   Problem 16.11

- Top branch will end in a cost control circle C1 with two event branches.
- Top branch will be success terminating with a cost (initial cost + 1 day fishing cost = 125K).
- Bottom branch will end up in failure terminating with the cost control circle MC1 branching out into two.
- Bottom branch will be sidetrack with a cost of (initial cost + side track cost + 1 day fishing cost = 325K).
- Top branch will be failure with the control node C2 and emanating two branches – failure and success.
- Top branch will terminate with success with the total cost of (initial cost + 2 days fishing cost = 150K).
- Bottom branch will be a failure with the cost of (initial cost + side track cost + 2 days fishing cost = 350K).
- Terminate the tree and estimate the control point costs for pruning the tree for decision making after 2 days.

Solution:

- The calculation proceeds from backwards from the second day. The cost at the control circle C2 is (350 × 0.5 + 150 × 0.5 = 250K)
- The cost at the control point MC1 is lowest of 250K and the option of the "sidetracking the well", after 1 day which is 325K and the smaller value was chosen. So MC1 is 250K.
- The cost at the control point C1 based on the probability of success on the first day which is (100K + 25K) × 0.75 + 250 × 0.25 = 156K. This gives an indication as to which direction decision should be taken.
- At C1 – since 156K is less than the cost of sidetrack 300K, it is economical to attempt fishing on the first day.
- At MC1 – Since 250K is less than the cost of sidetrack 325K it is economical to continue fishing for the second day
- At C2 – since 250 is less than the 350K it is economical to continue fishing for the third day.

531. Problem 16.12

After 2 days of mobilizing services and initial attempt has resulted in $300,000, the cost of sidetracking the well is estimated to be $200,000. The daily fishing operation cost is $75,000.

The probability of success on the first day is 50% whereas on the second day it is 25%. Draw the decision tree and find out whether it is profitable to continue fishing after second day. (Fig. 16.3)

Solution:

- Start with a square depicting an amount of 300K.
- Bottom branch will terminate with a cost of (initial cost + side track cost = 500K).
- Top branch will end in a cost control circle C1 with two event branches.
- Top branch will be success terminating with a cost (initial cost + 1 day fishing cost = 375K).

Well Cost

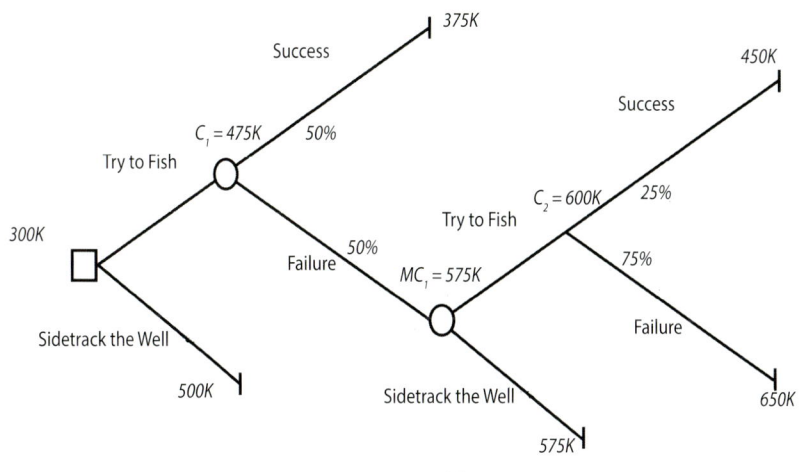

Figure 16.3 Problem 16.12

- Bottom branch will end up in failure terminating with the cost control circle MC1 branching out into two.
- Bottom branch will be sidetrack with a cost of (initial cost + side track cost + 1 day fishing cost = 575K).
- Top branch will be failure with the control node C2 and emanating two branches – failure and success.
- Top branch will terminate with success with the total cost of (initial cost + 2 days fishing cost = 450K).
- Bottom branch will be a failure with the cost of (initial cost + side track cost + 2 days fishing cost = 650K).
- Terminate the tree and estimate the control point costs for pruning the tree for decision making after 2 days.

Solution:

- The calculation proceeds from backwards from the second day. The cost at the control circle C2 is (650 × 0.75 + 450 × 0.25 = 600K)
- The cost at the control point MC1 is lowest of 575K and the option of the "sidetracking the well", after 1 day which is 600K and the smaller value was chosen. So MC1 is 575K.

- The cost at the control point C1 based on the probability of success on the first day which is (300K + 75K) × 0.50 + 575 × 0.50 = 475K. This gives an indication as to which direction decision should be taken.
- At C1 – since 475K is less than the cost of sidetrack 500K, it is economical to attempt fishing on the first day.
- At MC1 – Since 575K is less than the cost of sidetrack 600K it is economical to discontinue fishing for the second day.

532. Problem 16.13

After 2 days of mobilizing services and initial attempt has resulted in $300,000, the cost of sidetracking the well is estimated to be $200,000. The daily fishing operation cost $75,000. The probability of success on the first day is 50%

Using the data calculate the probability of success on the second day at which the engineer has to continue the fishing operation with the probability of success on the first day remaining the same.

Solution:

The task is to find the probability of success that will result in cost C2 less than the sidetrack cost of 575K.

Let the probability of success on the second day be P and the probability of failure be $(1-P)$ and so the equation can be formulated as below:

$$450 \times P + 650 \times (1 - P) \leq 575$$

Solving will result in P > 37.5 that will result in continuing of the fishing after second day.

533. Problem 16.14

Cost function of drillbits is given by the following equation:

$$R = 16{,}000\sqrt{D} + 24{,}000\sqrt{C} - 2000D - 1000C,$$

where
D = drilling cost.
C = capital cost

Well Cost

The cost constraint for drilling a particular well including the drilling cost and capital cost should not exceed $99,000. Calculate total cost with and without constraint.

Lagrangian

$$L(C, D, \lambda) = 16,000\sqrt{D} + 24,000\sqrt{C} - 2000D - 1000C + \lambda(99,000 - C - D)$$

Taking the derivatives

$$\frac{\partial L}{\partial C} = 0$$

$$\frac{\partial L}{\partial D} = 0 \text{ and}$$

$$\frac{\partial L}{\partial \lambda} = 0$$

Substituting, will yield $\dfrac{8000}{\sqrt{D}} - \dfrac{12,000}{\sqrt{C}} = 1000$

Substituting the condition from the third equation
C = 99,000 − D
and solving will result in the drillbit costs.
C = $86,768 D = $12,232

534. Problem 16.15

Assuming striking oil at one location is independent of another and probability of striking oil at an individual location is 0.3.
    a. What is the probability of striking oil exactly in two wells if drilling is carried out simultaneously at six different locations?
    b. What is the expected value of the number of locations having oil if drilling is carried out simultaneously at six different locations?
    c. What is the probability that the third time striking oil will occur at the sixth drilling if drilling is carried out one after another locations?

Solution:

a. $n = 6$, $p = 0.3$; $\binom{6}{2}(0.3)^2(1-0.3)^{6-2} = 0.324$

b. $np = 6 \times 0.3 = 1.8$

c. If third struck oil on the sixth drilling, then there are three success and three failures out of the sixth drilling. The sixth drilling is success and the remaining two success is anywhere in the previous five drilling. So the probability is

$$\binom{5}{2}(0.3)^3(1-0.3)^{5-2} = 0.092$$

Also Probability of two successes in five trials is

$$\binom{5}{2}(0.3)^2(1-0.3)^{5-2} = 0.309$$

Probability of success in six trials is 0.3
So probability that sixth drilling will find 3rd oil is
$0.309 \times 0.3 = 0.093$

APPENDIX

USEFUL CONVERSION FACTORS

Length

1 meter = 39.37 inch
1 inch = 2.54 centimeter 1 foot = 30.48 centimeter = 0.3048 meter
1 nautical mile = 6076 feet
1 meter = 1.0936 yards
1 centimeter = 0.39370 inch
1 inch = 2.54 centimeters
1 kilometer = 0.62137 mile
1 mile = 5280 feet
      = 1.6093 kilometers
1 angstrom = 10^{-10} meter

Area

SI Unit: meter-squared (m^2)
1 square kilometer = 0.3861 $mile^2$
1 square meter = 1.1960 $inch^2$
              = 10.763 $feet^2$
1 square centimeter = 0.155 $inch^2$
1 square mile = 2.590 km^2
1 square yard = 0.8361 m^2
1 square foot = 0.0929 m^2
1 square inch = 6.452 cm^2

Mass

SI Unit: Kilogram (kg)
1 lbm = 453.6 grams = 0.4536 kg = 7000 gr. (grain)
1 kg = 1000 grams = 2.204 6 lbm
1 slug = 1 lbf s^2/ft = 32.174 lbm
1 U.S. ton = 2000 lbm (also called short ton)
1 long ton = 2240 lbm (also called British ton)
1 ton = 1000 kilograms (also called metric ton) = 2204.6 lbm
1 kip = 1000 lb
1 kilogram = 1000 grams
          = 2.2046 pounds

Useful Conversion Factors

1 pound = 453.59 grams
= 0.45359 kilogram
= 16 ounces
1 ton = 2000 pounds
= 907.185 kilograms
1 metric ton = 1000 kilograms
= 2204.6 pounds
1 amu = 1.66056×10^{-27} kilograms

Force

1 lbf = 4.448 N = 4.448×10^5 dynes
1 lbf = 32.174 poundals = 32.174 lbm ft/s^2
1 US Short ton = 2000 lb
1 Metric ton = 2205 lbm

Gravitational Acceleration

$g = 32.2$ ft/s^2 = 9.81 m/s^2

$$g_\lambda = 9.7803267714 \left(\frac{1+0.00193185138639\sin^2\lambda}{\sqrt{1-0.00669437999013\sin^2\lambda}} \right) \text{ m/s}^2,$$

where
λ = geographic latitude of the earth ellipsoid measured from the equator in degrees.

Pressure

SI Unit: pascal (Pa)
1 atm = 14.69595 psia = 2118 lbf/ft^2
= 29.92 in. Hg = 760 mm = 1.013 bars
= 33.93 ft H$_2$O = 1.013×10 Pa = 101.3 kPa
1 Pa = 1 N/m^2 = 10^{-5} bars

1 pascal = $\dfrac{1 \text{ N}}{\text{m}^2}$

$$= \frac{1 \text{ kg}}{\text{m s}^2}$$

1 atmosphere = 101.325 kilopascals
= 760 torr (mmHg)
= 14.70 pounds per square inch
1 bar = 10^5 pascals

Volume

SI Unit: cubic meter (m^3)
1 ft^3 = 7.481 U.S. gal = 6.31 Imperial gal = 28.316 L
1 m^3 = 1000 L = 10^6 cm^3 = 264.2 U.S. gal = 35.31 ft = 264.2 gal = 35.31 ft^3
1 bbl = 42 U.S. gal = 5.61 ft^3
1 bbl = 9694.08 in^3
1 liter = 10^{-3} m^3
= 1 dm^3
= 1.0567 quarts
1 gallon = 4 quarts
= 8 pints
= 3.7854 liters
1 quart = 32 fluid ounces
= 0.94633 liter

Density

Water = 62.4 lbm/ft^3 = 1000 kg/m^3 = 1 g/cm^3 = 8.33 lbm/U.S. gal
°API, 60° F = (141.5/SG, 60°F) − 131.5
SG, 60° F = (141.5)/(°API, 60°F + 131.5)
1 lb/ft^3 = 0,01602 gm/cm^3
1 $slug/ft^3$ = 0,5154 gm/cm^3

Velocity

1 knot = 1 nautical mile/hr
60 mph = 88 ft/s

$$\left(\frac{\text{lb} \times s^n}{ft^2} \right) = 0.002088543 \times \text{eq.cP}$$

Useful Conversion Factors

Temperature

°F = 1.8 (°C) + 32
°R = °F + 459.67 = 1.8 (K)
0 K = −273.15°C
    = −459.67°F
K = °C + 273.15
°C = $\frac{5}{9}$(°F − 32)
°F = $\frac{9}{5}$(°C) + 32

Energy

SI Unit: joule (J)
1 J = 1 kg m²/s² = 10⁷ dyne
1 BTU = 777 ft lbf = 252 cal = 1055 J
1 hp hr = 2545 btu
1 kW hr = 3412 Btu = 1.341 hp hr
1 joule = 1 $\frac{kg\ m^2}{s^2}$
        = 0.23901 calorie
        = 0.4781 × 10⁻⁴ btu
        (British thermal unit)
1 calorie = 4.184 joules
          = 3.965 × 10⁻³ btu
1 btu = 1055.06 joules
      = 252.2 calories

Power

1 hp = 550 ft lbf/s = 33,000 ft lbf/min
Watt (W) = Joule (J) / s
Joule (J) = N-m
Coulomb (C) = Amp − second (As)

Gas Constant

R = 1.987 btu/lb mol °R = 1.987 cal/g mol K
  = 0.7302 atm ft³/lb mol °R = 1545 ft lbf/lb mol °R

= 0.08206 L atm/g mol K
= 82.06 atm cm³ mol K
= 8314 Pa m³ mol K or J/kg mol K
= 8.314 kJ/kgmol K

Viscosity

1 cP = 0.01 Poise = 0.01 g/cm s = 0.01 dyne s/cm
= 0.001 kg/ms = mPa-s = 0.001 N s/m²
= 2.42 lbm/ft hr = 0.0752 slug/ft hr
= 6.72×10^{-4} lbm/ft-s
= 2.09×10^{-5} lbf s/ft²
1 Pa s = 0.0209 lbf s/ft² = 0.672 lbm/ft s

$$\left(\frac{lb \times s^n}{ft^2}\right) = 0.002088543 \times eq.cP$$

$$\left(\frac{lb \times s}{ft^2}\right) = 4.79 \times 10^4 cP$$

1 ST = 100 cSt
1 ST = 0.0001 m² s⁻²
1 Calorie = 4.1868 joule (J)
Kinematic viscosity (cSt) = viscosity (cP)/density (g1cm⁻³)

Trigonometric Relationships

$$\csc\theta + \cot\theta = \frac{1+\cos\theta}{\sin\theta}$$

cscθ + cotθ = cot(θ/2)
sinθ = 2sin(θ/2)cos(θ/2)
cosθ = cos²(θ/2) − sin²(θ/2)
sin($a \pm b$) = sin$a$cos$b$ ± cos$a$sin$b$
cos($a \pm b$) = cos$a$cos$b$ ∓ sin$a$sin$b$

Useful Conversion Factors

$$\sin\theta = \frac{2e^t}{1+e^t} = \frac{2}{e^t + e^{-t}}$$

$$\cos\theta = \frac{1-e^{2t}}{1+e^{2t}} = \frac{e^{-t}-e^t}{e^t + e^{-t}}$$

$\sin\theta = \operatorname{sech} t$
$\cos\theta = \tanh t$
$\cosh^2 t - \sinh^2 t = 1$
$\tanh^2 t + \operatorname{sech}^2 t = 1$

$$\cosh t = \frac{e^t + e^{-t}}{2}$$

$$\sinh t = \frac{e^t - e^{-t}}{2}$$

$\sinh^2\theta + \cosh^2\theta = \cosh 2\theta$
$\cosh(a \pm b) = \cosh a \cosh b \mp \sinh a \sinh b$

$$\sin\theta = \frac{2\tan\frac{\theta}{2}}{1+\tan^2\frac{\theta}{2}}$$

$$\cos\theta = \frac{1-\tan^2\frac{\theta}{2}}{1+\tan^2\frac{\theta}{2}}$$

Trigonometric Approximations

$\sin\theta \cong \theta$
$\cos\theta \cong 1$
$\tan\theta \cong \theta$

For higher order approximation,

$$\cos\theta \cong 1 - \frac{\theta^2}{2}$$

Useful Relation

$a^x = e^{x \ln a}$

Integration Techniques

Integration by parts,

$$\int_a^b u\,dv = uv\Big|_a^b - \int_a^b v\,du$$

Taylor Series

If $f(c)$ is a continuous function in an open interval, its value at neighboring points can be expressed in terms of Taylor Series as

$$\sum_{n=0}^{\infty} \frac{f^{(n)}(c)}{n!}(x-c) = f(c) + f'(c)(x-c) + f''(c)\frac{(x-c)^2}{2!}$$

$$+ \ldots + f^{(n)}(c)\frac{(x-c)^n}{n!} + \ldots$$

where f^n denotes n^{th} derivative.
If $c = 0$, then the series is Maclaurin series for f.

Conversion

SI Metric Conversion Factors

bbl/ft × 5.216	E–01 = m³/m
cP × 1.0	E–03 = Pa.s
ft × 3.048	E–03 = m
gal × 7.460 43	E–03 = m³
hp × 7.460 43	E–01 = kW
in. × 2.54	E+00 = cm
lbf × 9.869 233	E–00 = Nm
lbf/ft² × 4.788 026	E–02 = kPa
lbm × 4.535 924	E–01 = kg

Useful Conversion Factors

lbm/ft³ × 1.601 846   E+01 = kg/m³
lbm/gal × 1.198 264   E+02 = kg/m³
md × 6.894 757   E−04 = μm²
psi × 6.894 757   E+00 = kPa
psi/ft × 2.262 059   E+01 = kPa/m
sq.in × 6.451   E+00 = cm²

Metric Prefix Conversions

pico-	$1p(x)$	$= 10^{-12}(x)$
nano-	$1n(x)$	$= 10^{-9}(x)$
micro-	$1u(x)$	$= 10^{-6}(x)$
milli-	$1m(x)$	$= 10^{-3}(x)$
centi-	$1c(x)$	$= 10^{-2}(x)$
deci-	$1d(x)$	$= 10^{-1}(x)$
Base Unit (x)		$= m, L, g$
Deca-	$1D(x)$	$= 10^{1}(x)$
Hecto-	$1H(x)$	$= 10^{2}(x)$
Kilo-	$1K(x)$	$= 10^{3}(x)$
Mega-	$1M(x)$	$= 10^{6}(x)$
Giga-	$1G(x)$	$= 10^{9}(x)$
Tera-	$1T(x)$	$= 10^{12}(x)$

BIBLIOGRPAHY

1. Robello Samuel and Dali Gao, "Horizontal Drilling Engineering – Theory, Methods and Applications." Sigma Quadrant Publications, Houston, 2014.
2. Robello Samuel, Kaiwan Bharucha, and Yuejin Luo, "Tortuosity Factors for Highly Tortuous Wells: A Practical Approach," SPE 92565, SPE/IADC Drilling Conference, 23–25 February, 2005, Amsterdam, Netherlands.
3. Robello Samuel, "Downhole Drilling Tools – Theory and Practice for Students and Engineers," Gulf Publishing, 2007.
4. Robello Samuel, "Formulas and Calculations for Drilling Operations," Wiley-Scrivener, 2010.
5. Robello Samuel and Xiushan Liu, "Advanced Drilling Engineering – Principles and Designs," Pennwell Publishers, 2007.
6. Azar JJ and G. Robello Samuel, "Drilling Engineering," Pennwell Publishers, 2006.
7. Gao DL, "Down-Hole Tubular Mechanics and Its Applications," Dongying, China University of Petroleum Press, 2006 (in Chinese, with English Abstr.).
8. "API Bulletin on Performance Properties of Casing, Tubing and Drill Pipe," API Bul. 5C2, 18th ed., March 1982.
9. "API Recommended Practice for Drill Stem Design and Operation Limits," API RP7G.
10. Akers TJ, Improving hole quality and casing-running performance in riserless top holes: deepwater Angola, "Proceedings of the IADC/SPE Drilling Conference," Orlando, Society of Petroleum Engineers, 4–6 March 2008.
11. Archard JF, Wear Theory and mechanism: In: "Wear Control Handbook," Edited by Peterson MB and Winer WO, New York, American Society of Mechanical Engineers, 35–88, 1980.
12. Balen R. Mark, Mens H-Z, Economides, Michael J, "Application for the Net Present Value (NPV) in the Optimization of Hydraulic Fractures," SPE paper 18541, 1988.
13. Begg S, Bratvold R, and Campbell J, "Improving Investment Decisions Using a Stochastic Integrated Asset Model", paper SPE 71414

SPE Annual Technical Conference and Exhibition, New Orleans, LA, 30 Sept.–3 Oct., 2001.
14. Bol GM, "Effect of Mud Composition on Wear and Friction of Casing and Tool Joints," SPE Drilling Engineering 1(5), 369–376, 1986.
15. Bourgoyne AT, Chenevert ME, Millheim KK, and Young FS, "Applied Drilling Engineering," Society of Petroleum Engineers, Richardson, TX, 1986.
16. Bradley WB and Fontenot JE, "The Prediction and Control of Casing Wear," SPE Journal of Petroleum Technology 27(2): 233–245 (Paper SPE 5122), 1975.
17. Bronshtein IN, Semendyayev KA, Musiol G, Muehlig H, and Mühlig H. "Handbook of Mathematics," 4th ed, Springer-Verlag Berlin Heidelberg, Germany, 2004.
18. Brooks GA and Harry W, "An Improved Method for Computing Wellbore Position Uncertainty and Its Application to Collision and Target Intersection Probability Analysis," SPE 36863, 1996.
19. Bruist EH, "A New Approach in Relief Well Drilling," SPE 3511, 1971.
20. Calhoun WM, Langdon SP, Wu J, et al., Casing Wear Prediction and Management in Deepwater Wells. SPE Deepwater Drilling and Completions Conference, Galveston, Texas, USA, 5–6 October, 2010.
21. Chen XB and Zhao GZ, Study on the Transmitting Mechanism of CSELF Waves: Response of the Alternating Current Point Source in the Uniform Space. Chinese, 2009.
22. Chen, Yu-Che, Lin, Yu-Hsu, and Cheatham, John B, "Tubing and Casing Buckling in Horizontal Wells," Journal of Petroleum Technology February 1990.
23. Chu S, Fan J, and Zhang L, Influence of the Weighting Material on Casing Wear in Impact-Sliding Test Conditions. Paper IJTC2008-71273, pp. 735–737, STLE/ASME International Joint Tribology Conference, Miami, Florida, 2008.
24. Contributed by Roberto Aguilera and adapted from his book "Horizontal Wells," Gulf Pub. Co., 1 Nov, 1991.

25. Cooper RE and Troncoso JC, "An Overview of Horizontal Well - Completion Technology," SPE 17582 SPE International Meeting held in Tianjin, China, 1988.
26. Cooper RE, "Coiled Tubing in Horizontal Wells," SPE paper 17581 SPE International Meeting in Tianjin, China, 1-4 November, 1988.
27. Cuthbert A and Russell R, A matched suite of drilling tools improves borehole quality and increases drilling efficiency, 2008, Unpublished.
28. Dawson, Rapier, and Paslay PR, "Drillpipe Buckling in Inclined Holes," Journal of Petroleum Technology October 1984.
29. Deli Gao, "Down-Hole Tubular Mechanics and Its Applications," China University of Petroleum at Beijing, China Copyright © 2006 by Petroleum University Press, China.
30. Deli Gao, "Down Hole Tubular Mechanics and Its Applications," China University of Petroleum Press, Dongying, Shandong Province, China, 2006.
31. Deli Gao, "Modeling & Simulation in Drilling and Completion for Oil & Gas," Tech Science Press, Duluth, USA, 2012.
32. Deli Gao, Chengjin Tan, and Wenyong Li, Research on Numerical Analysis of Drag and Torque for Xijiang Exteded-reach Wells in South-China Sea," Oil Drilling & Production Technology 25(5): 7–12, 2003.
33. Deli Gao, Xisheng Liu, and Binye Xu, "Prediction and Control of Wellbore Trajectory," Petroleum University Press, ISBN 7-5636-0584-3/TE 95, Dongying, Shandong Province, China, 1994.
34. Dellinger T, Gravley W, and Tolle GC, "Directional Technology will Extend Drilling Reach," Oil and Gas Journal 15, 153–169, 1980.
35. Deng JG, "Calculation Method of Mud Density to Control Borehole Closure Rate," Chinese Journal of Rock Mechanics and Engineering 16(6), 522–528, 1997.
36. Deng JG, Guo DX, and Zhou JL, "Mechanics - Chemistry Coupling Calculation Model of Borehole Stressing in Shale Formation and Its Numerical Solving Method," Chinese Journal of Rock Mechanics and Engineering 22 (Supp. 1), 2250–2253, 2003.

37. Deng JG, Wang JF, and Luo JS, "A New Experimental Method to Measure Diffusion Coefficient of Shale Hydration," Rock and Soil Mechanics 23, 40–42, 2002.
38. Dezen F and Morooka C, "Real Options Applied to Selection of Technological Alternative for Offshore Oilfield Development," paper SPE 77587 SPE Annual Technical Conference and Exhibition, San Antonio, Texas, 29 Sept–2 Oct, 2002.
39. Diao BB and Gao DL, "Solenoid Ranging System while Drilling," Acta Petrolei Sinica 32, 1061–1066, 2011a.
40. Diao BB and Gao DL, Adjacent well distance detection technology of the relief well and the blowout well. The sixteenth national academic annual meeting of prospecting engineering technology (rock and soil drilling engineering), 192–196, 2011b.
41. Diao Binbin and Gao Deli," Calculation Method of Adjacent Well Oriented Separation Factors," Petroleum Drilling Techniques 40 (1): 22–27, 2012.
42. Do Carmo MP, "Differential Geometry of Curves and Surfaces," Prentice-Hall, Inc., New Jersey, NJ, 1976.
43. Edwards Jack W, "Engineering Design of Drilling Operations," API Meeting, March, 1964.
44. El Rabaa AW and Meadows DL, "Laboratory and Field Application of the Strain Relaxation Method," SPE paper 15072 SPE California Regional Meeting, Oakland, CA, April 2–4, 1986.
45. Eric VO, "On the Physical and Chemical Stability of Shales," Journal of Petroleum Science and Engineering 38, 213–235, 2003.
46. Eric VO, Hale AH, and Mody FK, Critical Parameters in Modeling the Chemical Aspects of Borehole Stability in Shales and in Designing Improved Water-based Shale Drilling Fluids," SPE 171–186, 1994.
47. Fan J, Chen Q, and Zhang L, The Design and Application of a New Type of Instrument for Testing the Worn Surface of Casing Wear. Paper IJTC2008-71258, STLE/ASME International Joint Tribology Conference, Miami, Florida, 2008.

48. Fang Min, Lu Hong Kong, Wang Li-bo, "Constant Angle Curve Semi-analytical Calculation," Tongji University, Natural Science 3(63): 844–848, 2008.
49. Feenstra R and Zijsling DH, The Effect of Bit Hydraulics on Bit Performance in Relation to the Rock Destruction Mechanism at Depth, SPE 13025, 1984.
50. Fontenot JE and McEver JW, "Tripping is Not a Key Cause of Casing Wear," The Oil and Gas Journal 27: 148–163, 1975.
51. Fontoura CR and Rosana FTL, Characterization of shales for drilling purposes. SPE/ISRM, 78218, 2002.
52. Franca LFP, "Drilling Action of Roller-Cone Bits: Modeling and Experimental Validation," J Energy Resour Technology 132(4): 043101, 2010.
53. Galle EM and Woods HB, "Variable Weight and Rotary Speed for Lowest Cost Drilling," Annual Meeting of AAODC, New Orleans, LA, September 25–27, 1960.
54. Gao DL, Tan CJ, and Tang HX, "Limit Analysis of Extended Reach Drilling in South China Sea," Petroleum Science 6(2): 166–171, 2009
55. Gao DL, "Down-hole Tubular Mechanics and Its Applications," Dongying, China University of Petroleum Press, 80–82, 2006 (in Chinese).
56. Gao Deli et al., "Deep and Ultra-deep Drilling Technologies under the Complicated Geological Conditions," Petroleum Industry Press, Beijing, China, ISBN 7-5021-4775-6, 2004.
57. Gao Deli et al., "Optimized Design and Control Techniques for Drilling & Completion of Complex-Trajectory Wells," China University of Petroleum Press, Dongying, Shandong Province, China, ISBN 978-7-5636-3598-6, 2011.
58. Gao Deli the Liuxi Sheng, Xu Bing industry. Well trajectory control [M]. Shandong: Petroleum University Press, 207–218, 1994.
59. Gao Deli and Han Zhi-Dong, "Scan Calculation of the Distance of Adjacent Wells and Drawing Principle [J]," Oil Drilling & Production Technology 15(5): 21–29, 1993.

60. Gao Deli, "Predicting and Scanning of Wellbore Trajectory in Horizontal Well Using Advanced Models," SPE 29982, 1995.
61. Gao Deli and Han Zhi-Dong, "The Distance Adjacent Wells Scan Calculation and Drawing Principle," Oil Drilling & Production Technology 15(5): 21–29, 1993.
62. Gao Deli, "Deep and Ultradeep Well Drilling Technology under the Complicated Geological Conditions," Beijing, Petroleum Industry Press, 2004.
63. Gao DL, Liu FW, and Xu BY, An analysis of helical buckling of long tubular strings in horizontal wells, Proceedings of the International Oil & Gas Conference and Exhibition in China (IOGCEC), SPE 50931, 517–523, 1998.
64. Gao DL, Sun LZ, and Lian JH, "Prediction Method of Casing Wear in Extended-Reach Drilling," Petroleum Science 7(4): 494–501, 2010.
65. Gao DL and Sun LZ, "New Method for Predicting Casing Wear in Horizontal Drilling," Petroleum Science and Technology, 2012.
66. Gao DL, Tan CJ, and Tang HX, "Limit Analysis of Extended Reach Drilling in South China Sea," Petroleum Science 6(2), 166–171, 2009.
67. Garkasi AY, Xiang Y, and Liu G, "Casing Wear in Extended Reach and Multilateral Wells," World Oil 231(6): 35–37, 2010.
68. Ghassemi A, Tao Q, and Diek A, "Influence of Coupled Chemo-Poro-Thermoelastic Processes on Pore Pressure and Stress Distributions around a Wellbore in Swelling Shale," Journal of Petroleum Science and Engineering 67(1/2): 57–64, 2009.
69. Grace RD, Kuckes AF, and Branton J, "Operations at a Deep Relief Well: The TXO Marshall," SPE 18059, 1988.
70. Guo YF, Ji SJ, and Tang CQ, "The Relief Well - The Terminator of the Gulf Oil Leak," Foreign Oilfield Engineering 26: 64–65, 2010.
71. Karlsson H, Cobbley R, and Jaques GE, "New Developments in Short-, Medium-, and Long-Radius Lateral Drilling," SPE 18706, 1989.
72. Haas RC and Stokley CO, "Drilling and Completing a Horizontal Well in Fractured Carbonate," World Oil 39–45, 1989.

73. Hale AH and Mody FK, "Borehole-Stability Model To Couple the Mechanics and Chemistry of Drilling-Fluid/Shale Interactions," Journal of Petroleum Technology 45(11): 1093–1101, 1993.
74. Hale AH, Mody FK, and Salisbury DP, "The Influence of Chemical Potential on Wellbore Stability," SPE 23885-PA, 1993.
75. Hall RW, Garkasi A, Deskins G, et al., "Recent Advances in Casing Wear Technology," SPE 27532 SPE/IADC Drilling Conference, 1994, Dallas, Texas, 1993.
76. Hall RW and Malloy KP, "Contact Pressure Threshold: An Important New Aspect of Casing Wear," SPE Production Operations Symposium. Society of Petroleum Engineers, Oklahoma City, Oklahoma, 2005.
77. Han J, "There is Value in Operational Flexibility: an Intelligent Well Application," paper SPE 82018 SPE Hydrocarbon Economics and Evaluation Symposium, Dallas, Texas, April 5–8, 2003.
78. Han Zhiyong, "Design of Two-buildup Trajectory for Horizontal Wells," Journal of the University of Petroleum 15(4): 15–16, 1991.
80. Han Y, Ji B, Ouyang C, et al., "Experiments Illuminate Reasons for Casing Wear," The Oil and Gas Journal 108(17): 35–41, 2010.
81. Hirose S, "Biologically Inspired Robot: Snake-like Locomotors and Manipulators," Oxford, New York, Oxford University Press, 1993.
82. Holmquist JL and Nadai A, "A Theoretical and Experimental Approach to the Problem of the Collapse of Deep Well Casing," API Drilling Production Practices, API, Dallas, TX, 1939.
83. Horn BKP, "The Curve of Least Energy," A.I. Memo No. 612, Artificial Intelligence Laboratory, Massachusetts Institute of Technology, March 1983.
84. Hovda S, Haugland T, Norsk Hydro, Waddell K, Halliburton, Leknes R, and Weatherford, "World's First Application of a Multilateral System Combining a Cased and Cemented Junction with Fullbore Access to both Laterals," SPE 36488, SPE Annual Tech Conf, Denver, Colorado, Oct 1996.
85. Hsiao C, "A Study of Horizontal Wellbore Failure," SPE Paper 16927 Annual SPE Technical Conference and Exhibition held in Dallas, TX, September 27–20, 1987.

86. Huang NC and Pattillo PD, "Helical Buckling of a Tubular String in an Inclined Wellbore," International Journal of Non-linear Mechanics 35: 911–923, 2000.
87. Huang RZ, Chen M, and Deng JG, "Study on Shale Stability of Wellbore by Mechanics Coupling with Chemistry Method," Drilling Fluid & Completion Fluid 12(3): 15–21, 25, 1995.
88. Huq MZ and Celis JP, "Expressing Wear Rate in Sliding Contacts based on Dissipated Energy," Wear 252(5–6): 375–383, 2002.
89. Inglis TA, "Petroleum Engineering Development Studies, Volume 2: Directional Drilling," Graham & Trotman, London, 1987.
90. Jackson JD, "Classical Electrodynamics," John Wiley and Sons Inc., New York City, 1962.
91. Jiang W, "Discussion on Casing Wear Problems Occurred in Bohai Area in Recent Years," China Offshore Oil and Gas (Engineering) 14(1): 31–34, 2002 (in Chinese, with English Abstr.).
92. JinY, Chen M, and Liu GH, "Wellbore Stability Analysis of Extended Reach Wells," Journal of Geomechanics 5(1): 4–11, 1999.
93. Johancsik CA, Friesen DB, and Dawson, Rapier, "Torque and Drag in Directional Wells – Prediction and Measurement," Journal of Petroleum Technology 987–992, 1984.
94. Joly EL, et al., "New Production Logging Technique for Horizontal Wells," SPE paper 14463 SPE Annual Technical Conference Las Vegas, Nevada, 1987.
95. Juvkam-Wold HC and Wu J, "Casing Deflection and Centralizer Spacing Calculations," SPE Drilling Engineering 7(4): 268–274, 1992 (Paper SPE 21282).
96. Kai Sun, Robello Samuel, and Boyun Guo, "Effect of Stress Concentration Factors due to Corrosion on Production-String Design," SPE Journal of Production & Facilities 20(4): 334-339SPE 90094.
97. Kevin P. McCoy and Brian Vlasko, "Drilling and Gravel-packing Sinusoidal Wells," World Oil June, 2011.
98. Kral E, et al., "Fracture Mechanics Estimates Drill Pipe Fatigue," Oil Gas Journal, Tulsa 82(32): 50–55, 1984.

99. Kuckes AF, Plural sensor magnetometer for extended lateral range electrical conductivity logging. United States Patent, US4323848, April, 1982.
100. Kuckes AF, "Method for Determining the Location of a Deep-well Casing by Magnetic Field Sensing," United States Patent, US4700142, October 1987.
101. Kuckes AF, Hay RT, Joseph McMahon, Nord AG, Schilling DA, "New Electromagnetic Surveying/Ranging Method for Drilling Parallel Horizontal Twin Wells," 27466, SPE Drilling & Completion, June 1996.
102. Kuckes AF, Cornell U, and Ritch HJ, "Successful ELREC Logging for Casing Proximity in an Offshore Louisiana Blowout," SPE 11996, 1983.
103. Kuckes AF, Hay RT, and Mcmahon J, "An Electromagnetic Survey Method or Directionally Drilling a Relief Well into a Blowout Oil or Gas Well," Society of Petroleum Engineers SPE 10946, 1984.
104. Kuriyama Y, Tsukano Y, Mimaki T, et al., "Effect of Wear and Bending on Casing Collapse Strength, in SPE Annual Technical Conference and Exhibition," Society of Petroleum Engineers, Washington, D.C., 1992.
105. Leraand F, Wright JW, and Zachary MB, "Relief-well Planning and Drilling for a North Sea Underground Blowout," SPE 20420, 1990.
106. Levien R and Carlos S, "Interpolating Splines: Which is the Fairest of Them All?", Computer-Aided Design & Applications 6: 1–4, 2009.
107. Li Chen, Shugen Ma, Yuechao Wang, and Bin Li, "Design and Modeling of a Snake Robot in Traveling Wave Locomotion," Mechanism and Machine Theory 42: 1632–1642, 2007.
108. Li Chen, Yuechao Wang, Shugen Ma, and Bin Li, "Analysis of Travelling Wave Locomotion of Snake Robot," Proc. International Conference on Robotics Intelligent Systems and Signal Processing, Changsha, China, October 2003.
109. Liang QJ, "Trajectory Risk Index – An Engineering Method to Measure Risks of Multiple-Well Complex Trajectories," Proceedings of the Offshore Technology Conference, Houston, Offshore Technology Conference, 30 April–3 May 2007.

110. Liu Qi, "String Footwork to Calculate the Actual Track of the Borehole," Natural Gas Industry 6(4): 40–46, 1986.
111. Liu Xiushan Daqian and Gu Lingdi, "How to Simulate the Actual Well Trajectory Using Spline Functions," Daqing Petroleum Institute 15(1): 45–51, 1991.
112. Liu Xiushan, "Wellbore Trajectory Geometry," [M] Beijing, Petroleum Industry Press, 245–260, 2006.
113. Liu FW, Gao DL, and Xu BY, "An Analysis of Sinusoidal Buckling of Long Tubular Strings in Horizontal Holes," Engineering Mechanics 19(6): 44–48, 2002.
114. Liu FW, Xu BY, and Gao DL, "Packer Effect on Helical Buckling of Well Tubing," Journal of Tsinghua University 39(8): 104–107, 1999.
115. Liu FW, Xu BY, and Gao DL, "An Analytic Solution for Buckling and Post-buckling of Rubular Strings Subjected to Axial and Torsional Loading in Horizontal Circular Cylinders," Acta Mechanica Sinica 30(2): 238–242, 1999.
116. Liu X and Samuel Robello, "Actual 3D Shape Wellbore Trajectory and Objective Description for Complex Steered Wells," Proceedings of the SPE Annual Technical Conference and Exhibition, Denver, Society of Petroleum Engineers, 21–24 September 2008.
117. Liu X, "New Technique Calculates Borehole Curvature, Torsion," Oil & Gas Journal 104(40): 41–49, 2006.
118. Lu Hong Kong, Wang Gang, and Sun Zhongguo, "Directional Drilling Space Arc Orbital Calculations," Petroleum Geology and Engineering 20(6): 53–55, 2006.
119. Lu YH, Chen M, Jin Y, Teng XQ, Wu W, and Liu XQ, "Experimental Study of Strength Properties of Deep Mudstone under Drilling Fluid Soaking," Chinese Journal of Rock Mechanics and Engineering 31(7): 1399–1405, 2012.
120. Lubinski A, "Fatigue of Range 3 Drillpipe," Rev Inst Fr Pet Paris 32(2): 209–231, 1977.
121. Lubinski A, Althouse WS, and Logan JL, "Helical Buckling of Tubing Sealed in Packer," JPT 14(6): 655–667, 1962.

122. Lubinski A, "Developments in Petroleum Engineering, Volume II," edited by S. Miska, Gulf Publishing Co., 1988.
123. Lubinski A, "Developments in Petroleum Engineering, Volume I," edited by S. Miska, Gulf Publishing Co., Tulsa, 1987.
124. Lubinski A, "Maximum Permissible Dog Legs in Rotary Boreholes," J Pet Technol Dallas 13(2): 256–275, 1961.
125. Lyons KD, Honeygan S, and Mroz T, "NETL Extreme Drilling Laboratory Studies High Pressure High Temperature Drilling Phenomena," Journal of Energy Resources Technology 130(4): 102, 2008.
126. Economides M, McLennan J, Marcinew RP, and Brown JE, Fracturing of Highly Deviated and Horizontal Wells paper 89-40-39 CIM Annual Technical, Banff, AB, May 28–31, 1989.
127. Wiggins ML and Juvkam-Wold HC, "Simplified Equations for Planning Directional and Horizontal Wells," SPE 21261, 1990.
128. de Zwart MTW, "Theory on Loading, Torque, Buckling, Stresses, Strain, Twist, Burst and Collapse," Shell Doc. 1990, FS920101.
129. MacDonald GC and Lubinski A, "Straight-hole Drilling in Crookedhole Country," Proceedings of the Spring Meeting of Mid-Continent District, Division of Production, Amarillo, American Petroleum Institute, March 1951.
130. Markle RD, "Drilling Engineering Considerations in Designing a Shallow, Horizontal Well at Norman Wells, N.W.T., Canada, SPEI IADC paper, 16148. SPE/IADC Conference, New Orleans, LA, March 15–18, 1987.
131. McNair GA, Codling J, and Watson R, "Implementation of a New Risk Based Well Collision Avoidance Method," SPE/IADC 92554, 2005.
132. Millich E, et al., "New Technologies for the Exploration and Exploitation of Oil and Gas Resources," Volume 2, Proceeding of the 3rd E.C Symposium held in Luxemburg, March 22–24, 1988.
133. Miska S, Qiu WY, and Volk L, An improved analysis of axial force along coiled tubing in inclined/horizontal wellbores. SPE 37056 SPE International Conference on Horizontal Well Technology, Calgary, 207–214, 18–20 Nov 1996.

134. Miska S and Cunha JC, An analysis of helical buckling of tubular strings subjected to axial and torsional loading in inclined wellbore. Paper SPE 29460, Oklahoma City, OK, 173–178, 2–4 April 1995.
135. Mitchell R and Robello S, "How Good is the Torque Drag Model," SPE 105068. SPE/IADC Conference, February 2006.
136. Moll M and Kavaaki LE, "Path Planning for Deformable Objects," Journal of IEEE Trans Robotics 22(4): 625–636, 2006.
137. Moore PL, "Drilling Practices Manual," Petroleum Publishing Company, Tulsa, 1974.
138. Moreton H and Sequin C, Surface design with minimum energy networks. Proceedings of the First ACM Symposium on Solid Modeling Foundations and CAD/CAM Applications, New York, Association for Computing Machinery, 1991.
139. Morris FJ, Walters RL, and Costa JP, "A New Method of Determining Range and Direction from a Relief Well to a Blowout," SPE 6781, 1977.
140. Mukherjee H and Economides MJ, "A Parametric Comparison of Horizontal and Vertical Well Performance," SPE paper 18303 Annual SPE Technical Conference and Exhibition, Houston, TX, 2–5 October, 1988.
141. Oag AW and Williams M, The directional difficulty index – a new approach to performance benchmarking. Proceedings of the IADC/SPE Drilling Conference, New Orleans, Society of Petroleum Engineers, 23–25 February 2000.
142. Paslay PR and Bogy DB, "The Stability of a Circular Rod Laterally Constrained to Be in Contact with an Inclined Circular Cylinder," ASME Journal Applied Mechanics 31: 605–610, 1964.
143. Pastusek P and Brackin V, A model for borehole oscillations. Proceedings of the SPE Annual Technical Conference and Exhibition, Denver, Society of Petroleum Engineers, 5–8 October 2003.
144. Patil AP and Teodoriu C, "Model Development of Torsional Drillstring and Investigating Parametrically the Stick Slips Influencing Factors," J Energy Resour Technology 135(1): 013103, 2013.

145. Payne J, Ackerman W, Aretz R, Al Khodori S, Al-Farsi N, Ojulari B, Earl J, "Controlling Production Using Intelligent Completion Technology in Multilateral Wells", Oil and Gas Journal High Tech Wells Conference, Galveston, Texas, Feb. 2003.
146. Payne ML, Cocking DA, and Hatch AJ, "Critical Technologies for Success in Extended Reach Drilling," SPE 28293, 1994.
147. Poedjono B, Akinniranye G, Conran G, et al., "Minimizing the Risk of Well Collisions in Land and Offshore Drilling," SPE 108279, 2007.
148. Prince P and Tarmidi R, Hou TW, "Drilling and Completing Wells Uphill" SPE 96832, SPE/IADC Middle East Drilling Technology Conference and Exhibition, 12-14 September 2005, Dubai, United Arab Emirates
149. Rasmussen B, Sorheim JO, Seiffert E, Angeltvadt O, Gjedrem T, Statoil, World Record in Extended Reach Drilling, Well 33/9-C10, Statfjord Field, Norway, SPE/IADC Drilling Conference, 11-14 March, Amsterdam, Netherlands.
150. Reiber F, Vos BE, and Eide SE, On-line torque & drag: A real-time drilling performance optimization tool. Paper SPE 52836, 1999, Amsterdam, Netherlands.
151. Rezmer-Cooper I, Chau M, Hendricks A, et al., "Field Data Supports the Use of Stiffness and Tortuosity in Solving Complex Well Design Problems," SPE 52189 SPE/IADC Drilling Conference, Amsterdam, Netherlands, 9–11 March 1999.
152. Robello Samuel and Liu X, Wellbore tortuosity, drilling indices and energy: what do they do with well path design. Proceedings of the IADC/SPE Drilling Conference. 4–7 Oct. 2009, New Orleans, Society of Petroleum Engineers, 2009b.
153. Robello Samuel, "Modeling and Analysis of Drillstring Vibration in Riserless Environment," J Energy Resour Technology 135(1): 013101, 2013.
154. Robello Samuel and Liu X, "Wellbore Tortuosity, Torsion, Drilling Indices, and Energy: What do They Have to Do with Well Path Design?" SPE 124710 SPE Annual Technical Conference and Exhibition, New Orleans, Louisiana, 2009.

155. Schoenmakers JM, "Prediction of Casing Wear Due to Drilllstring Rotation: Field Validation of Laboratory Simulations," SPE Drilling Engineering 2(4): 375–381 (Paper SPE 14761), 1987.
156. Schuh F, Engineering Essentials of Modern Drilling, Energy Publications Division of HBJ. Dallas, 1982.
157. Schuh Frank J, "Horizontal Well Planning-Build Curve Design," Centennial Symposium Petroleum Technology into the Century, New Mexico Tech., N.M., 1989.
158. Sextro W, "Dynamical Contact Problems with Friction- Models, Methods, Experiments and Applications," Berlin, Springer, 2002.
159. Soliman MY, Hunt JL, and El Rabaa AWM, "On Fracturing Horizontal Wells," SPE paper 18542 SPE Eastern Regional Meeting Charleston, WV, 1–4 November, 1988.
160. Song JS, Bowen J, and Klementich F, "The Internal Pressure Capacity of Crescent-Shaped Wear Casing," SPE/IADC Drilling Conference. IADC/SPE Drilling Conference, New Orleans, Louisiana, 1992.
161. Stroud D, Peach S, and Johnston I, "Optimization of Rotary Steerable System Bottomhole Assemblies Minimizes Wellbore Tortuosity and Increases Directional Drilling Efficiency," SPE 90396, Houston, Texas, USA, and September 2004.
162. Sun Tengfei and Gao Deli, "Design of Horizontal Well Path under the Condition of the Uncertainty of Vertical Depth of Target," Fault-Block Oil & Gas Field 19(4): 526–528, 2012.
163. Tan CJ, Gao DL, Tang HX, et al., "The Method of Casing Wear Prediction for Liuhua Mega-Extended-Reach Wells in South China Sea," Oil Drilling & Production Technology 28(3): 1–3, 2006 (in Chinese).
164. Tikhonov VS and Safronov AI, "Analysis of Postbuckling Drillstring Vibrations in Rotary Drilling of Extended Reach Wells," J Energy Resour Technology 133(1): 043102, 2011.
165. Timoshenko SP and Goodier JN, "Theory of Elasticity, 3rd Edition," McGraw-Hill, New York, 1970.
166. Torgeir T, Havardstein TS, Weston LJ, et al. "Prediction of Wellbore Position Accuracy when Surveyed with Gyroscopic Tools," SPE Drilling & Completion 23(1): 5–12, 2008.

167. Walstrom JE, Brown AA, and Harvey RP, "An Analysis of Uncertainty in Directional Surveying," Journal of Petroleum Technology 21(4): 515–523, 1969.
168. Wang Zhenying, Contribution of Casing Wear Worked Problems.
169. Wang CJ, "Research of Friction and Wear between Drill Pipe and Casing. M.S. Thesis, Yanshan Yniversity, China - Qin Huangdao, 2007, (in Chinese, with English Abstr.).
170. Wang DG and Gao DL, "Study of Magnetic Vector Guide System in Tubular Magnet Source Space," Acta Petrolei Sinica 20: 608–611, 2008.
171. Wang Q, Zhou YC, Tan YL, and Jiang ZB, "Analysis of Effect Factor in Shale Wellbore Stability," Chinese Journal of Rock Mechanics and Engineering 31(1): 171–179, 2012a.
172. Weirich JB, Zaleski TE, and Mulcahy PM, "Perforating the Horizontal Well: Designs and Techniques Prove Successful," SPE paper lOO2l3, Dallas, TX, 1987.
173. White JP and Dawson R, "Casing Wear: Laboratory Measurements and Field Predictions," SPE Drilling Engineering 2(1): 56–62, 1987.
174. Williamson HS, "Accuracy Prediction for Directional Measurement while Drilling," SPE Drilling & Completion 15(4): 221–233, 2000.
175. Williamson JS, "Casing Wear: The Effect of Contact Pressure," SPE Journal of Petroleum Technology 33(12): 2382–2388, 1981.
176. Williamson JS and Bolton J, "Performance of Drill String Hardfacings," Journal of Energy Resources Technology 106(2): 278, 1984.
177. Wolff CJM and De Wardt JP, "Borehole Position Uncertainty - Analysis of Measuring Methods and Derivation of Systematic Error Model," SPE Journal of Petroleum Technology 33(12): 2339–2350, 1981.
178. Wu J and Hans C, and Juvkam-Wold, "Drag and Torque Calculations for Horizontal Wells Simplified for Field Use," Oil & Gas Journal 49–56, 1991.
179. Wu J and Zhang M, "Casing Burst Strength after Casing Wear," SPE Production Operations Symposium. Society of Petroleum Engineers, Oklahoma City, Oklahoma, 2005.

180. Wu J and Juvkam-Wold HC, "Buckling and Lockup of Tubular Strings in Inclined Wellbores," American Society of Mechanical Engineers, Petroleum Division (Publication), 56: 7–15, 1994.
181. Xiaojun He and Age Kyllingstad, "Helical Buckling and Lock-Up Conditions for Coiled Tubing in Curved Wells," SPE Drilling and Completions, March 1995.
182. Xingquan S and Wei'an G, Mathematical description of the spatial configuration of deviated wells and a new approach to calculate borehole curvature. Proceedings of the SPE International Oil and Gas Conference and Exhibition in China, Beijing, Society of Petroleum Engineers Inc., 2–6 November 1998.
183. Xiushan L, "Catenary Profile Design Method Study," Natural Gas Industry 27(7): 74–75, 2007.
184. Xiushan L, "Geometry of Wellbore Trajectory," Petroleum Industry Press, Beijing, China, 2006.
185. Xiushan L and Zaihong S, "Numerical Approximation Improves Well Survey Calculation," Oil & Gas Journal 15(2001): 50–54, 2007.
186. Yang LJ, "A Methodology for the Prediction of Standard Steady-State Wear Factor in an Aluminium-based Matrix Composite Reinforced with Alumina Particles," Journal of Materials Processing Technology (pp. 162–163): 139–148, 2005
187. Yew CH and Chenevert ME, "Wellbore Stress Distribution Produced by Moisture Adsorption," SPE 19536, 1990.
188. Yost II, AB and Overbey WK Jr, "Production and Stimulation - Analysis of Multiple Hydraulic Fracturing of a 2,000-ft Horizontal Well," SPE paper 19090 SPE Gas Technology Symposium held in Dallas, TX, June 7–9, 1989.
189. Young Jr. FS, "Computerized Drilling Control," SPE Paper No. 2241, 1968.
190. Yu L, Zhang L B, Fan J C, et al., "The Research on Calculation Method of Casing Wear Caused by Drill- Pipe Revolution," Drilling & Production Technology. 27(4): 66–69, 2004 (in Chinese).

191. Yuejin Luo, Kaiwan Bharucha, Robello Samuel, and Faris Bajwa, "Simple Practical Approach Provides a Technique for Calibrating Tortuosity Factors," Oil & Gas Journal 15, 2003.
192. Zaleski TE Jr. and Spatz E, "Horizontal Completions Challenge for Industry," Oil and Gas Journal 59–70, 1988.
193. Zhang L and Fan J, On the Casing Wear Mechanism in Deep and Ultra-Deep Well Drilling. Paper WTC2005-64064, Vol. 1, pp. 887–888, World Tribology Congress III, Washington, D.C., 12–16 September, 2005.
194. Zhang LW, Qiu DH, and Cheng YF, "Research on the Wellbore Stability Model Coupled Mechanics and Chemistry," Journal of Shandong University: Engineering Science 39(3): 111–114, 2009.
195. Zhiyong H, "Method of Non-Dimensional Design of Cautionary Shape Profile of Directional Well," Oil Drilling & Production Technology 19(4): 14, 16, 1997.
196. Zimmerman JC, Winslow DW, Hinkie RL, and Lockman RR, "Selection Tools for Stimulation in Horizontal Cased Hole," SPE paper 18995 SPE Joint Rocky Mountain Low Permeability Reservoirs Symposium, Denver, Colorado, 1989.

INDEX

A

Abnormal Pressure 138
Absolute 132
Absolute Tortuosity 110
Acceleration - Gravity 133
Acidity 147,202
Acoustic Velocity 420
Actual Flowrate 60
Air Weight 26
Alkalinity 147,202,213
Amplifying Factor 116
Angle Bending 79
Angle Fleet 461
Angle Toolface 79
Angle-Bend 75
Angle-BRA 88
Angle-Build Rate 88
Angle-Riser 40-41
Angle-Toolface 82-83
Angle-Torsion 85
Annular Velocity 320
Annular Velocity 219
Annular Volume 44
Annulus Flowrate 276
API Fluid Loss 199
Apparent viscosity 334
Appendix: Useful Conversion Factors 481
Arc Length 85
Astern 453
Average Curvature 75
Average DLS 75
Average DLS 75
Average Weight 149
Azimuth 75

B

Balance Volume 158
Balanced Plug 436
Ball Joint 461
Barite 151
Barite Cost 141
Barite Sacks 157
Base Fluid 127,151178
Basic elements 2
Beam Length 457
Bending Angle 78
Bending Angle 82
Bending Energy 307
Bending moment 304
Bending sign 307
Bending Stiffness 370
Bending stress 304,307
Bentonite Mass 151
Bernoulli 270
Bibliography 491
Bingham Plastic 184
Bingham Plastic 152
Bit HHP 223
Bit Hydraulics 272
Bit Optimization 277
Bit Pressure 271
Bit Pressure drop 223,275
BLF 422
Block 8
Block efficiency 11
Block Efficiency Factor 11
Block Line 9
Block Line Strength 12
Blocks 9
BML 450

Index

BOEMRE 450
Borehole Curvature 75
Borehole Curvature 75
Borehole curvature 311
Borehole Curvature 77
Borehole Torsion 83
Bottomhole Pressure 141-146
Buoyancy Factor 351
Bow 453
Bow Force 457
Bow Force 39
bottomhole pressure 211
Buckling Force 426
Buckling Limit Factor 422
Build Rate Angle 88
Build Section 74

C

Cake Thickness 145
Capacity 44
Capacity drillpipe 57
Capacity Linear 44
Capacity Volume 44
Casing TM 22
Casson 185
Cement Plug 350
Cement Slurry Requirements 341
Cement Yield 434
Cementing 341,430
Cementing (Problem 474 to 496) 429
Cementing Inner 436
Cementing Stage 436
Centrifuge 151
Chapter 2 43
Chapter 3 73
Chapter 4 131

Chapter 5 147
Chapter 6 181
Chapter 8 269
Chien Correlation 334
Clay-Water 152
Closed End Displacement 45
Coefficient of Friction 375
Coefficients Height 38
Collar Float 438
Collins-Graves 185
Composite Material 382
Concentration Cuttings 326
Concentration Factor 403
Conditioning Mud 438
Consistency Index 153,185
Constant Toolface 93
Constant Toolface Angle 99
Contact Time 342,399
Convergence Grid 115-116
Coring Costs 472
Coring TM 22
Correction Factor 399
Correction-Total 116
Cost Barite 141
Cost Function 474
Cost Per Foot 469
Critical Flowrate 221
Critical Transport Velocity 321
Crown Block 26
Crown Block Capacity 21-26
Critical Velocity 219
Curvature Borehole 75
Current Force 35,454
Curvature 79-83,304
Curvature-Borehole 75
Curvature-Horizontal 79-83

511

Curvature-Radius 75
Curvature-Radius 89
Curvature-Vertical 79-83
Cuttings 167
Cuttings Concentration 326
Cuttings Generated 167

D

Dead Line 11
Deadline 11
Decision Tree 473
Declination 115-116
Declination-Easterly 116
Declination-Westerly 116
Density 134,148
Density Final 149
Density Slurry 434
Derrick 10,15
Derrick Efficiency 16
Derrick load 11,17
Derrick-Dynamic Load 10,17
Derrick-Efficiency 11
Derrick-Load 10,16
Derrick-Maximum 11
Derrick-Maximum Equivalent Load 11
Derrick-Static 11
Desander 151
Desilter 151
Diameter Hole 169
Diesel 7,152
Diesel Engine 7
Diluting Mud 127,155
Discarded Volume 168
Discharge Coefficient 270
Displacement 45

Displacement Close End 45
Displacement Volume 45
Displacement-Horizontal 103
Displacement-Horizontal 105
Distance-Horizontal 88
Distance-Vertical 88
Diverter 461
DLS 75,79-81
Dogleg 79-83
Dogleg Severity 79-83
Draft 453,457
Drawworks 29
Drawworks Output 32
Drill bit Hydraulics (Problem 298 to 325) 269
Drilling Costs 469
Drilling Line 8,9
Drilling Operations (Problem 459 to 473) 413
Drilling Tools (Problem 326 to 344) 295
Drum Reel 28
Duplex Pump 51

E

Easterly Declination 116
ECD 214,215,303
Efficiency Factor 27
Effective Tension 389
Effective Weight 26
Efficiency 5,29-30,49
Efficiency - Line Pull 27
Efficiency Block 11
Efficiency Derrick 16
Efficiency Factor 18,27
Efficiency Hoisting 7,13

Index

Efficiency Mechanical 49
Efficiency Motor 49
Efficiency Overall 49
Efficiency Volumetric 49
Efficiency-Drawworks 31
Efficiency-Electric Motor 31
Efficiency-Engine 31
Efficiency-Hoisting 31
Efficiency-Overall 30-31
Efficiency-Volumetric 48
Elastic beam 307
Ellis Model 185
EMW 137-139,214-215
Endurance Limit 368
Energy 6
Energy Bending 307
Energy Mechanical 414
Energy Specific 414
Energy Transfer 6,31,49
Energy-Well profile 114
Engine Efficiency 49
Engine Horsepower 71
Equivalent Mud Weight 119,137-139
ESD 207,214
Excitation Harmonic 383

F

Factor Buoyancy 351
Factor Buckling 422
Factor Buckling Limit 422
Factor Concentration 403
Factor Correction 399
Factor Efficiency 18,27
Factor Safety 12,351
Factor Wear 399
Factor-Amplifying 116

Fast Line 10,11,14
Fast line speed 14
Fatigue Endurance Limit 368
Fatigue limit 368
Filtrate 145
Filtration 190
Final Density 149
Fishing Cost 473
Fleet Angle 461
Float Collar 438
Float Shoe 441
Flow Exponent 280
Flow Index 278
Flow Velocity 218
Flowrate 2
Flowrate Actual 60
Flowrate Optimum 279
Flowrate Theoretical 60
Fluid Base 178
Fluid Loss 145
Fluid Newtonian 183
Fluid Water Ratio 179
Fluids (Problem 220 to 258) 181
Force Current 454
Force Impact 272
Force Maximum Impact 279
Force Wave 454
Force Wind 454
Forces 35
Forces-Current 35
Forces-Wind 35
Force-Wave 35
Formation Porosity 167
Four Parameter 185
Fraction of Oil 171
Fraction of Water 171

Fraction Solid 171
Fraction Volume 382
Fraction Weight 150
Fracture Gradient 139
Friction Factor 11
Friction Static 376
Free vibration 383
Freeboard 453
Friction Factor 12
Frictional Pressure 271
Friction-COE 375-378
Friction-Coefficient of Friction 375-378
Friction-Friction Factor 375-378
Friction-Kinetic 375-378
Friction-Rolling 375-378
Friction-Rotational Speed 375-378
Friction-Sliding 375-378
Friction-Static 375-378
Forced vibration 383
Fuel 7
Fuel Consumption 7,8,71
Fuel Cost 32
Fuel Heating 8
Functions of drilling fluid 318

G

Galena 152
Gas Migration Potential 347
Gauge 132
Gel Strength 188
Gel Strength 213
Gel Strength Calculation 188
Gell Control 188
Generalized Casson Model 187
Gradient Fracture 139

Gradient Mud 133
Grid Convergence 116

H

Harmonic excitation 383
Heating Value 71
Heave 450
Hedstrom Number 266
Height Coefficients 38
Helical Method 85
Helical Method 111
Herschel Bulkley 184
Herschel Bulkley 150
Hertzian Model 396
HHP 46,2721,272,415
HMSE 415
Hoisting Efficiency 7,13
Hole Cleaning 326
Hole Cleaning (Problem 345 to 380) 317
Hole Diameter 169
Hook HP 26
Horsepower 4
Horizontal Curvature 75
Horizontal Displacement 103,105
Horizontal Distance 88
Horizontal Wellpath 99
Horsepower 9
Horsepower 46
Horsepower 278
Horsepower Engine 71
Horsepower Hydraulic 272
Horsepower-HP 9
Horsepower-Hydraulic 9
Horsepower-Mechanical 9
Horsepower-Rotating 9

Index

HP 4,5,46
HP Hook 26
HSI_ 274
Hull 37
Hybrid 457
Hydration 430
Hydraulic Diameter 322
Hydraulic Horsepower 46,272
Hydraulic Horsepower per square inch 274
Hydraulics (Problem 259 to 297) 209
Hydromechanical Specific Energy 415
Hydrostatic Pressure 132
Hydrostatic Pressure Reduction 342

I

Impact Wear 396
Impact Force 272
Inclination 79-83
Index Consistency 185
Index Flow 278
Index Power Law 185
Inertia Moment 306
Inline 29
Inline Reel 29
Inner String Cementing 436
Input 31
Input Power 9

J

J-Type 99

K

Kelly 2
Keyseating 339
KOD 74

L

Lag Time 330
Laser Specific Energy 415
Lateral Strain 362
Lazy S 457
Lazy Wave 457
Lazy Wave 457
Limitation 1 – Available Pump 278
Limitation 2 – Surface 278
Line Block 9
Line Dead 11
Line Pull 27
Line Pull Efficiency Factor 21-25
Linear Capacity 44
Linear Wear 396
Liner cementing 436
Liner Diameter 61,72
LMRP 450
Loading 425
LSE 415

M

Magnetic Declination 115-116
Manometer 136
Manometer 135
Marine Risers 458
Marsh Funnel 205
Marsh Funnel 149
Mass of Bentonite 151
Maximum Impact Force 279
MD 206
MD 214
Mechanical Efficiency 49
Mechanical Specific Energy 414
Minimum Curvature 88
Minimum Yield strength 367

515

Mix Water 434
Model - Four 185
Model - Three 185
Model - Two 185
MODU 450
Modulus of Elasticity 306, 362
Moment Bending 304
Moment of Inertia 306
Motor Efficiency 49
MSL 450
Mud 20-21
Mud Conditioning 438
Mud density 148
Mud Dilution 155
Mud Gradient 133
Mud Original 149
Mud pump 47
Mud Pump - Volumetric Efficiency 51
Mud Pump-Duplex 57
Mud Pump-Factor 48
Mud Pumps 47
Mud Pumps (Problem 51 to 85) 43
Mud Pumps-Double Acting 47
Mud Pumps-Liner 51
Mud Pumps-Single Acting 47
Mud Pump-Triplex 48,49
Mud Rheology 150,183
Mud weight 133,148,213
Mud Weight Equivalent 137-139
Mud Weight Increase Due to Cuttings 331
Mud Weight-EMW 119
Mud Weighting 151
Mud Weighting 121
Mud Weighting (Problem 176 to 219) 147

Mud-Dilution 155
Mud-Dilution 127
Mud-Gradient 20-21
Mud-Weighting
Mud-Weighting 92,121,151

N

Natural Curve
Natural Curve Method 92
Neutral 213
Newtonian Fluid 183
Normal Pressure 138
Nozzle Optimum 279
Nozzle Selector 292-293
Nozzle Velocity 272
Nozzle Selection 270
Number Hedstrom 243
Number of slips 31
Number Reynolds 243
Number ROP 326

O

Obstruction 74
Offset Reel 30
Offshore 35
Offshore (Problem 497 to 519) 449
offshore vessel motion 450
Oil Fraction 171
Oil Water Ratio 172
Oil Water Ratio 138
Operating Pressure 278
Optimal Nozzle 280
Optimization Bit 278
Optimization Calculations 277
Optimum Flowrate 279
Optimum Nozzle Size 279

Index

Original Mud 149
Output 32
Output Drawworks 32
Output Power 8
Output Volumetric 63
Overall Efficiency 49
OWR 172

P

Parallel Pressure drop 312
Peden Correlation 335
Percentage Bit Pressure 243
pH 206
pH 213
Pipe Velocity 219
Pipe Volume 45
Pitch 450
Plastic Real 183
Plastic Viscosity 184
Plastic Viscosity 182
Plastic Viscosity 152
Polar Moment 356
Porosity Formation 167
Port 453
Power Law 153,184
Power Law Exponent 153,185
Power Law Index 185
Power Rotary 27
Power-Input 31
Power-Output 31
Pre flush 436
Pressure 20
Pressure Abnormal 138
Pressure Absolute 132
Pressure Bit 271
Pressure Bottomhole 141-146

Pressure Calculations (Problem 153 to 175) 131
Pressure Frictional 271
Pressure Gauge 132
Pressure Hydrostatic 132
Pressure Normal 138
Pressure Pump 69
Pressure Subnormal 138
Pressure-Absolute 20
Pressure-Gauge 20
Pressure-Hydrostatic 20
Pressure-Measurement 20-21
Pressure-Mud 20
Primary Cementing 430
Pull Line 27
Pump - Volumetric Efficiency 47
Pump Factor 48
Pump Pressure 69
Pump-Factor 48
Pumps 47-72
Pumps-Double Acting 47
Pumps-Liner 47
Pumps-Single Acting 47
PV 182

R

Radius of Curvature 75,89
Radius of Curvature 78
Ratio-Oil Water 138,172
Real Plastic 183
Reel Drum 28,29
Reel Inline 29
Reel Offset 30
Relative Tortuosity 110
Resistivity 206
Resistivity 214

Reynolods Number 227
Rheology 150,183
Rig Equipment (Problem 1 to 50) 1
Riser Angle 40-41
Riser Angle 461
Riser-Angle 40-41
Robertson-Stiff 185
Roll 450
ROP 167
ROP Number 326
Rotary Power 27
Round trip TM 25
RPM 2

S

Sacks Barite 157
Sacks of Cement 434
Safety Factor 351
Safety factor 12
SCF 403
SCR 457
Sea Water Temperature 432
Secondary Cementing 430
Self-excited vibration 383
Series pressure drop 312
Shape Coefficients 37
Shear Rate 152,183
Shear Stress 152,183
Shoe Float 441
Shoe Track 438
Sign Convention 307
Sine wave 111
sine Wave Method 111
Slack-off Test 420
Slip velocity 318,321
Slip Velocity Calculations 333

Slurry Density 342,434
Solid Fraction 171
Solids 178,191
Solids Volume 191
Specific Energy 396,414
Spurt Loss 190,214
Stage Cementing 436
Starboard 453
Static Density 134
Static Friction 376
Steady Vibration 383
Steel Catenary 457
Steep S 457
Steep Wave 457
Steiner Model 188
Stern 453
Stiffness Bending 370
Stiffness Torsional 370
Strain 362
Stress 362
Stress Concentration 403
Strokes 59
Strong Gel 190
S-Type 99
Subnormal Pressure 138
Surge 450
Sway 450
Sway 450
Swivel 3
System Block 9
System Tackle 9

T

Tackle System 9
Tank Trip 67
Temperature 23

Index

Temperature Sea Water 432
Temperature-Centigrade 23
Temperature-Kelvin 23
Temperature-Rankine 23
Tension Calculation 388-394
Tension Effective 389
Tension True 389
Terminal Velocity 320
The Chien Correlation 333
The Moore Correlation 334
The Walker Mays Correlation 335
Theoretical Displacement 50
Theoretical flowrate 60
Three Parameter 185
Time Lag 330
Time Total 56
TM casing 22
TM coring 22
Ton Mile 21,44
Ton-Mile - Casing 21-25
Ton-Mile - Coring 21-25
Ton-Mile - Drilling 21-25
Ton-Mile - Reaming 21-25
Ton-Mile-Round Trip 21-25
Toolface 79
Toolface Angle 79,82
Toolface Angle Change 99
Tooljoint 300
Torque 5,47
Torsion 85
Torsional Stiffness 370
Torsion-Helical Method 86
Tortuosity 110
Tortuosity-Absolute 110
Tortuosity-Helical 111
Tortuosity-Random Inc Azm 111

Tortuosity-Random Inc Dependent Azm 112
Tortuosity-Relative 110
Tortuosity-Sine Wave 111
Total Correction 116
Total Input Energy 416
Total pressure 311
Total Strokes 56
Total Time 56
Total Weight 151
Track Shoe 438
Trajectory 89
Transport Cuttings 326
Transport ratio 320
Transport Velocity 321,335
Trip Tank 67
Tripping in 391
Tripping Out 391
True Tension 389
Tubular wear 396
Tubular Wear (Problem 450 to 458) 395
Tubulars (Problem 381 to 449) 347
TVD 206
TVD 214
TVDBML 450
Twist 356,369
Two Parameter 185

U

Ultimate Tensile strength 368
Uncertainty 116
Undercut 99
Unloading 425
UTS 368
U-Tube 210

V

Velocity Critical 321
Velocity Nozzle 272
Velocity Slip 318,321
Velocity Transport 321
Vertical Curvature 75
Vertical Distance 88
Vessel Length 457
Vibration Forced 383
Vibration Free 383
Vibration Self-excited 383
Vibration Steady 383
Viscosity 182,213
Viscosity Apparent 334
Volume 148
Volume Annular 44
Volume Balance 158
Volume Capacity 44
Volume Dilution 191
Volume Displaced 64
Volume Discarded 168
Volume Displacement 45
Volume Fraction 382
Volume per Stroke 53
Volume Pipe 45
Volumetric Efficiency 49
Volumetric Output 63

W

Wall Thickness 362
Water 178
Water Fraction 171
Wave Force 35,41,454
Weak Gel 190
Wear 396
Wear Factors 396

Weight Effective 26
Weight Fraction 150
Weight Mud 133
Weight Total 151
Weighting Material 123,151
Well Cost 469
Well Cost (Problem 520 to 534) 467
Well Path Design (Problem 86 to 152) 73
Well profile Energy 114
Wellbore curvature 307
Wellbore energy 396
Wellpath 74
Wellpath Design 99
Wellpath- Natural Curve 92
Wellpath-Length 87
Wellpath-Minimum Curvature 88
Wellpath-Trajectory 88
Wellpath-Uncertainty 116
Westerly-declination 116
Whipstock 81
Wind Force 35,41,454
Wiper Plug 443
Walker Mayes 334
WOB 2

Y

Yaw 450
Yield 434
Yield of Cement 341
Yield Point 152,184,213
Yield Stress 185
Yield Value 185

Z

Zero-Sec-Gel 152,185,192

Check out the forthcoming titles from SigmaQuadrant.com!

Learn for the Quadrant not for the Coordinate

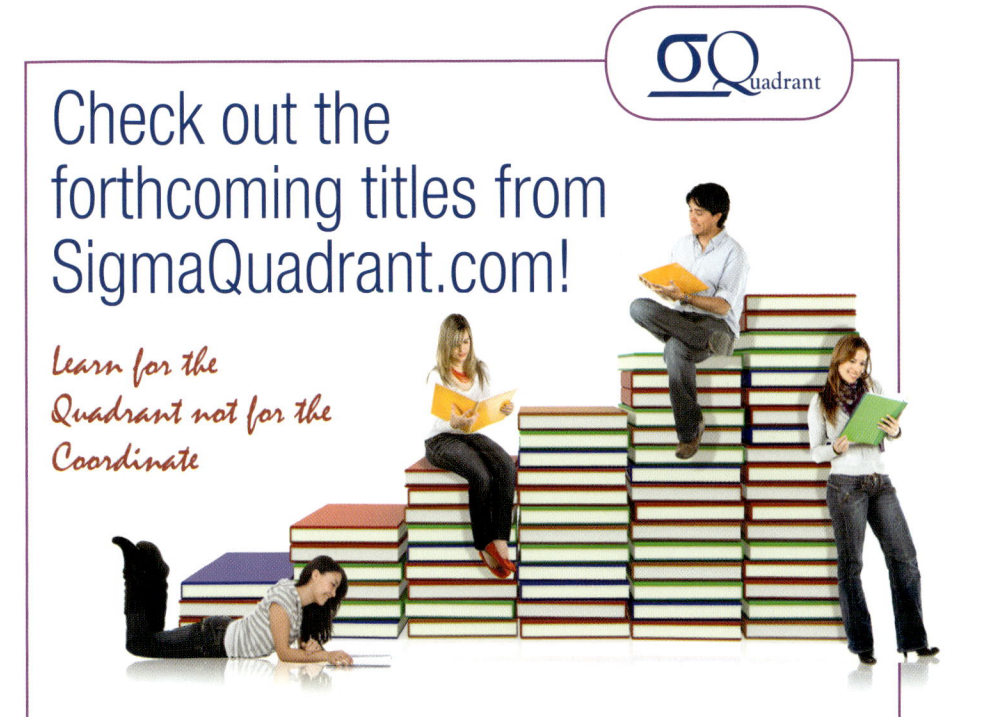

- **Horizontal Drilling Engineering** - *Theory Methods and Application*
  Robello Samuel, Dali Gao
- **Heuristics in Drilling Engineering**
  Prosper Aideyan
- **Positive Displacement Motors** - *Theory Design and Applications*
  Robello Samuel, Dmitry D. Baldenko and Dimtry F. Baldenko
- **Applied Drilling Engineering Optimization**
  Robello Samuel, J.J. Azar
- **Deepwater Drilling Engineering**
  Otto L. Santos, Robello Samuel

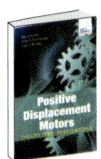

Are you looking for a book publisher?

Tell us about your book or Submit a Proposal today!

We accept manuscripts for a complete publishing from copy editing to printing. You will be working with experienced professionals. So whether you are looking to publish a small book or a complex engineering book we would love to work with you.

Visit and order at **SigmaQuadrant.com** or send e-mail to **info@sigmaquadrant.com**